T0192417

Naive Decision Making

How should one choose the best restaurant to eat in? Or whom to marry? Can one really make money at gambling? Or predict the future?

Naive Decision Making discusses how mathematics can help us make decisions when the outcome is uncertain or the interests of others need to be taken into consideration. Professor Körner provides the reader with an enjoyable journey through many aspects of mathematical decision making with pithy observations, anecdotes and quotations. Topics include probability, statistics, Arrow's theorem, Game Theory and Nash equilibrium. Readers will also gain insight into mathematics in general and the role it can play within society.

Intended for those with elementary calculus, this book is ideal as a supplementary text for undergraduate courses in probability, game theory and decision making. This engaging book will appeal to all those of a mathematical mind. To aid understanding, many exercises are included with solutions available online.

T. W. KÖRNER is Professor of Fourier Analysis at the University of Cambridge. He is the author of *Fourier Analysis* and *The Pleasures of Counting*.

Naive Decision Making
Mathematics Applied to the Social World

T. W. KÖRNER
Trinity Hall
Cambridge

CAMBRIDGE
UNIVERSITY PRESS

Shaftesbury Road, Cambridge CB2 8EA, United Kingdom

One Liberty Plaza, 20th Floor, New York, NY 10006, USA

477 Williamstown Road, Port Melbourne, VIC 3207, Australia

314–321, 3rd Floor, Plot 3, Splendor Forum, Jasola District Centre, New Delhi – 110025, India

103 Penang Road, #05–06/07, Visioncrest Commercial, Singapore 238467

Cambridge University Press is part of Cambridge University Press & Assessment,
a department of the University of Cambridge.

We share the University's mission to contribute to society through the pursuit of
education, learning and research at the highest international levels of excellence.

www.cambridge.org
Information on this title: www.cambridge.org/9780521731638

First published 2008

A catalogue record for this publication is available from the British Library

ISBN   978-0-521-51616-7   Hardback
ISBN   978-0-521-73163-8   Paperback

This world is a comedy to those that think, a tragedy to those that feel.

(*Walpole* Letter to Sir Horace Mann, December 31, 1760 *[68]*)

Perhaps someone will quite rightly ask whether the same people who know these rules also play well or not. For it seems to be a different thing to know and to execute, and many who play well are very unlucky. The same question arises in other discussions. Is a learned physician also a skilled one? In those matters which give time for reflection, the same man is both learned and successful, as in mathematics, jurisprudence and also medicine, for it is very rare for a patient not to pose any problem.

But in those matters in which no time is given and guile prevails, it is one thing to know and another to exercise one's knowledge successfully, as in gambling, war, duelling and commerce. For although acumen depends on both knowledge and practice, still practice and experience can do more than knowledge.

(*Cardano* The Book of Games of Chance *[48] (Translation modified)*)

Studies serve for delight, for ornament, and for ability. Their chief use for delight is in privateness and retiring; for ornament, is in discourse; and for ability, is in the judgement and disposition of business. For expert men can execute, and perhaps judge of particulars, one by one; but the general counsels, and the plots and marshaling of affairs, come best from those that are learned. To spend too much time in studies is sloth; to use them too much for ornament, is affectation; to make judgement wholly by their rules, is the humour of a scholar. They perfect nature, and are perfected by experience: for natural abilities are like natural plants, that need [cultivating] by study; and studies themselves do give forth directions too much at large, except they be bounded in by experience. Crafty men con[d]emn studies, simple men admire them, and wise men use them; for they teach not their own use; but that is a wisdom without them, and above them, won by observation.

(*Bacon* Essays; Of Studies *[3]*)

Such a tract as this may be useful to several ends; the first of which is, that there being in this world several inquisitive persons, who are desirous to know what foundation they go upon, when they engage in play, whether from motive of gain, or barely diversion, they may, by the help of this or the like tract, gratify their curiosity, either by taking pains to understand what is here demonstrated, or else making use of the conclusions, and taking it for granted that the conclusions are right.

Another use to be made of this *Doctrine of Chances* is, that it may serve in conjunction with the other parts of mathematics, as a fit introduction to the art of reasoning; it being known from long experience that nothing can contribute more to the attaining of that art than the consideration of a long train of consequences, rightly deduced from undoubted principles, of which this book affords many examples. To this may be added that some of the problems about chance having a great appearance of simplicity, the mind is easily drawn into a belief, that their solution may be attained by mere strength of good sense; which generally proving otherwise, and the mistakes occasioned thereby being not infrequent, it is presumed that a book of this kind, which teaches to distinguish truth from what seems so nearly to resemble it, will be looked upon as a help to good reasoning.

(*de Moivre* The Doctrine of Chances *[16]*)

When you study mathematics you will be fully enriched, if you keep away from it, you will find yourself intellectually lacking. If you study mathematics readily like a youth with an open mind, you will be instantly enlightened. However, if you approach it like an old man with an obstinate attitude, you will not become skillful in it.

(Sun Zi Sun Zi's Mathematical Manual *[72])*

The trouble ... is that many of the real situations which are apt to arise are so complicated that they cannot be fully represented by one mathematical model. With structures there are often several alternative possible modes of failure. Naturally the structure breaks in whichever of these ways turns out to be the weakest – which is too often the one which nobody had happened to think of, let alone do sums about.

(Gordon Structures *[25])*

I told Canada Bill the game he was playing in was crooked, he said, 'I know it is, but it's the only game in town'.

(Old gambler's story)

Contents

Introduction

A well known newspaper columnist once wrote:

> I studied maths to 16. I could sword-fight with a slide rule and consort with logarithms. As in Ronald Searle's St Trinian's,[1] I could stalk the square on the hypotenuse and drop a surd at fifty paces. I ate quadratic equations for breakfast and lunched on differential calculus. It was completely pointless. For all the good it did me, I could have been parsing Mongolian verbs.
>
> (The Times *April 25, 2003*)

This book requires the knowledge and skill which the columnist believes he once had, together with a rather more open mind. Roughly speaking, it requires the tools available after two years of school or one year of university calculus. I assume the reader can use those tools readily and without too much effort.

The level aimed for is a year or so higher than that I envisaged for *The Pleasures of Counting*. In that book, I tried to offer something to those readers who skipped the mathematical details but, in this book, the mathematics is the message.

Students who have been trained to think of mathematics as being about finding right answers often find it hard to adjust to subjects like statistics which are about making decisions that may turn out to be right or wrong. The object of this book is to help readers think about how decisions are made and how mathematics can help the process. I have called it *Naive Decision Making* partly as a tribute to Halmos's *Naive Set Theory* but mainly to warn the reader that the real world is a much more complicated place than the shadow of a shadow which appears in this book.

A mathematician consulting the text *Mathematics: A Simple Tool for Geologists* [69] would expect to learn more about geology than about mathematics. Although I hope that this book will be found amusing and instructive by many

---

[1] I suspect he means St Custard's. TWK.

outside its intended audience, prospective readers should expect to learn more about mathematics than about decision making. They should note that I shall make no serious attempt either to describe or prescribe human behaviour and I will not investigate the meaning of rationality.[2]

Classical applied mathematics uses mathematics to study the physical world. This book deals with what is sometimes called 'applicable mathematics', that is mathematics applied to the social world of mankind.

Applied mathematicians have learnt that the real world is too complex to be studied as a whole and that even such simple things as the shimmer on a beetle's wing or the behaviour of a tippy top present major challenges to our understanding. Experience incites the applicable mathematician to even greater modesty than her applied colleague. The grand themes of philosophers, economists and historians have proved remarkably resistant to mathematical treatment. Mankind demands simple answers to complex questions and a multitude of confident people minister to that demand. Mathematicians have discovered that even simple questions may have complicated answers and are debarred from this useful trade.

Mathematicians cannot explain the rise and fall of societies, or prescribe rules to produce the best of all possible human worlds. Instead, applicable mathematicians indulge a taste for the low company of gamblers, stage magicians and financial speculators. These low-minded individuals have low and simple desires.[3] They wish to make as much money as possible or to make a certain sum of money with as little work as possible or just to find the best place to park. Mathematics which remains silent before great questions has much to say about maximisation and low questions such as these.

In our first chapter, we discover that, even amidst the noise and confusion of the race-track, some sorts of betting are more sensible than others. We give different advice to those who know about horses and those who do not. In Chapter 2, we take a step back from the hubbub to produce a simple formal theory of probability. The high point of this chapter is the deep result known as the 'weak law of large numbers'. In Section 2.6, we use the law of large numbers to produce Kelly's rule which tells us not only how to bet, but also how much to bet. Our discussion of Kelly's rule emphasises that different gamblers

---

[2] Life somewhat better might content him,
   But for the gleam of heavenly light which Thou hast lent him:
   He calls it Reason – thence his power's increased,
   To be far beastlier than any beast.
                                            (Goethe, Faust, *Bayard's translation)*

[3] One of my readers noted that 'artificial decision-making entities, which, like it or not, are going to be more and more a part of our societies', may also be considered as having simple desires.

may have different goals and that their appropriate strategies will also differ. In the next chapter, we use the mathematics of the race-track to discuss insurance, pensions and a matter of life and death.

In Chapter 4, we introduce the notion of an algorithm. This is the mathematical equivalent of a cooking recipe and, just as with recipes, the skill lies not in following the instructions, but in producing the instructions in the first place. Only the first two sections of this chapter will be referred to later, but the reader who perseveres with the remainder of the chapter will learn how ideas of a Greek mathematician from 2500 years ago, a Chinese mathematician from 1500 years ago and a French mathematician from 400 years ago combine to give one of the most modern secret codes.

The next chapter uses the model of a shuffled pack of cards to illustrate topics as disparate as athletic records, choosing a restaurant and finding the shortest way from *A* to *B*. Since the object of the book is to help form educated consumers and, perhaps, producers of mathematical ideas, I have, throughout this book and particularly in this chapter, chosen methods and results which illustrate the chosen topics rather than those which are *best possible*. The reader is strongly warned that, if she wishes to apply the ideas in practice, there are often better and sometimes far better ways of doing things than those described here.[4]

So far we have dealt with single individuals following single-minded goals. Much of the rest of the book deals with what happens when several such single-minded individuals interact. The optimist will be disappointed to learn that matters become much more complicated, but the pessimist will be surprised to learn that there is still much we can do. In Chapter 6, we produce the Gale–Shapely marriage algorithm to comfort the optimist by showing that, under certain limited circumstances, we can produce a reasonable outcome when faced with conflicting interests. We then prove Arrow's theorem to confirm the pessimist by showing that, in general, there is no 'fair' system for combining the preferences of individuals into a single list of preferences for a group of individuals.

The simplest examples of conflict between individuals occur in games. Games are important to mathematicians because they have well-defined rules, carry no ethical overtones and have well-defined outcomes. The statement '*A* beat *B* at a game of chess' is either true or false, whilst statements like 'The opening of Japan was ultimately to the great benefit of both Japan and the

---

[4] 'I may say that a good many of these papers have come to him through me, and I need not add are thoroughly untrustworthy. It would brighten my declining years to see a German cruiser navigating the Solent according to the mine-field plans which I have furnished.' Conan Doyle, *His Last Bow.*

United States' resist such classification. In Chapter 7, we show how to play games like 'Scissors, Paper, Stone' in which the players have to choose their moves simultaneously.

In simple games, one player's loss is the other player's gain. What happens if the players can gain by cooperating? Even if the players involved trust each other to carry out their share of the bargain, it is not clear how the benefits of cooperation should be shared. Chapter 8 gives a clever argument of Nash in favour of one particular method. The second half of the chapter gives another of Nash's arguments which shows that, even if the players do not trust each other, they may be able to derive some of the benefits of cooperation. However, the game of 'Prisoner's Dilemma' shows that this is not always the case.

So far we have only discussed single decisions, but we often make a series of decisions in which each decision depends on the outcome of previous decisions. Chapter 9 considers this in the context of various types of duel and the next chapter extends our discussion by looking at casinos both from the point of view of the customers and from the point of view of the owners. Both from a philosophical and a practical point of view, a casino is a much simpler place than a race-track. We take advantage of this simplicity to use more complicated mathematics to answer more complicated questions. In the final part of this chapter, we introduce the notion of a lottery in which you have a very small chance of winning a very large sum. It turns out that not only is this situation amenable to a rather elegant mathematical treatment but that both the ideas and the mathematics are relevant to a large number of practical problems.

When we bet in a casino, we know the appropriate odds and when we bet on the race-track, we believe we know the appropriate odds. If we do not know the appropriate odds, how should we try to find them? This is the concern of statistics, and in Chapter 11 we illustrate the ideas involved by looking at the search for new drugs.[5]

The last chapter illustrates the tendency of mathematics to complicate simple matters by seeking a calculus proof of the statement that if something becomes cheaper then people will buy more of it. We conclude with a brief meditation on the limits of the kind of naive decision making discussed in this book.

Although this is not a text book, I have used the standard theorem–proof format. Theorems make it clear what the author actually claims and proofs enable the reader to check that those claims have been supported by correct reasoning.

---

[5] Even here we act in a low-minded way. Instead of seeking to treat the whole person within the context of an overarching theory of heath and illness, we simply look for a particular remedy for a particular disease.

I am tempted to say that the ideal reader will read the book three times: the first time skipping both proofs and exercises, the second time studying the proofs and skipping the exercises and the third time doing the exercises. In fact, my ideal reader is anybody who enjoys any part of the book. Some of the exercises are little more than remarks, some extend the ideas of the main text and some are intended to intrigue. None is intended to prey on the reader's conscience. Sketch solutions to most of the exercises will be found on my home page at http://www.dpmms.cam.ac.uk/~twk/ together with a list of corrections.

Borges often cites

> Valéry's project: to write a history of literature without proper names. A history that would present all the books of the world as though they were written by a single person.
>
> (German Literature in the Age of Bach *[7]*)

It is all too easy to present an account of mathematics without proper names. I am particularly conscious that my references fail to reflect the influence of the books of Knuth, the articles of Martin Gardner, the lectures and conversation of Conway and my teenage reading of the English version of Kraitchik's *Mathematical Recreations* [36]. I have rarely consulted *Wikipedia* on a mathematical topic without finding something of interest.

I should like to thank Terry Gagen for allowing me to conduct the experiment described on page 153. I owe particular thanks to Dennis Courtney, Andrew Colman, Tadashi Tokieda (who finds Mongolian verbs rather interesting) and two anonymous reviewers, but many other people have also contributed help, corrections and advice.

I dedicate this book to Dennis and Sally Avery, friends of mathematics, of Trinity Hall and of the Körners.

1

A day at the races

1.1 Money for nothing

Horatio Bottomley was a flamboyant Edwardian journalist, financier and crook. It is fitting that he is now chiefly remembered by a story which ought to be true but, apparently, is not.

According to legend, Bottomley arranged a race at a Belgian seaside course in such a way that he owned all six competing horses and could instruct the jockeys as to the precise order in which they should finish. The bets he laid should have made a fortune but, unfortunately, half way through the race, a thick sea fog swept in and all ended in confusion.

On arriving at a race course, the first question that occurs to a mathematician is 'Can I make money without risk?'. This suggests the harder question 'If I bet Y, what is the maximum sum L that I can guarantee to get back?'.[1] If $L > Y$, then I can guarantee a profit. If $Y > L$, I cannot.

In the old days, when two gentlemen A and B differed in their views on the ability of a certain horse to win a certain race, A would offer to back his judgement by wagering y at odds of a to b. If B accepted the wager, then B would pay A the amount y if the horse won and A would pay B the amount ya/b if the horse lost.

Exercise 1.1.1 *If $c > 0$, show that offering odds of ca to cb is the same as offering odds of a to b.*

In view of Exercise 1.1.1, odds are most usually quoted as v to 1.

---

[1] Should we say that 'a particle travels a distance x centimetres in a time t seconds' or 'a particle travels a distance x in a time t'? Most advanced texts on physics follow the second convention with the implicit assumption that it does not matter whether we measure distance in centimetres or kilometres. We shall often follow the same convention for money and assume that it does not matter whether we use dollars or euros. We will return to this point in Section 3.5.

Sometimes the two gentlemen would settle the bet after the race. Sometimes, when it was not clear what sort of gentlemen were involved, A and B would pay over the sums for which they might be liable to a 'stakeholder' who would pay back the total sum to the winner of the bet.

On modern British racecourses you will find 'bookmakers' who will offer to bet with you at appropriate odds. However, because they do not fully trust you, the arrangements for making a bet are rather different.[2] If the bookmaker offers odds of v to 1 on a horse and you wish to wager y on that horse, you give the bookmaker y which vanishes into her pocket. If the horse wins, she gives you $y(v + 1)$ (that is to say, she returns your winnings plus the original stake), if it loses, you get nothing.

Exercise 1.1.2 *Show that, provided both sides are honest, the new arrangement (you pay y before the race and get back $y(v + 1)$ if your horse wins) is equivalent to the old (after the race you pay y if your horse loses and get yv if your horse wins).*

From our point of view, we pay the bookmaker y for a promise to pay $(v + 1)y$ if our horse wins. Mathematically, it is slightly simpler to deal with the quantity $u = v + 1$ than with v itself. We then pay y for a piece of paper which is worth uy if our horse wins and nothing otherwise. We shall call u the *payout multiplier*.

We shall suppose that there are n horses running and that the bookmaker offers a payout multiplier of u_j on the jth horse. If I bet y_1 on the first horse, y_2 on the second horse and so on, then I shall have paid the bookmaker

$$y_1 + y_2 + \cdots + y_n = Y$$

in return for a promise to pay $y_j u_j$ if the jth horse wins. Thus the least sum I may get back is

$$\min_j u_j y_j,$$

the minimum of $u_1 y_1, u_2 y_2, \ldots, u_n y_n$.

My problem thus becomes one of choosing $y_1, y_2, \ldots, y_n$ in such a way that

$$y_1 + y_2 + \cdots + y_n = Y$$

and $\min_j u_j y_j$ is as large as possible. It is worthwhile noting explicitly that $u_j > 0$ and that, at least for the moment, we take $Y > 0$.

---

[2] If you ask what reason you have to trust them, they will reply, often making use of vivid figures of speech, that those who do not trust them need not bet with them. For the purposes of simplicity rather than verisimilitude, we shall assume throughout the book that, unless otherwise stated, everybody trusts everybody and that their trust is justified.

If we cannot solve this general problem, it makes sense to consider the simpler case when $n = 2$. We now have to choose $y_1 = s$ and $y_2 = Y - s$ with $0 \le s \le Y$ to make

$$\min \{u_1 y_1, u_2 y_2\} = \min \{u_1 s, u_2(Y - s)\}$$

as large as possible. The next two exercises give two different approaches to the problem.

Exercise 1.1.3 *Let u_1, u_2, $Y > 0$. Sketch the graph of*

$$f(s) = \min \{u_1 s, u_2(Y - s)\}$$

as a function of s. (Your choice of u_1, u_2 and Y should not affect the general look of the graph.)

Exercise 1.1.4 *Let u_1, u_2, $Y > 0$. Find $\min \{u_1 s, u_2(Y - s)\}$ when $u_1 s < u_2(Y - s)$ and show that, in this case, if t is small and strictly positive,*

$$\min \{u_1(s + t), u_2(Y - (s + t))\} > \min \{u_1 s, u_2(Y - s)\}.$$

What can you say if $u_1 s > u_2(Y - s)$?

Both exercises suggest (or, if considered carefully, prove) the following result.

Lemma 1.1.5 *Let u_1, u_2, $Y > 0$. If $u_1 s^* = u_2(Y - s^*)$, then*

$$\min \{u_1 s^*, u_2(Y - s^*)\} \ge \min \{u_1 s, u_2(Y - s)\}$$

for all s.

Proof If $s \ge s^*$, then $Y - s^* \ge Y - s$ so

$$\min \{u_1 s^*, u_2(Y - s^*)\} = u_2(Y - s^*) \ge u_2(Y - s) \ge \min \{u_1 s, u_2(Y - s)\}.$$

A similar argument works if $s^* \ge s$. ∎

Exercise 1.1.6 *(i) Fill in the details of the 'similar argument' referred to in the proof of Lemma 1.1.5.*
  (ii) Let s^* be as in the statement of Lemma 1.1.5. Show that, if $s \ne s^*$, then

$$\min \{u_1 s^*, u_2(Y - s^*)\} > \min \{u_1 s, u_2(Y - s)\}.$$

We have seen that the two-horse problem is solved by taking $y_1 u_1 = y_2 u_2$. It is not hard to guess the solution of the n horse problem and to adapt the proof of Lemma 1.1.5 to the more general case.

Theorem 1.1.7 *Let $u_j > 0$ for all j and let $Y > 0$. If*

$$u_1 y_1^* = u_2 y_2^* = \cdots = u_n y_n^* \text{ and } y_1^* + y_2^* + \cdots + y_n^* = Y,$$

while

$$y_1 + y_2 + \cdots + y_n = Y,$$

then

$$\min_j u_j y_j^* \geq \min_j u_j y_j.$$

Exercise 1.1.8 *Let $u_j > 0$ for all j and let $Y > 0$. Show that the equations*

$$u_1 y_1^* = u_2 y_2^* = \cdots = u_n y_n^* \text{ and } y_1^* + y_2^* + \cdots + y_n^* = Y$$

have the unique solution given by

$$y_j^* = \frac{Y u_j^{-1}}{u_1^{-1} + u_2^{-1} + \cdots + u_n^{-1}}.$$

[Recall that we write $u^{-1} = 1/u$.]

Exercise 1.1.9 *(i) Prove Theorem 1.1.7.*
(ii) Show that, under the conditions of Theorem 1.1.7,

$$\min_j u_j y_j^* = \min_j u_j y_j$$

only if $y_j = y_j^*$ for all j.

We now know how we should bet in order to be sure that the bookmaker returns us L. Will we make a profit or a loss? We need to bet $L u_j^{-1}$ on the jth horse, and so we must bet a total of

$$\frac{L}{u_1} + \frac{L}{u_2} + \cdots + \frac{L}{u_n} = \left(\frac{1}{u_1} + \frac{1}{u_2} + \cdots + \frac{1}{u_n} \right) L,$$

so we shall make a loss unless the amount we get back from the bookmaker is at least as large as what we paid, in other words,

$$L \geq \left(\frac{1}{u_1} + \frac{1}{u_2} + \cdots + \frac{1}{u_n} \right) L.$$

We will lose if

$$1 > \frac{1}{u_1} + \frac{1}{u_2} + \cdots + \frac{1}{u_n},$$

we will break even if

$$1 = \frac{1}{u_1} + \frac{1}{u_2} + \cdots + \frac{1}{u_n},$$

and we will make guaranteed profit if

$$1 < \frac{1}{u_1} + \frac{1}{u_2} + \cdots + \frac{1}{u_n}.$$

Inspection of the odds given on any race will reveal that this result is also known to the bookmakers.[3]

1.2 The ideal bookmaker

In mechanics we often consider a particle sliding over a frictionless surface. We do this, not because we think that frictionless surfaces exist, but because we believe that understanding the simpler idealised problem will help us understand the more complicated real one.

In a similar way we introduce the concept of 'an ideal bookmaker' who is so confident in her choice of odds that she is willing to reverse roles with you so that she will *take* or *make* a bet on each horse at the same odds. The next exercise shows several ways in which this could be done.

Exercise 1.2.1 *Let $u = v + 1$. Show that the following different ways of betting are equivalent.*

(1) You agree with the bookmaker that, after the race, you will give her vy if the horse wins and she will give you y if it loses.

(2) Before the race the bookmaker gives you y which vanishes into your pocket. If the horse wins, you give her yu, otherwise you give her nothing.

(3) Before the race you give the bookmaker yv. If the horse loses, she gives you $y(v + 1)$, otherwise she gives you nothing.

To fix ideas, let us suppose that you bet in the manner described in (2), that is to say, you ask the bookmaker to give you the sums $y_1, y_2, \ldots, y_n$ before the race, but promise to return $u_j y_j$ if the jth horse wins. Your object is now to *minimise* the amount that you might have to give the bookmaker. If you wish to bet a total of Y your problem thus becomes one of choosing $y_1, y_2, \ldots, y_n$ in such a way that

$$y_1 + y_2 + \cdots + y_n = Y$$

and

$$\max_j u_j y_j,$$

---

[3] Sometimes the first horse will be such a favourite that, although $1 < 1/u_1 + 1/u_2 + \ldots + 1/u_n$, we have $1 > 1/u_2 + \cdots + 1/u_n$. In such circumstances anyone who knows (how, we shall not enquire) that the first horse will lose can make a guaranteed profit.

the maximum of the $u_j y_j$, is as small as possible. As before, we suppose that $u_j > 0$ and $Y > 0$.

As might be expected, similar ideas to those in the previous section give a complete solution.

Exercise 1.2.2 *Let $u_j > 0$ for all j and let $Y > 0$.*
   (i) If

$$u_1 y_1^* = u_2 y_2^* = \cdots = u_n y_n^* \text{ and } y_1^* + y_2^* + \cdots + y_n^* = Y,$$

while

$$y_1 + y_2 + \cdots + y_n = Y,$$

show that

$$\max_j u_j y_j^* \le \max_j u_j y_j.$$

   (ii) Show that

$$\max_j u_j y_j^* = \max_j u_j y_j$$

only if $y_j = y_j^*$ for all j.
   (iii) Show that, if we bet in accordance with the discussion above so as to minimise our maximum possible loss, we will lose if

$$1 < \frac{1}{u_1} + \frac{1}{u_2} + \cdots + \frac{1}{u_n},$$

we will break even if

$$1 = \frac{1}{u_1} + \frac{1}{u_2} + \cdots + \frac{1}{u_n},$$

and we will make guaranteed profit if

$$1 > \frac{1}{u_1} + \frac{1}{u_2} + \cdots + \frac{1}{u_n}.$$

Thus a bookmaker who is prepared to take or make bets at the same odds must choose them so that

$$\frac{1}{u_1} + \frac{1}{u_2} + \cdots + \frac{1}{u_n} = 1.$$

If she sets such odds, there is no way we can place bets so that we are certain of making money.

However, our dreams of easy money are not completely dashed. Suppose that there are two such bookmakers the first of whom uses a payout multiplier of a_j for the horse j and the second who uses a payout multiplier of b_j. They must set their odds so that

$$\frac{1}{a_1} + \frac{1}{a_2} + \cdots + \frac{1}{a_n} = 1 \text{ and } \frac{1}{b_1} + \frac{1}{b_2} + \cdots + \frac{1}{b_n} = 1,$$

but this does not imply that $a_j = b_j$.

Exercise 1.2.3 *Suppose that a_j, $b_j > 0$,*

$$\frac{1}{a_1} + \frac{1}{a_2} + \cdots + \frac{1}{a_n} = 1 \text{ and } \frac{1}{b_1} + \frac{1}{b_2} + \cdots + \frac{1}{b_n} = 1.$$

Set $c_j = a_j$ if $a_j > b_j$ and $c_j = b_j$ otherwise. Show that

$$\frac{1}{c_1} + \frac{1}{c_2} + \cdots + \frac{1}{c_n} < 1$$

unless $a_j = b_j$ for all j. Explain why this means that (unless the two bookmakers set exactly the same odds) you can bet on the horses in such a way as to guarantee a strictly positive profit.

I said earlier that the study of frictionless motion may be a useful prelude to the study of motion in general, but, at some stage, we must consider the effects of friction. In practice, bookmakers do not allow you to bet on or against a horse at the same odds. One simple way of introducing a frictional effect is for our more realistic bookmaker to take a proportion $1 - \alpha$ of every pound that an ideal bookmaker would give you 'to cover expenses' so that, if you bet y on a horse with payout multiplier u, you get back $\alpha u y$ if the horse wins and nothing if the horse loses.

Exercise 1.2.4 *(i) Suppose that a bookmaker sets a payout multiplier of u_j for the horse j but takes a proportion $1 - \alpha$ (with $1 \geq \alpha > 0$) of every pound you bet 'to cover expenses'. Show that you can only make a guaranteed profit by making appropriate bets if*

$$\frac{1}{u_1} + \frac{1}{u_2} + \cdots + \frac{1}{u_n} < \alpha.$$

Show that you can only make a guaranteed profit by taking appropriate bets if

$$\frac{1}{u_1} + \frac{1}{u_2} + \cdots + \frac{1}{u_n} > \frac{1}{\alpha}.$$

(ii) Suppose that one bookmaker sets a payout multiplier of a_j on a horse j and another sets a multiplier b_j. Both take a proportion $1 - \alpha$ of every pound

you bet 'to cover expenses'. Show that you can only make a guaranteed profit by making *appropriate bets if*

$$\frac{1}{\max\{a_1, b_1\}} + \frac{1}{\max\{a_2, b_2\}} + \cdots + \frac{1}{\max\{a_n, b_n\}} < \alpha.$$

Show that you can only make a guaranteed profit by taking *appropriate bets if*

$$\frac{1}{\min\{a_1, b_1\}} + \frac{1}{\min\{a_2, b_2\}} + \cdots + \frac{1}{\min\{a_n, b_n\}} > \frac{1}{\alpha}.$$

(iii) Explain why

$$\frac{1}{\min\{a, b\}} - \frac{1}{\max\{a, b\}} = \left| \frac{1}{a} - \frac{1}{b} \right|.$$

(Recall that $|x|$ is the absolute value of x, that is to say, $|x| = x$ if $x \geq 0$ and $|x| = -x$ if $x < 0$.)

Show that, if, under the conditions of (ii), it is possible to make a guaranteed profit by making *bets and also possible to make a guaranteed profit* taking *bets, then*

$$\left| \frac{1}{a_1} - \frac{1}{b_1} \right| + \left| \frac{1}{a_2} - \frac{1}{b_2} \right| + \cdots + \left| \frac{1}{a_n} - \frac{1}{b_n} \right| > \frac{1 - \alpha^2}{\alpha}.$$

(iv) In parts (i) to (iii), we have supposed that we either make or take all our bets. Suppose that we are allowed to take and make bets with both bookmakers. Explain why, if

$$\max_j \frac{a_j}{b_j} > \frac{1}{\alpha^2} \quad or \quad \max_j \frac{b_j}{a_j} > \frac{1}{\alpha^2},$$

we can make and take bets in such a way that we are guaranteed to make money.

When we can make a guaranteed profit by betting with different bookmakers we are said to be exercising 'arbitrage'. Financiers and mathematicians dream about arbitrage in the same way that explorers dreamt about Eldorado. By their nature, opportunities for arbitrage are rare and fleeting, and no one who knows of such an opportunity will tell anyone else about it.[4] The present author does not deny that opportunities for arbitrage may occasionally occur at the races, but points out that spending day after day on the race-track doing nothing but wait for such opportunities may be neither interesting nor profitable. After a

---

[4] Just as people come up to you in the street and offer to sell you gold bars, so their more smartly dressed friends will come to your office and offer to let you in on fabulous arbitrage opportunities. Watch out for the weasel words 'statistical arbitrage'.

brief detour to discuss negative money we shall discuss what to do if we cannot exercise arbitrage.

1.3 Negative money

In the previous sections we considered the following bets.

(A) Before the race you give the bookmaker y which vanishes into her pocket. If the horse wins she gives you yu, otherwise she gives you nothing.

(B) Before the race the bookmaker gives you y which vanishes into your pocket. If the horse wins you give her yu, otherwise you give her nothing.

Naturally we supposed that $y > 0$ but, if we drop this condition, we see that bet (B) can be written as follows.

(A^*) Before the race you give the bookmaker $-y$ which vanishes into her pocket. If the horse wins she gives you $-yu$, otherwise she gives you nothing.

Once we allow negative money, we can combine the results of the previous two sections into a single theorem.

***Theorem* 1.3.1** *Let $u_j > 0$ for all j and $Y \neq 0$. (Observe that these are just the hypotheses of Theorem 1.1.7 with the condition $Y > 0$ replaced by $Y \neq 0$.)*
  (i) The equations

$$u_1 y_1^* = u_2 y_2^* = \cdots = u_n y_n^* \text{ and } y_1^* + y_2^* + \cdots + y_n^* = Y$$

have the unique solution given by

$$y_j^* = \frac{Y u_j^{-1}}{u_1^{-1} + u_2^{-1} + \cdots + u_n^{-1}}.$$

  (ii) If y_j^* is as in (i) and

$$y_1 + y_2 + \cdots + y_n = Y,$$

then

$$\min_j u_j y_j^* \geq \min_j u_j y_j.$$

If $\min_j u_j y_j^* = \min_j u_j y_j$ then $y_j^* = y_j$.
  (iii) If $Y > 0$, then

$$\min_j u_j y_j^* \geq Y$$

if and only if

$$1 \geq \frac{1}{u_1} + \frac{1}{u_2} + \cdots + \frac{1}{u_n}.$$

If Y < 0, then

$$\min_j u_j y_j^* \geq Y$$

if and only if

$$1 \leq \frac{1}{u_1} + \frac{1}{u_2} + \cdots + \frac{1}{u_n}.$$

The reader may check that the same proof as was suggested for Theorem 1.1.7 works for this more general theorem.

At first sight, the notion of negative money seems rather odd. We do not expect news items to tell us that 'Thieves broke into a bank last night and deposited a large sum of negative currency'. However, first impressions are misleading.

If A receives n from B and gives B a piece of paper saying 'I promise to pay who ever owns this piece of paper n' then (provided that A is trustworthy) B can use the promise to purchase n of goods from C who can then use the promise to purchase n of goods from D and so on. The realisation that a 'promise to pay' can be treated as money is the foundation of the modern economy. But, so long as the promise to pay remains in circulation, it represents $-n$ for A.

In 1985 the Bank of England withdrew £1 notes from circulation and replaced them with coins. Fourteen years later, 55 million of the old notes had still not been exchanged for coins and presumably never would be. To the ordinary citizen, the notes were positive currency but, to the Bank, the 55 million old notes represented 55 million pounds worth of negative currency. The disappearance of -55 million pounds represented a 55 million pound profit.

1.4 Probability and expectation

Suppose that there are three horses running in a race and that an ideal bookmaker (that is to say, a bookmaker who is prepared to allow you to bet on or against any horse at the same odds) offers a payout multiplier of 3 on each horse. Since

$$\frac{1}{3} + \frac{1}{3} + \frac{1}{3} = 1,$$

there is no way that you can be guaranteed a profit. However, if you know that the first horse usually (though not always) beats the other two horses, you would be well advised to bet on it. Although you know that there is some risk involved, you believe that the 'promise to pay' 3 if the first horse wins is worth more than its cost of 1.

You now meet another ideal bookmaker who offers a payout multiplier of 100/98 for the first horse and 100 for both the other horses. You know that, although the first horse is good, it is not that good and it is therefore worth betting against the first horse. Although you know that you will usually lose, you believe that a 'promise to pay' 50 if the first horse loses is worth more than its cost of 1.

You believe that both bookmakers have chosen the wrong odds. Presumably correct odds must exist. If n horses are running, we say that the odds $u_1, u_2, \ldots, u_n$ offered are correct if there is no advantage in choosing to *make* or to *take* any bet at those odds. Thus, since it costs

$$y_1 + y_2 + \cdots + y_n$$

to make a bet which pays $u_j y_j$ if horse j wins for $j = 1, 2, \ldots, n$, it must be equally advantageous to accept a sum of

$$y_1 + y_2 + \cdots + y_n$$

or a promise to pay $u_j y_j$ if horse j wins for $j = 1, 2, \ldots, n$. Writing $x_j = y_j u_j$, we obtain the equivalent statement that it must be equally advantageous to accept a sum of

$$\frac{x_1}{u_1} + \frac{x_2}{u_2} + \cdots + \frac{x_n}{u_n}$$

or a promise to pay x_j if horse j wins for $j = 1, 2, \ldots, n$.

We note that, by the arguments of the previous section, the correct odds u_j must satisfy the conditions $u_j > 0$ and

$$\frac{1}{u_1} + \frac{1}{u_2} + \cdots + \frac{1}{u_n} = 1.$$

We write $p_j = 1/u_j$ and say that p_j is the *probability* that horse j will win. Observe that $p_j > 0$ and

$$p_1 + p_2 + \cdots + p_n = 1.$$

The sentence which concludes the previous paragraph now tells us that it must be equally advantageous to accept a sum of

$$p_1 x_1 + p_2 x_2 + \cdots + p_n x_n$$

or a promise to pay x_j if horse j wins for $j = 1, 2, \ldots, n$. We call $p_1x_1 + p_2x_2 + \cdots + p_nx_n$ the *expected value* of our winnings.[5]

Much of the rest of this book will be concerned with probability and expectation. To help fix our ideas we will look at the game of Crown and Anchor.[6]

To play Crown and Anchor, the banker needs a rectangular piece of canvas divided into six squares distinguished by the symbols of a club, diamond, heart, spade, crown and anchor. In addition, he needs three dice with their faces marked with the same symbols. The banker invites the onlookers to place their money on whichever of the squares they fancy. He then throws the three dice. Suppose you back a particular symbol. If it does not show on any of the dice, you lose. If it shows up on one of them, you get double your money back. If it shows up on two, you get three times your money back and if it shows on all three dice, you get back four times your money. (In other words, if your symbol shows at all, you get your stake back plus the same sum multiplied by the number of times it shows on the three dice.) If you place a pound on the crown square, what is the expected value of your bet?

We can only *guess* the probability that a horse will win a race. However, if the dice are fair, we can use symmetry to *calculate* the exact probability that a particular throw will appear. There are 6 ways in which the first die can be thrown, 6 ways in which the second die can be thrown, and 6 ways in which the third can be thrown, There are thus $6 \times 6 \times 6 = 216$ different throws. By symmetry, they must all have the same probability and, since the probabilities must add up to 1, this probability must be $1/216$.

How many throws are there of the form $*C*$ where $*$ means any throw not a crown and C means a crown? There are 5 ways of throwing the first die so that it is not a crown, 1 way of throwing the second so it is a crown and 5 ways of throwing the third so it is not a crown. Thus there are $5 \times 1 \times 5 = 25$ ways of obtaining $*C*$. We tabulate the various possibilities in Table 1.1. The information we need can now be summarised in Table 1.2. Since each possible throw has probability $1/216$, we can now see that the expected value of our bet is

$$125 \times \frac{1}{216} \times 0 + 75 \times \frac{1}{216} \times 2 + 15 \times \frac{1}{216} \times 3 + 1 \times \frac{1}{216} \times 4 = \frac{199}{216}.$$

---

[5] There is no difficulty in extending our definitions to include the possibility $p_j = 0$. If $p_j = 0$ we will not be prepared to bet on the horse at any odds.

[6] This example has stuck in my mind ever since I read it as a beginning student in the splendid book *Facts from Figures* [44].

Table 1.1. *Possibilities in Crown and Anchor*

Type	Number
$***$	$5 \times 5 \times 5 = 125$
$C**$	$1 \times 5 \times 5 = 25$
$*C*$	$5 \times 1 \times 5 = 25$
$**C$	$5 \times 5 \times 1 = 25$
$*CC$	$5 \times 1 \times 1 = 5$
$C*C$	$1 \times 5 \times 1 = 5$
$CC*$	$1 \times 1 \times 5 = 5$
CCC	$1 \times 1 \times 1 = 1$
Total	216

Table 1.2. *Payouts in Crown and Anchor*

Type	Number	Payout multiplier
no crown	125	0
1 crown	75	2
2 crowns	15	3
3 crowns	1	4

Thus we have spent 1 on a bet with expected value 199/216. By symmetry, the same is true whichever symbol we choose. Unsurprisingly, the advantage rests with the banker.

There is another way of obtaining the same result which is psychologically illuminating. Let us look at the expected value E of our bet to the banker. By symmetry, the value is the same whichever symbol we choose. Thus, if there are six players who each choose a different symbol and wager 1, the expected value of their combined bets to the banker is $6E$.

How many throws are there of the form abc where a, b and c are distinct? We can choose a in 6 ways, b in 5 ways (since it cannot be a) and c in 4 ways giving $6 \times 5 \times 4 = 120$ ways in all. We tabulate the various possibilities in Table 1.3. The information we need can now be summarised in Table 1.4. The banker's expected gain if 1 is placed on each square is

$$6E = 120 \times \frac{1}{216} \times 0 + 90 \times \frac{1}{216} \times 1 + 6 \times \frac{1}{216} \times 2 = \frac{102}{216}.$$

Thus $E = 17/216$. Since the banker's expected gain is our expected loss, we see again that the value of our bet is $1 - 17/216 = 199/216$.

Moroney writes with unrestrained enthusiasm.

Table 1.3. *The banker looks at Crown and Anchor*

Type	Number
abc	$6 \times 5 \times 4 = 120$
abb	$6 \times 5 \times 1 = 30$
aba	$6 \times 5 \times 1 = 30$
bba	$5 \times 1 \times 6 = 30$
aaa	$6 \times 1 \times 1 = 6$
Total	216

Table 1.4. *Payouts in Crown and Anchor*

Type	Number	Banker's gain
3 distinct	120	0
2 the same	90	1
3 the same	6	2

The game is beautifully designed. In over half the throws the banker sees nothing for himself. Whenever he makes a profit, he pays out more bountifully to other people, so that the losers' eyes turn to the lucky winner, rather than to the banker in suspicion. Spectacular wins [for the banker] are kept to the minimum, but when they do fall they are softened by apparent generosity. [[44], Chapter 7]

The game is not risk-free for the banker. If the board is not evenly covered, he may have to pay out more than he takes in. Apparently, the most popular bets are the crown and anchor themselves. If, in the first game, six people bet a pound each on the crown and the 1 in 216 chance of three crowns comes up, the banker will have to find 18 pounds at once. The game was particularly popular among British soldiers and sailors and they would not have dealt particularly kindly with a banker who failed to pay out. Expectations tell us a great deal about a game, but not everything that we wish to know.

Exercise 1.4.1 *(i) What happens if we keep the rules of Crown and Anchor except that we return 9/4 times the stake to anyone whose sign appears on exactly one die?*

(ii) What happens if we keep the rules of Crown and Anchor except that we return 10 times the stake when a symbol appears on all 3 dice?

(iii) *What happens (assuming that the players play sensibly) if we keep the rules of Crown and Anchor except that we return* 20 *times the stake when a heart appears on all* 3 *dice and* 5 *times the stake when any other symbol appears on all* 3 *dice?*

1.5 Back to the races

We have seen that we cannot expect a bookmaker to offer odds which enable us to make a guaranteed profit. However, if we know the correct odds and the bookmaker does not, we might be able to place a bet whose expected value is greater than the cost of making it. Is this possible, and if it is possible what is the best bet to make?

We shall suppose that there are n horses running, that we can only make and not take bets and that the bookmaker offers a payout multiplier of u_j on the jth horse. We shall suppose that

$$1 \geq \frac{1}{u_1} + \frac{1}{u_2} + \cdots + \frac{1}{u_n},$$

so that it is impossible to guarantee a strictly positive profit. We take the probability that the jth horse wins to be p_j.

If I bet y_1 on the first horse, y_2 on the second horse and so on, then I shall have paid the bookmaker

$$y_1 + y_2 + \cdots + y_n = Y$$

in return for a promise to pay $y_j u_j$ if the jth horse wins. The expected value of my bet is thus

$$p_1 y_1 u_1 + p_2 y_2 u_2 + \cdots + p_n y_n u_n.$$

My problem thus becomes one of choosing non-negative $y_1, y_2, \ldots, y_n$ in such a way that

$$y_1 + y_2 + \cdots + y_n = Y$$

and

$$p_1 u_1 y_1 + p_2 u_2 y_2 + \cdots + p_n u_n y_n$$

is as large as possible. We can simplify the problem slightly by setting $a_j = p_j u_j$ and seeking to maximise

$$a_1 y_1 + a_2 y_2 + \cdots + a_n y_n.$$

Since $p_j, u_j > 0$, we have $a_j > 0$. We also suppose $Y > 0$.

If we cannot solve this general problem, it makes sense to consider the simpler case when $n = 2$. We now have to choose $y_1 = s$ and $y_2 = Y - s$ with $0 \leq s \leq Y$ so as to make

$$a_1 s + a_2(Y - s)$$

as large as possible.

Exercise 1.5.1 *Let a_1, a_2, $Y > 0$. Sketch the graph of $f(s) = a_1 s + a_2(Y - s)$ as a function of s. (Your choice of a_1, a_2 and Y should not affect the general look of the graph.)*

Once we have looked at Exercise 1.5.1 the full answer is clear.

Lemma 1.5.2 *Suppose that $a_1 \geq a_2 \geq \cdots \geq a_n > 0$. Then, if $Y > 0$, $y_j \geq 0$ and $y_1 + y_2 + \cdots + y_n = Y$, we have*

$$a_1 Y \geq a_1 y_1 + a_2 y_2 + \cdots + a_n y_n.$$

Proof Observe that

$$a_1 Y - (a_1 y_1 + a_2 y_2 + \cdots + a_n y_n)$$
$$= (a_1 - a_2)y_2 + (a_1 - a_3)y_3 + \cdots + (a_1 - a_n)y_n \geq 0.$$

∎

It will turn out to be useful to say slightly more.

Lemma 1.5.3 *Suppose that*

$$a_1 = a_2 = \cdots = a_r > a_{r+1} \geq a_{r+2} \geq \cdots \geq a_n > 0.$$

Then, if $Y > 0$, $y_j \geq 0$ and $y_1 + y_2 + \cdots + y_n = Y$, we have

$$a_1 Y \geq a_1 y_1 + a_2 y_2 + \cdots + a_n y_n.$$

Moreover

$$a_1 Y = a_1 y_1 + a_2 y_2 + \cdots + a_n y_n$$

if and only if $y_{r+1} = y_{r+2} = \cdots = y_n = 0$.

Proof Left to the reader.   ∎

Thus we maximise the expected value of our bet by betting on the horse (or horses) for which $p_j u_j$ is a maximum and the expected value of a bet which costs Y will then be $Y \times \max_j p_j u_j$. Since we should only bet if the expected value of our bet is not less than its cost, we should not bet if $\max_j p_j u_j < 1$.

Exercise 1.5.4 *Suppose that*

$$a_1 \geq a_2 \geq \cdots \geq a_{r-1} > a_r = a_{r+1} = \cdots = a_{n-1} = a_n > 0.$$

Show that, if $Y < 0$, $y_j \leq 0$ and $y_1 + y_2 + \cdots + y_n = Y$, we have

$$a_n Y \leq a_1 y_1 + a_2 y_2 + \cdots + a_n y_n.$$

Moreover

$$a_n Y = a_1 y_1 + a_2 y_2 + \cdots + a_n y_n$$

if and only if $y_1 = y_2 = \cdots = y_{r-1} = 0$.

Exercise 1.5.5 *As usual, suppose that n horses are running. Suppose that an ideal bookmaker offers a payout multiplier of u_j on the jth horse and is prepared to take or make bets on each horse. We write $q_j = 1/u_j$ and recall that $q_1 + q_2 + \cdots + q_n = 1$. We take the probability that the jth horse wins to be p_j.*

(i) Show that if you decide to make *bets and wish to maximise the expected value of your bet, you should bet on the horse (or horses) such that p_j/q_j is a maximum.*

(ii) Show (using Exercise 1.5.4 if you need it) that, if you decide to take *bets and wish to maximise the expected value of your bet, you take a bet on the horse (or horses) such that q_j/p_j is a maximum.*

Thus, if faced with an ideal bookmaker, you should take or make a bet on the horse for which the true probability of winning differs most (in ratio) from the probability implied by the odds.

Now suppose that two ideal bookmakers A and B meet. A believes that the correct probability (as shown by her odds) that the jth horse will win is p_j [$1 \leq j \leq n$]. B believes that correct probability (as shown by her odds) that the jth horse will win is q_j. Without loss of generality, we may suppose that $\max_j p_j/q_j \geq \max_k q_k/p_k$ and $p_1/q_1 = \max_j p_j/q_j$. Bookmaker A will wish to bet with bookmaker B on the first horse (giving the maximum expectation if her probabilities are correct) and bookmaker B will wish to bet with bookmaker A against the first horse (giving the maximum expectation if her probabilities are correct). Thus A and B can bet with each other and both will believe that they have made the best bet possible! 'It is difference of opinion that makes horse races'[7] and the bigger the difference of opinion, the better the race.

[7] *Pudd'nhead Wilson*, Mark Twain.

Exercise 1.5.6 *Suppose that each bookmaker takes a proportion* $1 - \alpha$ *of each pound bet so that bookmaker A will either take bets on the jth horse with multiplier* αp_j^{-1} *or make bets on the jth with multiplier* $\alpha^{-1} p_j^{-1}$ *and bookmaker B will either take bets on the jth horse with multiplier* αq_j^{-1} *or make bets on the jth with multiplier* $\alpha^{-1} q_j^{-1}$.

Show that the bookmakers can bet with each other if

$$\max_j \frac{p_j}{q_j} \geq \frac{1}{\alpha^2} \text{ or } \max_j \frac{q_j}{p_j} \geq \frac{1}{\alpha^2}$$

but not otherwise.

Of course, no one knows the true probability of a horse winning a race. It is only worth betting if you believe that you know more about the horses involved (that is, can make a better guess at the true odds) than your opponent. If you have no confidence in your ability at judging horses, you should not be betting, and, if you do have confidence, you will have no difficulty in treating your guessed probabilities as true probabilities.

Exercise 1.5.7 *In the game of* High Dice *a player throws n dice (usually n = 5). She must then declare one and may declare more dice to be 'fixed'. If any dice remain unfixed, she then throws those dice and must then fix one or more dice. She continues throwing until all of the dice are fixed. The total shown on the dice after her final throw represents her score.*

In order to help us understand the game, we introduce a second game High Dice with Free Turn. *This is exactly the same as* High Dice *except that, after the first throw (and only then), the player can, if she chooses, throw all the dice again.*

(i) Show that the expected score in High Dice *with one die is* $7/2$.

(ii) Explain why, if we wish to maximise our expected score in High Dice with Free Turn *with one die, we should throw again if the die shows* 1, 2 *or* 3.

(iii) Show that, if we follow the prescribed tactics in (ii), our expected score is $17/4$.

Exercise 1.5.8 *(i) Now suppose we play* High Dice *with two dice. Show, using Exercise 1.5.7, that we should fix both dice if the lowest die shows* 4 *or more and throw the lowest die otherwise. Find our expected score with these tactics.*

(ii) How should we play if we wish to maximise our expected score in High Dice with Free Turn *with two dice?*

(iii) Explain how (given sufficient incentive) you would work out how to maximise our expected score in High Dice *with five dice.*

Exercise 1.5.9 *Suppose that we play* High Dice *with a very large number of dice. What tactics should we employ in the early turns and why? (An informal argument is sufficient.)*

Exercise 1.5.10 *Consider a game of* High Dice *with two dice between two players. The second player knows the final score of the first player before she begins her throws. The lower scorer pays the higher scorer €1. If both players have the same score, no money changes hands. What strategy should the second player follow?*

1.6 Betting last at the tote

So far, we have dealt with a single bet (possibly covering several horses) between two individuals. If many people wish to bet, there is another way of betting which, in England, is called 'tote betting' and, in the USA, 'parimutuel betting'. In its simplest form, all the money bet is placed in the care of some trustworthy individual or body and returned to those who have backed the winner in proportion to the bet placed. If there are n horses and a total sum s_j has been bet on the jth horse [$1 \leq j \leq n$], then, if the kth horse wins, someone who has bet y on that horse will get back

$$\frac{y}{s_k} S \text{ where } S = s_1 + s_2 + \cdots + s_n.$$

How should you bet if you are the *last* person to make a bet under such a system? We shall assume that there are n horses, that other members of the pool have placed t_j on the jth horse and that $t_j > 0$.

Exercise 1.6.1 *What should you do if $t_k = 0$ for some k?*

Suppose that you bet y_j on the jth horse. Then anyone who has placed z on the kth horse will get back

$$\frac{z}{t_k + y_k}(T + Y) \text{ where } T = t_1 + t_2 + \cdots + t_n \text{ and } Y = y_1 + y_2 + \cdots + y_n.$$

Thus the payout multipliers u_j will be given by

$$u_j = \frac{T + Y}{t_j + y_j}$$

and, since

$$\frac{1}{u_1} + \frac{1}{u_2} + \cdots + \frac{1}{u_n} = \frac{t_1 + y_1}{T + Y} + \frac{t_2 + y_2}{T + Y} + \cdots + \frac{t_n + y_n}{T + Y} = 1,$$

they correspond to the payout multipliers given by an ideal bookmaker. We could say that the $q_j = 1/u_j$ are 'the probabilities implied by the weight of money bet', but there is no need to take this sort of statement too seriously. More important, from our point of view, is that changing y_j changes the payout multipliers. The act of betting changes the odds we receive!

This does not matter if we bet only a small amount. Since this will affect the odds only very slightly, we can use the rules established earlier for fixed odds. If we wish to maximise the expected value of our bet, then those rules tell us to bet on the horse or horses such that the probability of it winning multiplied by the payout multiplier is greatest.

We thus know what to do if we bet a *small* amount *last*. How should we bet a *large* amount *last*? Since the mathematics involved now becomes more complicated, we shall consider the case when there are only two horses. (If the reader does Exercise 1.7.8, which deals with the general case of n horses, she can check that, although the calculations are more intricate, the general picture remains the same.)

We start with a couple of algebraic simplifications.

Lemma 1.6.2 *Suppose t_1, t_2, p_1, p_2, $Y > 0$, $t_1 + t_2 = T$ and we wish to maximise*

$$\frac{T+Y}{t_1+y_1}p_1y_1 + \frac{T+Y}{t_2+y_2}p_2y_2$$

subject to the conditions $y_1 + y_2 = Y$ and y_1, $y_2 \geq 0$.

(i) Our problem is equivalent to maximising

$$\frac{p_1y_1}{t_1+y_1} + \frac{p_2y_2}{t_2+y_2}$$

under the same conditions.

(ii) Our problem is equivalent to minimising

$$\frac{p_1t_1}{t_1+y_1} + \frac{p_2t_2}{t_2+y_2}$$

under the same conditions.

Proof (i) Observe that $T + Y$ is constant.

(ii) Observe that

$$\left(\frac{p_1y_1}{t_1+y_1} + \frac{p_2y_2}{t_2+y_2}\right) + \left(\frac{p_1t_1}{t_1+y_1} + \frac{p_2t_2}{t_2+y_2}\right) = p_1 + p_2.$$

∎

(Of course, in a two-horse race, $p_1 + p_2 = 1$, but we do not need this fact.)

Theorem 1.6.3 *Suppose t_1, t_2, p_1, p_2, $Y > 0$, $t_1 + t_2 = T$, $p_1/t_1 \geq p_2/t_2$ and we wish to maximise*

$$\frac{T + Y}{t_1 + y_1} p_1 y_1 + \frac{T + Y}{t_2 + y_2} p_2 y_2$$

subject to the conditions $y_1 + y_2 = Y$ and y_1, $y_2 \geq 0$.
If $\dfrac{p_1 t_1}{(t_1 + Y)^2} \geq \dfrac{p_2}{t_2}$, then the maximum occurs when $y_1 = Y$ and $y_2 = 0$.
If not, then the maximum occurs when

$$\frac{p_1 t_1}{(t_1 + y_1)^2} = \frac{p_2 t_2}{(t_2 + y_2)^2}.$$

Proof By Lemma 1.6.2 we need to minimise

$$\frac{p_1 t_1}{t_1 + y_1} + \frac{p_2 t_2}{t_2 + y_2}$$

subject to $y_1 + y_2 = Y$ and y_1, $y_2 \geq 0$. If we set $y = y_2$, our problem becomes one of minimising

$$f(y) = \frac{p_1 t_1}{t_1 + Y - y} + \frac{p_2 t_2}{t_2 + y}$$

subject to $Y \geq y \geq 0$. We can use the methods of the calculus.
Observe that

$$f'(y) = \frac{p_1 t_1}{(t_1 + Y - y)^2} - \frac{p_2 t_2}{(t_2 + y)^2} \quad \text{and}$$

$$f''(y) = \frac{2 p_1 t_1}{(t_1 + Y - y)^3} + \frac{2 p_2 t_2}{(t_2 + y)^3}.$$

Since $f''(y) > 0$ for $y \geq 0$, we know that $f'(y)$ is a strictly increasing[8] function of y. If

$$\frac{p_1 t_1}{(t_1 + Y)^2} \geq \frac{p_2}{t_2},$$

then $f'(0) \geq 0$, so $f'(y) > 0$ for all $y > 0$ and $f(y)$ is a strictly increasing function of y. Thus $f(y)$ attains its minimum at $y = 0$.
  If

$$\frac{p_1 t_1}{(t_1 + Y)^2} < \frac{p_2}{t_2},$$

---

[8] Recall that a function g is *increasing* if $g(x) \geq g(y)$ whenever $x \geq y$ and *strictly increasing* if $g(x) > g(y)$ whenever $x > y$.

then $f'(0) < 0$. We also know that

$$f'(Y) = \frac{p_1 t_1}{t_1^2} - \frac{p_2 t_2}{(t_2 + Y)^2}$$
$$> \frac{p_1 t_1}{t_1^2} - \frac{p_2 t_2}{t_2^2}$$
$$= \frac{p_1}{t_1} - \frac{p_2}{t_2} \geq 0.$$

Since f' is strictly increasing and $f'(0) < 0 < f'(Y)$, there must be a unique y^* with $0 < y^* < Y$ and $f'(y^*) = 0$. It must be a minimum for f. If we translate back into terms of y_1 and y_2, we see that we have proved our theorem.

∎

We asked how to bet last on a tote on a two-horse race in which there is a probability p_j of the jth horse winning and t_j has already been staked on that horse. We suppose $p_1/t_1 \geq p_2/t_2$. Theorem 1.6.3 tells us, as we already know, that when Y is small we should bet on the first horse. However, when Y is large we should split our bet, placing y_1 on the first horse and y_2 on the second in such a way that

$$\frac{p_1 t_1}{(t_1 + y_1)^2} = \frac{p_2 t_2}{(t_2 + y_2)^2}.$$

Let us look more closely at the case when Y is large. We observe that, using the previous formula,

$$\frac{t_1 + y_1}{t_2 + y_2} = \frac{(p_1 t_1)^{1/2}}{(p_2 t_2)^{1/2}}.$$

After we have bet, the payoff multipliers are given by

$$u_1 = \frac{T + Y}{t_1 + y_1} \quad \text{and} \quad u_2 = \frac{T + Y}{t_2 + y_2}$$

and so

$$\frac{u_1}{u_2} = \frac{t_2 + y_2}{t_1 + y_1} = \frac{(p_2 t_2)^{1/2}}{(p_1 t_1)^{1/2}}.$$

Since we know that

$$\frac{1}{u_1} + \frac{1}{u_2} = 1,$$

it follows that

$$\frac{1}{u_1} = \frac{(p_1 t_1)^{1/2}}{(p_1 t_1)^{1/2} + (p_2 t_2)^{1/2}} \quad \text{and} \quad \frac{1}{u_2} = \frac{(p_2 t_2)^{1/2}}{(p_1 t_1)^{1/2} + (p_2 t_2)^{1/2}}.$$

The expected value E of the total bet placed by other bettors is thus

$$p_1t_1u_1 + p_2t_2u_2 = p_1t_1\frac{(p_1t_1)^{1/2} + (p_2t_2)^{1/2}}{(p_1t_1)^{1/2}} + p_2t_2\frac{(p_1t_1)^{1/2} + (p_2t_2)^{1/2}}{(p_2t_2)^{1/2}}$$
$$= (p_1t_1)^{1/2}\left((p_1t_1)^{1/2} + (p_2t_2)^{1/2}\right) + (p_2t_2)^{1/2}\left((p_1t_1)^{1/2} + (p_2t_2)^{1/2}\right)$$
$$= \left((p_1t_1)^{1/2} + (p_2t_2)^{1/2}\right)^2.$$

Since our idealised tote pays back everything that is bet, the expected value of our bet is $T + Y - E$ and our expected profit (that is the expected value of our bet minus its cost) is

$$T - E = (t_1 + t_2) - \left((p_1t_1)^{1/2} + (p_2t_2)^{1/2}\right)^2.$$

Exercise 1.6.4 *Check the algebra above.*

Once Y is so large that we bet on both horses, our expected profit remains the same however much we bet. When we make a small bet, we are betting against the other bettors, but, if we make a bet which is so large that it forms a substantial part of the total bet, we are, in effect, betting against ourselves, a pastime which is unlikely to prove profitable. The following exercises reinforce the moral.

Exercise 1.6.5 *Consider betting on a tote in a two-horse race. We give p_j, t_j and T their usual meanings. We assume that $p_1/t_1 \geq p_2/t_2$.*

(i) Show that if we bet y on the first horse and nothing on the second, our expected profit is

$$f(y) = T - \left(\frac{p_1t_1(T + y)}{t_1 + y} + p_2(T + y)\right).$$

(ii) Show that

$$f(y) = p_1t_2 - p_2y - \frac{p_1t_1t_2}{t_1 + y}.$$

By considering the first and second derivatives, show that $f(y)$ increases as y increases from 0 to Y_0 where

$$\frac{p_1t_1}{(t_1 + Y_0)^2} = \frac{p_2}{t_2},$$

but that the rate of increase $f'(y)$ of f decreases as y increases.

(iii) Show that, when $y > Y_0$, $f(y)$ decreases as y increases. Show also that $f(y) \to -\infty$ as $y \to \infty$.

(iv) Explain the meaning of results (ii) and (iii) to a non-mathematical bettor. Explain also why we might expect them to be true.

Exercise 1.6.6 *We consider our usual two-horse race. Suppose that, instead of one person betting y last at the tote, a large number of people (who agree on the value of p_1 and p_2) bet small sums, one after another, following the rule 'bet on the horse such that the probability of a winning multiplied by the payout multiplier is greatest'.*

(i) What will happen as the total y that they bet increases? (Argue informally.)

(ii) Show that, when y is large, unless a particular condition holds, the total amount bet on each horse by the group will be different from that bet by a single person following our advice.

(iii) Let y_1 and y_2 be the totals bet on each horse under the circumstances just sketched.

Show that the expected total gain of the new bettors remains the same once $y_2 > 0$.

(iv) (This just repeats earlier calculations.) Suppose that the new bettors form a syndicate and bet together following our advice. Suppose they bet z_1 and z_2 on the two horses.

Show that the expected total gain of the new bettors remains the same once $z_2 > 0$ but, unless a particular condition holds, is higher than the expected gain in (iii).

(v) Explain what is happening in (iii) and (iv) to a non-mathematical bettor.

Here is another way of looking at the matter. The reader should certainly do this exercise.[9] The calculations are very similar to those used in the proof of Theorem 1.6.3 and the discussion which followed.

Exercise 1.6.7 *(i) Let A, B > 0 and set*

$$f(q) = \frac{A}{q} + \frac{B}{1-q}.$$

Show that $f(q)$ has a unique minimum with $0 < q < 1$ and find the value of f at that point.

(ii) In a two-horse race, the first horse has probability p_1 of winning and the second p_2. A (rather naive) bettor announces his intention of placing t_1 on the first horse and t_2 on the second. He then asks a bookmaker to name appropriate multipliers u_1 and u_2 such that u_1, $u_2 > 0$ and

$$\frac{1}{u_1} + \frac{1}{u_2} = 1.$$

---

[9] If you cannot do it, remember that sketch solutions to most of the exercises can be found at the internet address given in the Introduction.

Show that, if the bookmaker wishes to minimise her expected payout, she should choose

$$\frac{1}{u_1} = \frac{(p_1 t_1)^{1/2}}{(p_1 t_1)^{1/2} + (p_2 t_2)^{1/2}} \quad and \quad \frac{1}{u_2} = \frac{(p_2 t_2)^{1/2}}{(p_1 t_1)^{1/2} + (p_2 t_2)^{1/2}}.$$

Thus, if we are betting a total of Y and $Y \geq Y_0$ (with Y_0 as in Exercise 1.6.5), we can bet y_1 and y_2 on the two horses in such a way as to ensure the worst possible payout multipliers for the total bet placed by the other bettors. Increasing Y does not enable us to make things any worse.

1.7 Betting first

The discussion of the previous section depended on our being able to bet last at the tote. If someone manages to bet later, they can exploit our choices in the same way as we exploited the choices of earlier bettors.

What should we do if we cannot bet last?

At first sight, this seems an impossible question. The bets of those who bet after us will change the payout ratios in ways that we cannot control.

However,we can extract one very useful fact from the discussion, in the previous section, of what happens if we make the last bet.

Lemma 1.7.1 *Suppose p_1, $p_2 > 0$, $p_1 + p_2 = 1$ and $t_1 + t_2 = T$, $p_1/t_1 = p_2/t_2$. Then, if u_1, $u_2 > 0$ and $u_1^{-1} + u_2^{-1} = 1$, it follows that*

$$p_1 t_1 u_1 + p_2 t_2 u_2 \geq T.$$

Proof Observe first that the conditions of the first sentence of the lemma tell us that $t_1 = p_1 T$ and $t_2 = p_2 T$

By Exercise 1.6.7, the minimum value of

$$p_1 t_1 u_1 + p_2 t_2 u_2$$

occurs when

$$\frac{1}{u_1} = \frac{(p_1 t_1)^{1/2}}{(p_1 t_1)^{1/2} + (p_2 t_2)^{1/2}} \quad and \quad \frac{1}{u_2} = \frac{(p_2 t_2)^{1/2}}{(p_1 t_1)^{1/2} + (p_2 t_2)^{1/2}}.$$

For these values of u_1 and u_2,

$$\begin{aligned} p_1 t_1 u_1 + p_2 t_2 u_2 &= p_1 t_1 \frac{(p_1 t_1)^{1/2} + (p_2 t_2)^{1/2}}{(p_1 t_1)^{1/2}} + p_2 t_2 \frac{(p_1 t_1)^{1/2} + (p_2 t_2)^{1/2}}{(p_2 t_2)^{1/2}} \\ &= \left((p_1 t_1)^{1/2} + (p_2 t_2)^{1/2}\right)^2 \\ &= \left(p_1 T^{1/2} + p_2 T^{1/2}\right)^2 = \left(T^{1/2}\right)^2 = T. \end{aligned}$$

We have shown that the minimum value of $p_1t_1u_1 + p_2t_2u_2$ is T and so proved our lemma.   ∎

Thus, if we bet t_1 on the first horse and t_2 on the second, in such a way that $p_1/t_1 = p_2/t_2$ (that is, if we take $t_1 = p_1T$ and $t_2 = p_2T$ where T is our total bet), then, no matter what payout multipliers we are given (by an ideal bookmaker), the expected value of our bet will be at least as large as its cost. Moreover, unless payout multipliers take particular values, the expected value of our bet will be greater than its cost.

Exercise 1.7.2 *Suppose that* p_1, $p_2 > 0$, $p_1 + p_2 = 1$ *and* $t_1 + t_2 = T$, $p_1/t_1 = p_2/t_2$. *Show that, if* u_1, $u_2 > 0$ *and* $u_1^{-1} + u_2^{-1} = 1$,

$$p_1t_1u_1 + p_2t_2u_2 > T$$

unless $u_1 = 1/p_1$ *and* $u_2 = 1/p_2$.

Since, no matter what the other bettors do, the tote payout multipliers u_1 and u_2 must satisfy the condition $u_1^{-1} + u_2^{-1} = 1$, it follows that, if we bet t_1 on the first horse and t_2 on the second, in such a way that $p_1/t_1 = p_2/t_2$, then the expected value of our bet will be at least as large as its cost. Moreover, unless the other bettors place their bets in a particular manner, the expected value of our bet is strictly greater than its cost.

Exercise 1.7.3 *Consider a tote on a two-horse race in which the jth horse has probability* p_j *of winning. If* $T > 0$ *and we bet* $t_j = Tp_j$ *while the other bettors bet* y_j *on the jth horse, show that the expected value of our bet is strictly greater than* T *unless* $y_j = Yp_j$ *for some* Y.

It is not hard to extend the results of this section to a race with n horses.

Lemma 1.7.4 *(i) If* A, B, x, y, $c > 0$ *and* $x + y = c$, *then*

$$\frac{A}{x} + \frac{B}{y} \geq \frac{\left(A^{1/2} + B^{1/2}\right)^2}{c}.$$

If

$$\frac{A}{x} + \frac{B}{y} = \frac{\left(A^{1/2} + B^{1/2}\right)^2}{c},$$

then

$$\frac{x}{y} = \frac{A^{1/2}}{B^{1/2}}.$$

*(ii) Suppose that $A_j > 0$, $w_j > 0$ $[1 \leq j \leq n]$ and $W = w_1+w_2+\cdots+w_n$.
If*

$$\frac{w_i}{w_k} \neq \frac{A_i^{1/2}}{A_k^{1/2}}$$

*for some i and k with $1 \leq i < k \leq n$, then we can find $w_j^* > 0$, $[1 \leq j \leq n]$
with $W = w_1^* + w_2^* + \cdots + w_n^*$ such that*

$$\frac{A_1}{w_1} + \frac{A_2}{w_2} + \cdots + \frac{A_n}{w_n} > \frac{A_1}{w_1^*} + \frac{A_2}{w_2^*} + \cdots + \frac{A_n}{w_n^*}.$$

(iii) If $A_j > 0$, $w_j > 0$, $[1 \leq j \leq n]$ and $W = w_1 + w_2 + \cdots + w_n$, then

$$\frac{A_1}{w_1} + \frac{A_2}{w_2} + \cdots + \frac{A_n}{w_n} \geq \frac{(A_1^{1/2} + A_2^{1/2} + \cdots + A_n^{1/2})^2}{W}$$

with equality if and only if

$$w_j = \frac{A_j^{1/2} W}{A_1^{1/2} + A_2^{1/2} + \cdots + A_n^{1/2}}.$$

*(iv) Suppose that p_1, p_2, $\ldots$, $p_n > 0$, $p_1 + p_2 + \cdots + p_n = 1$, $T > 0$
and $t_1 = p_1 T$, $t_2 = p_2 T$, $\ldots$, $t_n = p_n T$. Then, if u_1, $u_2 \ldots u_n > 0$ and
$u_1^{-1} + u_2^{-1} + \cdots + u_n^{-1} = 1$,*

$$p_1 t_1 u_1 + p_2 t_2 u_2 + \cdots + p_n t_n u_n \geq T$$

with equality only if $u_j = p_j^{-1}$ for all $1 \leq j \leq n$.

Proof (i) The argument follows a familiar form. Set

$$f(x) = \frac{A}{x} + \frac{B}{c - x}.$$

Then

$$f'(x) = -\frac{A}{x^2} + \frac{B}{(c - x)^2} = \frac{Bx^2 - A(c - x)^2}{x^2(c - x)^2}$$

so, if $0 < x < c$, it follows that $f'(x) < 0$ for $Bx^2 < A(c - x)^2$ and
$f'(x) > 0$ for $Bx^2 > A(c - x)^2$. Thus f has a unique minimum for the range
considered, which is attained when $Bx^2 = A(c - x)^2$, that is to say, when
$B^{1/2}x = A^{1/2}(c - x)$. The required result follows.
  (ii) Without loss of generality, we may suppose $i = 1$ and $k = 2$, so

$$\frac{w_1}{w_2} \neq \frac{A_1^{1/2}}{A_2^{1/2}}.$$

If we set $A = A_1$, $B = A_2$, $x = w_1$, $y = w_2$, $c = w_1 + w_2$ and

$$w_1^* = \frac{A^{1/2}c}{A^{1/2} + B^{1/2}}, \quad w_2^* = \frac{B^{1/2}c}{A^{1/2} + B^{1/2}},$$

then w_1^*, $w_2^* > 0$, $w_1^* + w_2^* = c = w_1 + w_2$ and, by part (ii),

$$\frac{A_1}{w_1} + \frac{A_2}{w_2} > \frac{A_1}{w_1^*} + \frac{A_2}{w_2^*}.$$

If we now set $w_j^* = w_j$ for $j \geq 3$, the stated result follows.

(iii) Part (ii) shows that, if the quantity

$$\frac{A_1}{w_1} + \frac{A_2}{w_2} + \cdots + \frac{A_n}{w_n}$$

(with $W = w_1 + w_2 + \cdots + w_n$) is minimised by taking $w_j = w_j^*$, then

$$\frac{w_i^*}{w_k^*} = \frac{A_i^{1/2}}{A_k^{1/2}}$$

for all i and k and so

$$w_j^* = \frac{A_j^{1/2}W}{A_1^{1/2} + A_2^{1/2} + \cdots + A_n^{1/2}}.$$

Since a minimum must exist,[10] it is given by $w_j = w_j^*$ and we have

$$\frac{A_1}{w_1} + \frac{A_2}{w_2} + \cdots + \frac{A_n}{w_n} \geq \frac{A_1}{w_1^*} + \frac{A_2}{w_2^*} + \cdots + \frac{A_n}{w_n^*}$$
$$= \frac{\left(A_1^{1/2} + A_2^{1/2} + \cdots + A_n^{1/2}\right)^2}{W}$$

with equality if and only if $w_j = w_j^*$. This is the stated result.

(iv) Set $A_j = p_j^2 T$, $w_j = u_j^{-1}$ and apply part (iii). ∎

Thus, if we have to bet on an n-horse race and we do not know what the final odds will be, we should divide our bet T so that we bet $p_j T$ on the jth horse where p_j is our estimate of the true probability of that horse winning.

Exercise 1.7.5 *Explain why this follows from Lemma 1.7.4 if the payout multipliers u_j satisfy*

$$u_1^{-1} + u_2^{-1} + \cdots + u_n^{-1} = 1.$$

Explain why the advice holds good even if the equation is not satisfied.

---

[10] In very advanced mathematics we come across cases when such statements are doubtful, but this is not one of those cases. We shall pass over this point in silence when it occurs again.

If we have to bet early in a tote, without knowing the final odds, we should follow the advice in the previous paragraph.

Exercise 1.7.6 *(This exercise repeats results already obtained and should not require much work from the reader.) A bookmaker and a bettor both know that the probability of the jth horse winning an n-horse race is p_j. The bettor has announced her fixed intention of choosing her bets so that, if t_j is her bet on the jth horse,*

$$t_1 + t_2 + \cdots + t_n = T$$

and the bookmaker has announced her fixed intention of choosing her payout ratios so that, if u_j is her payout ratio on the jth horse,

$$u_1^{-1} + u_2^{-1} + \cdots + u_n^{-1} = K.$$

Naturally the bettor wishes to maximise the expected value of her bet and the bookmaker to minimise it.

(i) Show that if the bettor announces her bets before the bookmaker names her payout ratios, she should choose $t_j = p_j T$.

(ii) Show that if the bookmaker names her payout ratios before the bettor names her bets, she should choose $u_j = p_j^{-1} K$.

(iii) Show that in both cases, if bookmaker and bettor act wisely, the result will be the same and the expected value of the bet to the bettor is $T K$.
[The results of this exercise echo the view of W. C. Fields that 'You can't cheat an honest man'.]

Exercise 1.7.7 *(i) In a real tote the organisers take some money from the pot (that is to say, the total sum bet) to compensate them for their trouble.*[11]

The natural way of arranging things is for the organisers to take a fixed proportion $1 - \alpha$ of the pot. If there are n horses and a total sum s_j has been bet on the jth horse $[1 \le j \le n]$, then if the kth horse wins, someone who has bet y on that horse will get back

$$\frac{y}{s_k} \alpha S \text{ where } S = s_1 + s_2 + \cdots + s_n.$$

Show that, if you wish to bet Y and maximise the expected value of your bet, then our recommendations under the various conditions ('a small sum last', 'a large sum last', 'any sum first') are unaltered.

---

[11] The tote system was invented by Pierre Oller in 1865 when a bookmaker friend asked him produce a system which would be fair to bettors while guaranteeing a profit for the organiser.

Now suppose that you are betting last. Show that, under certain circumstances, to be stated, there is no small bet with a positive expected profit.[12] *Show that under all circumstances there is a Y_0 such that there is no value of Y with $Y > Y_0$ such that we can bet Y with a positive expected profit.*

(ii) Another way of organising things (which will only be attractive to bettors if they can be confident that other people will wager large sums) is for the organisers to pay themselves a fixed sum b. If there are n horses and a total sum s_j has been bet on the jth horse $[1 \le j \le n]$, then if the kth horse wins, someone who has bet y on that horse will get back

$$\frac{y}{s_k}(S - b) \text{ where } S = s_1 + s_2 + \cdots + s_n.$$

What should you do if you are betting a small sum last? What should you do if you are betting first?

Exercise 1.7.8 *(This exercise fulfils our promise to extend the treatment of betting last on the tote to races with n horses. Do this exercise only if you are interested.)*

(i) Use Theorem 1.6.3 to show that, if t_1, t_2, p_1, p_2, $Y > 0$ and $p_1/t_1 \ge p_1/t_2$, then taking $y_1 = y_1^*$, $y_2 = y_2^*$ maximises

$$\frac{p_1 y_1}{t_1 + y_1} + \frac{p_2 y_2}{t_2 + y_2},$$

for y_1, $y_2 \ge 0$ and $y_1 + y_2 = Y$, if and only if either

$$\frac{p_1 t_1}{(t_1 + Y)^2} \ge \frac{p_2}{t_2} \text{ and } y_1^* = Y, \ y_2^* = 0,$$

or

$$\frac{p_1 t_1}{(t_1 + Y)^2} < \frac{p_2}{t_2} \text{ and } \frac{p_1 t_1}{(t_1 + y_1^*)^2} = \frac{p_2 t_2}{(t_2 + y_2^*)^2}.$$

(ii) Suppose that t_j, $p_j > 0$ for $1 \le j \le n$ and $Y > 0$. Suppose further that

$$\frac{p_1}{t_1} \ge \frac{p_2}{t_2} \ge \cdots \ge \frac{p_n}{t_n}.$$

Show that, if $y_j = y_j^*$ maximises

$$\frac{p_1 y_1}{t_1 + y_1} + \frac{p_2 y_2}{t_2 + y_2} + \cdots + \frac{p_n y_n}{t_n + y_n},$$

---

[12] So, since small bets maximise the ratio of expected gain to amount bet, there is no bet with a positive expected profit.

for $y_j \geq 0$ and $y_1 + y_2 + \cdots + y_n = Y$, then, whenever $1 \leq j < k \leq n$, we will have either

$$\frac{p_j t_j}{(t_j + y_j^*)^2} \geq \frac{p_k}{t_k} \quad and \quad y_k^* = 0,$$

or

$$\frac{p_j t_j}{(t_j + y_j^* + y_k^*)^2} < \frac{p_k}{t_k} \quad and \quad \frac{p_j t_j}{(t_j + y_j^*)^2} = \frac{p_k t_k}{(t_k + y_k^*)^2}.$$

(iii) Show that, under the hypotheses of (ii), we can find an r with $1 \leq r \leq n$ such that

$$\frac{p_1 t_1}{(t_1 + y_1^*)^2} = \frac{p_2 t_2}{(t_2 + y_2^*)^2} = \cdots = \frac{p_r t_r}{(t_r + y_r^*)^2}$$

and, if $r \leq n - 1$,

$$y_{r+1}^* = y_{r+2}^* = \cdots = y_n^* = 0, \quad and \quad \frac{p_r t_r}{(t_r + y_r^*)^2} \geq \frac{p_{r+1}}{t_{r+1}}.$$

(iv) Use this result to describe the strategy for betting last in a tote on an n-horse race so as to maximise the expected profit. Show that there is no point in betting more than a certain amount.

1.8 Real race-tracks

There is a risk-free way of making money out of betting on horse races. It is to arrange that the bettors bet with each other through you, allowing you to take a proportion of each bet. This is the principle of tote betting and of the modern system of 'exchange betting'.

Bookmaking is not risk-free. If the sum that the bookmaker has to pay out on the winning horse exceeds the total bet on all the other horses, then the bookmaker will make a loss. In order to avoid this, the bookmaker will try to make her payout ratios as small as possible but, if her payout ratios are too low, no one will bet with her. Sometimes bookmakers can arrange things (possibly by laying bets with other bookmakers) so that they cannot lose. Usually they can arrange things so that their expected profit on a particular race is positive but, like the banker in Crown and Anchor, they run the risk of a loss.[13] Occasionally, no doubt, their expected profit on a particular race will be negative but, provided this does not happen very often, their trade will remain profitable.

[13] The improbable is merely improbable, not impossible. In 1997, Frankie Dettori rode all seven winners at Ascot. British bookmakers claim to have lost forty million pounds that day.

The payout ratios offered by bookmakers on a particular race change with time. If a lot of money is bet on a particular horse, then a bookmaker will reduce the payout ratio on that horse. Since bookmakers are in competition for the money of the bettors, if one bookmaker raises the payout ratio on a horse, other bookmakers will tend to do the same. Since any opportunity for arbitrage will be seized on by other bookmakers, the payout ratios will move more or less in line.

Suppose that, at some time before the race, a bookmaker offers a payout ratio on the jth horse of u_j and bettors believe that the probability that the jth horse winning is p_j $[1 \leq j \leq n]$. Our discussion suggests that the bettors should place their money on the horse k for which $u_j p_j$ is largest. The bookmaker will now lower the value of u_k so that $u_k p_k$ becomes smaller. If little money is placed on horse l because bettors believe that $u_l p_l$ is too small, then the bookmaker will raise u_l and so $u_l p_l$.

Let u_j^* be the payout ratio on the jth horse when the betting ends and let p_j^* be the true probability that the j horse will win. It is plausible that, as a result of the processes described in the previous paragraphs, all the $u_j^* p_j^*$ will be roughly equal and we will have

$$u_j^* p_j^* \approx \alpha$$

for all j. Here, as before, $1 - \alpha$ represents something like 'the fraction of each bet that the bookmakers hope to take to cover expenses and profit'. Our discussion of tote betting suggests that much the same will happen in the tote.

This observation, if correct, has important consequences. Ignorant bettors should either bet on the tote or bet just before the race. If they bet a total of Y, placing y_j on the jth horse then the expected value of their bet will be

$$y_1 u_1^* p_1^* + y_2 u_2^* p_2^* + \cdots + y_n u_n^* p_n^* = \alpha Y$$

and will not depend on the choice of y_j. Because the payout ratios reflect the true probabilities, ignorance is no disadvantage. If they follow this strategy the expected value of their bet will be less than its cost but only by the 'bookmaker's percentage' and they will avoid being exploited by more knowledgeable bettors.

What about knowledgeable bettors? They wish to bet when $u_j p_j$ is substantially larger than 1. They will not bet with the tote and will bet early with bookmakers. If u_j is small, it is likely that $u_j p_j$ will be small, so we may expect that most of their bets will be on horses with large u_j (outsiders). Finally, since it is part of the business of bookmakers to gather information on horses, even knowledgeable bettors will only rarely discover cases when $u_j p_j$

is substantially larger than 1. Thus the knowledgeable bettor will only bet on a few races, will mainly bet on long shots and will bet early.

So much for theory. What about practice? Of course, we can never know p_j^*. However, economists have looked at records of past races and worked out how bettors would have fared by following particular strategies.[14] Their most surprising finding, from our point of view, is that a bettor who consistently bet at final odds on strong favourites (that is horses with u_j^* very close to 1) would have lost very slowly indeed, while a bettor who consistently bet at final odds on long shots (that is horses with large u_j^*) would have lost money very rapidly.

Speaking very roughly, it would appear, at least historically, that if $u_j^* \leq 2$ the expected value $u_j^* p_j^*$ of a unit bet 'at starting odds' on the jth horse will be very close to the cost of the bet. However, if $u_j^* \geq 25$, the expected value of a bet will be less than half its original cost. We may conjecture that competition and greater knowledge force bookmakers to provide odds on favourites which are very close to the 'true odds'. It certainly appears that many bettors on long shots grossly overestimate the chance of their horses winning.[15]

What is true in the past may not continue to be true in the future,[16] but the evidence is strong enough to modify our advice to ignorant bettors. Not only should they bet as late as possible, but they should only bet on horses with $u_j^* \leq 2$. Our advice to knowledgeable bettors will remain the same, but we observe that the phenomenon described in the previous two paragraphs means that it will be even harder to find cases where $p_j u_j$ is large.

Bookmakers make money out of betting. Those who have knowledge about the merits of the horses which is not available to the general public make money out of betting. Is it possible for bettors with the same information as the general public to use the fact that not all the public bet wisely to make money out of betting? Possibly, but I would not bet on it.

[14] The book [71] discusses the results of these investigations in non-technical terms and provides references to the original literature.

[15] Even more oddly, this pattern does not appear in the betting on certain other sports. We discuss another reason why bookmakers may make 'long odds' bets more expensive when we talk about high variance bets on page 299.

[16] Two economists are walking down the street and see a $100 note in the gutter. The junior economist stoops down to pick it up but the senior economist stops him. 'If that note was worth anything someone else would already have picked it up.'

2

The long run

2.1 The laws of probability

The reader will have noted that, in the previous chapter, we moved uneasily between 'the true probability p that a certain horse will win a race', 'the number p that you believe is the probability that a certain horse will win a race' and 'the number p which the odds given imply is the probability that a certain horse will win a race'. The reader will also have observed that, even if there is such a thing as 'the true probability p that a certain horse will win a race', we have no means of measuring it. When the race is run, either the horse will win or it will not. Anyone tempted to mutter 'the probability p is the long run proportion of identical races that the horse will win' should reflect on the impossibility of arranging even two nearly identical races.[1] (If we could arrange two *identical* races the same horse would win both of them.)

Does this mean that we should dismiss the whole discussion as rubbish? That would be a pity. Even if the fundamental concepts used in the first chapter appear hazy on close inspection, they do seem to tell us something about how to bet. Experience seems to show that those who have some knowledge of probability, like bookmakers and casino owners, tend to do better than those, like some of their clients, who do not. Finally, if the reader dismisses the idea of probability out of hand, what alternative mode of dealing with uncertainty does she propose?

Fortunately, mathematicians have already met this difficulty in less controversial circumstances. Consider Newtonian mechanics. In the form that we first meet it, it concerns point masses which move under forces according to certain mathematical laws. To 'find the orbit of the earth round the sun' we consider a formula which gives 'the force due to a mass at the origin acting on another

---

[1] 'No man ever steps into the same river twice, for it is not the same river and he is not the same man.' (Paraphrase of a fragment attributed to Heraclitus.)

mass at a given point'. We then solve the resulting differential equations and find the resulting path of the moving particle. It turns out that, if we fix certain constants in the solution, the motion of our particle round the fixed point resembles the orbit of the earth round the sun.

We know that the sun is not a point mass but a large ball of gas and that it is not fixed in space. We know that the point mass earth must be replaced by a complex system in which the earth and moon conduct an intricate dance round each other. It is up to the physicist to judge what approximations are reasonable and, even more important, to decide whether the agreement between the model and reality is acceptable.

Thus we have a purely mathematical theory called Newtonian mechanics which can be studied for its own sake without considering whether or how it represents anything in the real world. The relevance or otherwise to the real world can be left as a separate study.

In the same way, we can set up certain rules for what we call 'probability theory' and work out their consequences. It is the separate job of statisticians to attempt to apply 'probability theory' to the real world and that of philosophers to worry why such an application can be made.

In this section we set out the rules for a simple theory of probability. We start with a finite non-empty set Ω called the *probability space*,[2] event space or sample space.[3] If we consider a horse race with n horses, then we might take

$$\Omega = \{\omega_1, \omega_2, \ldots \omega_n\}$$

with ω_j the point corresponding to the jth horse winning. We also have a function $p : \Omega \to \mathbb{R}$ such that $p(\omega) \geq 0$ for each $\omega \in \Omega$ and

$$\sum_{\omega \in \Omega} p(\omega) = 1.$$

(In other words, the sum of all the $p(\omega)$ with $\omega \in \Omega$ is 1.) In our horse-racing example, $p(\omega_j)$ would be the probability that the jth horse wins the race. The probability $p(\omega_j)$ that the jth horse wins the race must be positive (but we extend the ideas of the first chapter to allow $p(\omega_j) = 0$)[4] and the sum of the probabilities

$$p(\omega_1) + p(\omega_2) + \cdots + p(\omega_n) = 1.$$

---

[2] It is traditional to use the capital version Ω of the Greek letter omega for the probability space and the small version ω for its elements.
[3] In more advanced texts, authors may use these phrases to denote different things.
[4] Mathematicians say that x is *positive* if $x \geq 0$ and *strictly positive* if $x > 0$.

If A is a subset of Ω, we call A an *event* and write

$$\Pr(A) = \sum_{\omega \in A} p(\omega).$$

(in other words, $\Pr(A)$ is the sum of all the $p(\omega)$ with $\omega \in A$). We call $\Pr(A)$ the probability of the event A. In our horse race example, if the first k horses are brown and the rest black, then taking the event A to be the set of points corresponding to a brown horse winning we have

$$\Pr(A) = p(\omega_1) + p(\omega_2) + \cdots + p(\omega_k).$$

It is remarkable that such a simple set of rules gives rise to such a rich theory.

Exercise 2.1.1 *We use the notation above.*
(i) Show that, if $\omega \in \Omega$, then $\Pr(\{\omega\}) = p(\omega)$.
(ii) Show that, if A and B are events and $A \supseteq B$, then $\Pr(A) \geq \Pr(B)$.
(iii) Show that, if A is an event, then $1 \geq \Pr(A) \geq 0$.
(iv) If A and B are disjoint events (that is to say, disjoint sets) show that

$$\Pr(A \cup B) = \Pr(A) + \Pr(B).$$

The reader may ask if we could choose our probability space Ω to be infinite. In order to answer this question I shall need concepts not used elsewhere (so there is no harm in skipping directly to the next section).

There is no problem in extending our ideas to *countably infinite probability spaces*, that is to say, Ω which can written as:

$$\Omega = \{\omega_1, \omega_2, \ldots\}.$$

We then demand $p(\omega_j) \geq 0$ for each j and $\sum_{j=1}^{\infty} p(\omega_j) = 1$. As before, we assign probabilities to an event $A \subseteq \Omega$ by writing

$$\Pr(A) = \sum_{\omega_j \in A} p(\omega_j).$$

The development of the theory is slightly more technical because we have to consider the convergence of infinite sums, but there are no serious difficulties.

There are problems when we try to extend our ideas to *uncountable* probability spaces. The first problem, which turns out not to be very serious, is that, in the infinite case, expressions like

$$\Pr(A) = \sum_{\omega \in A} \Pr(\{\omega\}),$$

which we used in the finite case may not even make sense.

Exercise 2.1.2 *(You should treat the contents of this exercise as heuristic rather than rigorous.)*

In what follows we consider intervals

$$[a, b) = \{x \in \mathbb{R} : a \le x < b\}.$$

Suppose we take our probability space $\Omega = [0, 1)$, *the unit interval, and we want a probability 'which makes no part of the interval more likely than other'. Convince yourself that this means that we want*

$$\Pr\big([0, 1/2)\big) = \Pr\big([1/2, 1)\big)$$

and so $\Pr\big([0, 1/2)\big) = \Pr\big([1/2, 1)\big) = 1/2$. *More generally, convince yourself that we require*

$$\Pr\big([(r-1)2^{-n}, r2^{-n})\big) = 2^{-n}$$

for all integers r *and* n *with* $n \ge 0$ *and* $1 \le r \le 2^n$. *Observe that we want* $\Pr(A) \ge \Pr(B)$ *whenever* A *and* B *are events with* $A \supseteq B$ *and deduce that* $\Pr(\{\omega\}) = 0$ *for all points* $\omega \in [0, 1)$.

Informally, if we take a pencil and bend it till it breaks, it will break, but the probability that it breaks at any particular place is zero.

The second problem turns out to be more serious. The subsets of an uncountable space can be extremely complex and we now know that there are subsets so complex that there is no way of assigning them probabilities in any way that we would wish. We therefore need to divide the subsets of our space into two classes: the class $\mathcal{F}$ of 'well-behaved sets' to which we assign probabilities and which we call events, and the class consisting of all the other sets. It is not easy to do this in a satisfactory manner, but a marvellous theory now exists due to Borel, Lebesgue, Radon, Nikodym, Kolmogorov and others, which enables us to study probability on general spaces.

Bertrand Russell wrote that 'Pure Mathematics was discovered by Boole in a work that he called *The Laws of Thought*'.[5] In a similar vein, we could write that 'Probability Theory was discovered by Kolmogorov in a work that he called *The Foundations of the Theory of Probability*'. However, there was lots of pure mathematics before Boole and, as we shall see, there is lots of probability theory associated with finite probability spaces.

---

[5] Where he laid the foundation for modern mathematical logic.

2.2 Cards, dice and coin tossing

The research supervisor of the great probabilist Feller told him that the best mathematics consists of the general embedded in the concrete.[6] Probability becomes easier to grasp if we consider some simple but suggestive models.

Our first model is that of a pack of n distinct cards being dealt. The first card can be any of the n cards, the second can be any of the remaining $n - 1$ cards, the third any of the remaining $n - 2$ cards and so on. There are thus

$$n! = n \times (n - 1) \times (n - 2) \times \cdots \times 2 \times 1$$

different ways of dealing n cards.[7] (By convention $0! = 1$.)

Table 2.1 shows the $4! = 4 \times 3 \times 2 \times 1 = 24$ different ways of dealing a pack of 4 cards A, B, C and D.

Exercise 2.2.1 *Draw up a table like Table 2.1 for a pack of 3 cards. Draw up a sufficient part of a table for a pack of 5 cards to convince a reasonably intelligent student that there are, indeed, $5! = 120$ different ways of dealing 5 cards.*

By symmetry, we expect the probability of all deals to be the same, so we obtain a probability space Ω with $n!$ points each labelled by a particular order of cards and such that

$$\Pr(\{\omega\}) = 1/n!$$

for all $\omega \in \Omega$.

When we dealt with Crown and Anchor we could draw a diagram with every point of the probability space and just count points to get probabilities. Clearly, this will not be possible for large spaces, so we must introduce new tools.

Lemma 2.2.2 *If we have r red cards and $n - r$ blue cards, then there are*

$$\frac{n!}{(n - r)!r!}$$

different ways of arranging them if cards of the same colour are indistinguishable.

Proof Let $C(n, r)$ be the number of ways that we can arrange the cards if cards of the same colour are indistinguishable. Now suppose we number the n cards so that they are all distinguishable. The red cards can now be arranged in $r!$

---

[6] Feller claimed that it was some years before he realised this was not an anti-militarist slogan.

[7] We pronounce $n!$ as 'n factorial'. (In the nether regions of the British school system it is also called 'n bang' or 'n shriek'.)

Table 2.1. *Possible deals with 4 cards*

A	A	A	A	A	A	B	B	C	D	C	D	B	B	C	D	C	D	B	B	C	D	C	D
B	B	C	D	C	D	A	A	A	A	A	A	C	D	B	B	D	C	C	D	B	B	D	C
C	D	B	B	D	C	C	D	B	B	D	C	A	A	A	A	A	A	D	C	D	C	B	B
D	C	D	C	B	B	D	C	D	C	B	B	D	C	D	C	B	B	A	A	A	A	A	A

Table 2.2. *Two red cards and two blue*

a	a	a	a	a	a	b	b	C	D	C	D	b	b	C	D	C	D	b	b	C	D	C	D
b	b	C	D	C	D	a	a	a	a	a	a	C	D	b	b	D	C	C	D	b	b	D	C
C	D	b	b	D	C	C	D	b	b	D	C	a	a	a	a	a	a	D	C	D	C	b	b
D	C	D	C	b	b	D	C	D	C	b	b	D	C	D	C	b	b	a	a	a	a	a	a

Table 2.3. *One red card and three blue*

a	a	a	a	a	a	B	B	C	D	C	D	B	B	C	D	C	D	B	B	C	D	C	D
B	B	C	D	C	D	a	a	a	a	a	a	C	D	B	B	D	C	C	D	B	B	D	C
C	D	B	B	D	C	C	D	B	B	D	C	a	a	a	a	a	a	D	C	D	C	B	B
D	C	D	C	B	B	D	C	D	C	B	B	D	C	D	C	B	B	a	a	a	a	a	a

ways and the blue cards can be arranged in $(n - r)!$ ways without exchanging red cards for blue. Thus the total number of ways of arranging our cards without exchanging red cards for blue is $r!(n - r)!$ and the total number of ways of arranging our cards in any way we wish is $C(n, r)r!(n - r)!$. But we already know that the total number of ways of arranging our cards in any way we wish is $n!$ so

$$C(n, r)r!(n - r)! = n!$$

as required.   ■

Exercise 2.2.3 *Go through the proof of Lemma 2.2.2 in the cases illustrated in Tables 2.2 (with cards a and b red and cards C and D blue) and 2.3 (with card a red and cards B, C and D blue).*

Exercise 2.2.4 *(i) Show that, if we have r red cards, b blue cards and g green cards (so we have $n = r + b + g$ cards in all), then there are*

$$\frac{n!}{r!b!g!}$$

different ways of arranging them if cards of the same colour are indistinguishable.

(ii) State and prove the general result for cards of many colours.

The expression

$$\binom{n}{r} = \frac{n!}{r!(n-r)!}$$

is very important in probability theory.

Exercise 2.2.5 *(i) Show that the coefficient of $x^{n-r}y^r$ in the expansion of $(x+y)^n$ is $\binom{n}{r}$. In other words, show that*

$$(x+y)^n = \binom{n}{0}x^n + \binom{n}{1}x^{n-1}y + \binom{n}{2}x^{n-2}y^2 + \cdots + \binom{n}{n}y^n.$$

Because of this result, $\binom{n}{r}$ is called a binomial coefficient.

(ii) Find the coefficient of $x^r y^b z^g$ in the expansion of $(x+y+z)^n$.

We now consider a probability model corresponding to throwing an m-sided die n times. The first throw can show any of the m faces, the second can show any of the m faces, and so on. There are thus

$$\overbrace{m \times m \times m \times \cdots \times m}^{n \text{ times}} = m^n$$

different ways in which we can throw the die n times. By symmetry, we expect the probability of all sequences of throws to be the same, so we obtain a probability space Ω with m^n points each labelled by a particular sequence of k faces and such that

$$\Pr(\{\omega\}) = 1/(m^n) = m^{-n}$$

for all $\omega \in \Omega$.

Exercise 2.2.6 *Draw up a table like Table 2.4 to cover a two-sided die thrown 4 times and one for a four-sided die thrown 2 times. Draw up a sufficient part of a table to convince a reasonably intelligent student that there are, indeed, 4^3 different ways of throwing a four-sided die 3 times.*

The models of card dealing and dice throwing that we have set up are very attractive. We cannot imagine a horse race being run over and over again, but we can readily conceive of a pack of cards being shuffled and redealt or a die being thrown over and over again. It is not hard to persuade ourselves that if

Table 2.4. *Three-sided die thrown three times*

A	A	A	A	A	A	A	A	A	B	B	B	B	B	B	B	B	B	C	C	C	C	C	C	C	C	C
A	A	A	B	B	B	C	C	C	A	A	A	B	B	B	C	C	C	A	A	A	B	B	B	C	C	C
A	B	C	A	B	C	A	B	C	A	B	C	A	B	C	A	B	C	A	B	C	A	B	C	A	B	C

a pack of cards is well shuffled or a well constructed die is shaken and then thrown, the various possible outcomes must have equal probabilities. (On the other hand, people find it hard to agree on the probability that a horse will win a race.)

It might be thought that the ideas of this section will only produce probability spaces in which each singleton $\{\omega\}$ has the same probability, but this is not so.

Lemma 2.2.7 *Suppose that b sides of an m-sided die are coloured blue and the remaining sides are coloured green. If the die is thrown n times the probability of any particular sequence of blue and green sides with exactly r blues is $p^r(1-p)^{n-r}$ where $p = b/m$.*

If A_r is the event that a blue face appears exactly r times, then

$$\Pr(A_r) = \binom{n}{r} p^r (1-p)^{n-r}.$$

where $p = b/m$.

Thus if we are only interested in the number of times that a blue face appears, our probability space is

$$\{\omega_0, \omega_1, \omega_2, \ldots, \omega_n\}$$

with ω_r the event that a blue face appears exactly r times, and

$$\Pr(\{\omega_r\}) = \binom{n}{r} p^r (1-p)^{n-r}.$$

Proof Each blue face can arise in b ways and each green face in $m - b$ ways so the total number of ways in which we can get a particular sequence of n throws with exactly r blue faces is

$$\overbrace{b \times b \times \cdots \times b}^{r \text{ times}} \times \overbrace{(m - b) \times (m - b) \times \cdots \times (m - b)}^{n - r \text{ times}} = b^r (m - b)^{n-r}.$$

It follows that the probability of a particular sequence of n throws with exactly r blue faces is

$$b^r (m - b)^{n-r} \times m^{-n} = p^r (1-p)^{n-r}.$$

Table 2.5. *Three-sided die with two faces coloured blue*

| a a a a a a a a a B B B B B B B B B C C C C C C C C C |
| a a a B B B C C C a a a B B B C C C a a a B B B C C C |
| a B C a B C a B C a B C a B C a B C a B C a B C a B C |

We know that there are $\binom{n}{r}$ different ways in which we can arrange a sequence consisting of r blue faces and $n - r$ green faces, so

$$\Pr(A_r) = \binom{n}{r} p^r (1 - p)^{n-r}$$

as stated.   ∎

Exercise 2.2.8 *Go through the proof of Lemma 2.2.7 in the case illustrated in Table 2.5 (with face a blue and faces B and C green).*

In Lemma 2.2.7, p must be rational, but there is no reason why we should not *define* a corresponding probability space with p irrational.

Lemma 2.2.9 *Suppose $1 \geq p \geq 0$ and n is a positive integer. Let Ω be the space of all sequences of length n consisting of the letters H and T. If $\omega \in \Omega$ is a sequence containing the letter H exactly r times we take*

$$p(\omega) = p^r (1 - p)^{n-r}.$$

With these choices, $p : \Omega \to \mathbb{R}$ has the property that $p(\omega) \geq 0$ for all $\omega \in \Omega$ and $\sum_{\omega \in \Omega} p(\omega) = 1$. We thus have a probability space.

Proof Let A_r be the set of all sequences in Ω containing the letter H exactly r times. We know that A_r has $\binom{n}{r}$ elements and thus, with an obvious notation,

$$\sum_{\omega \in \Omega} p(\omega) = \sum_{0 \leq r \leq n} \sum_{\omega \in A_r} p(\omega) = \sum_{0 \leq r \leq n} \binom{n}{r} p^r (1 - p)^{n-r}$$
$$= \big(p + (1 - p)\big)^n = 1^n = 1.$$

                                                                                ∎

The reader will be unsurprised to learn that we refer to this model as 'tossing a coin n times' and that we refer to H as 'heads' and T as 'tails'.[8] If $p = 1/2$, we refer to a 'fair coin'.

---

[8] In monarchies the head of the sovereign appears on one side of a coin.

2.3 Random variables

This section continues the formal business of introducing and defining the various basic concepts of probability theory. However, at the end of the section, we shall give the solution due to Nicholas Bernoulli of the problem considered in the following exercise.

Exercise 2.3.1 *Two ordinary packs of 52 playing cards are each thoroughly shuffled. They are then dealt in such a way that the top card from each pack forms the first pair, the second card from each pack forms the second pair and so on until we have 52 pairs and the packs are exhausted. What do you think is the probability that at least one of the pairs consists of two identical cards?*

Suppose we have a finite probability space Ω with an associated probability Pr. We call a function $X : \Omega \to \mathbb{R}$ a *random variable*. Reverting to our horse race example, if $\Omega = \{\omega_1, \omega_2, \ldots, \omega_n\}$ where ω_j corresponds to a win by the jth horse, $X(\omega_j)$ could be the amount that I promise to pay you if the jth horse wins.

Every random variable X has an associated *expectation*[9] written $\mathbb{E}X$ and defined by

$$\mathbb{E}X = \sum_{\omega \in \Omega} p(\omega) X(\omega)$$

(where $p(\omega) = \Pr(\{\omega\})$). In the first chapter we saw that, for the horse-racing example of the previous paragraph, $\mathbb{E}X$ corresponds to the value of my promise to pay $X(\omega_j)$ if the jth horse wins.

The following results are trivial to prove but very important to remember and understand.

Lemma 2.3.2 *(i) If X and Y are random variables and a and b are real numbers, then*

$$\mathbb{E}(aX + bY) = a\mathbb{E}X + b\mathbb{E}Y.$$

(ii) If X and Y are random variables and $X(\omega) \geq Y(\omega)$ for all $\omega \in \Omega$, then

$$\mathbb{E}X \geq \mathbb{E}Y.$$

---

[9] Sometimes called *mathematical expectation* to distinguish it from the non-mathematical sense of the term. If I have a probability 10^{-7} of winning £10^7, then my *mathematical expectation* is £1 but I *expect* to win nothing.

Proof (i) Observe that, by definition,

$$\mathbb{E}(aX + bY) = \sum_{\omega \in \Omega} p(\omega)\big(aX(\omega) + bY(\omega)\big)$$

$$= a \sum_{\omega \in \Omega} p(\omega)X(\omega) + b \sum_{\omega \in \Omega} p(\omega)Y(\omega)$$

$$= a\mathbb{E}X + b\mathbb{E}Y.$$

(ii) Left as an exercise for the reader. ∎

Exercise 2.3.3 *Interpret the relations*

$$\mathbb{E}(aX) = a\mathbb{E}X \text{ and } \mathbb{E}(X + Y) = \mathbb{E}X + \mathbb{E}Y$$

in terms of the value of bets.
  Do the same for the statement that, if $X(\omega) \geq Y(\omega)$ for all $\omega \in \Omega$, then

$$\mathbb{E}X \geq \mathbb{E}Y.$$

We now introduce a key random variable.

Definition 2.3.4 *If Ω is a set and $A \subseteq \Omega$, we define the* indicator function $\mathbb{I}_A : \Omega \to \mathbb{R}$ *by*

$$\mathbb{I}_A(\omega) = \begin{cases} 1 & \text{if } \omega \in A, \\ 0 & \text{otherwise.} \end{cases}$$

Lemma 2.3.5 *Suppose Ω is a probability space with associated probability* Pr. *If A is an event, then*

$$\mathbb{E}\mathbb{I}_A = \Pr(A).$$

Proof Just observe that

$$\mathbb{E}\mathbb{I}_A = \sum_{\omega \in \Omega} p(\omega)\mathbb{I}_A(\omega) = \sum_{\omega \in A} p(\omega) = \Pr(A).$$

∎

Thus we can recover probabilities from expectations just as we can obtain expectations using probabilities.
  As might be expected, the algebraic properties of indicator functions are closely involved with the set theoretic relations of the underlying sets.

Lemma 2.3.6 *Let A and B be subsets of a set Ω.*
  (i) $\mathbb{I}_{A \cup B}(\omega) = \mathbb{I}_A(\omega) + \mathbb{I}_B(\omega) - \mathbb{I}_A(\omega)\mathbb{I}_B(\omega)$ for all $\omega \in \Omega$.
  (ii) $\mathbb{I}_{A \cap B}(\omega) = \mathbb{I}_A(\omega)\mathbb{I}_B(\omega)$ for all $\omega \in \Omega$.

(iii) We have $\mathbb{I}_A(\omega)^2 = \mathbb{I}_A(\omega)$ *for all* $\omega \in \Omega$.
(iv) If $\mathbb{I}_A(\omega) = \mathbb{I}_B(\omega)$ *for all* $\omega \in \Omega$, *then* $A = B$.

Proof (i) Observe that

$$\mathbb{I}_{A \cup B}(\omega) = \begin{cases} 1 & \text{if } \omega \in A \text{ and } \omega \in B \\ 1 & \text{if } \omega \in A \text{ and } \omega \notin B \\ 1 & \text{if } \omega \notin A \text{ and } \omega \in B \\ 0 & \text{if } \omega \notin A \text{ and } \omega \notin B \end{cases}$$

and

$$\mathbb{I}_A(\omega) + \mathbb{I}_B(\omega) - \mathbb{I}_A(\omega)\mathbb{I}_B(\omega) = \begin{cases} 1 + 1 - 1 = 1 & \text{if } \omega \in A \text{ and } \omega \in B \\ 1 + 0 - 0 = 1 & \text{if } \omega \in A \text{ and } \omega \notin B \\ 0 + 1 - 0 = 1 & \text{if } \omega \notin A \text{ and } \omega \in B \\ 0 + 0 + 0 = 0 & \text{if } \omega \notin A \text{ and } \omega \notin B \end{cases}$$

so

$$\mathbb{I}_{A \cup B}(\omega) = \mathbb{I}_A(\omega) + \mathbb{I}_B(\omega) - \mathbb{I}_A(\omega)\mathbb{I}_B(\omega)$$

for all $\omega \in \Omega$.

(ii), (iii) and (iv) are left to the reader. ∎

We can use indicator functions to derive set theoretic relations. For example,

$$\begin{aligned} \mathbb{I}_{(A \cup B) \cap C}(\omega) &= \mathbb{I}_{(A \cup B)}(\omega)\mathbb{I}_C(\omega) = \big(\mathbb{I}_A(\omega) + \mathbb{I}_B(\omega) - \mathbb{I}_A(\omega)\mathbb{I}_B(\omega)\big)\mathbb{I}_C(\omega) \\ &= \mathbb{I}_A(\omega)\mathbb{I}_C(\omega) + \mathbb{I}_B(\omega)\mathbb{I}_C(\omega) - \mathbb{I}_A(\omega)\mathbb{I}_B(\omega)\mathbb{I}_C(\omega) \\ &= \mathbb{I}_A(\omega)\mathbb{I}_C(\omega) + \mathbb{I}_B(\omega)\mathbb{I}_C(\omega) - \big(\mathbb{I}_A(\omega)\mathbb{I}_C(\omega)\big)\big(\mathbb{I}_B(\omega)\mathbb{I}_C(\omega)\big) \\ &= \mathbb{I}_{A \cap C}(\omega) + \mathbb{I}_{B \cap C}(\omega) - \mathbb{I}_{A \cap C}(\omega)\mathbb{I}_{B \cap C}(\omega) = \mathbb{I}_{(A \cap C) \cup (B \cap C)}(\omega) \end{aligned}$$

for all $\omega \in \Omega$. Thus

$$(A \cup B) \cap C = (A \cap C) \cup (B \cap C).$$

Exercise 2.3.7 *Recall that, if A is a subset of* Ω, *we write* A^c *for the complement of A in* Ω, *that is to say, we write*

$$A^c = \{\omega \in \Omega : \omega \notin A\}.$$

Show that

$$\mathbb{I}_{A^c} = 1 - \mathbb{I}_A.$$

Use indicator functions to show that, if A and B are subsets of Ω, *then*

$$(A \cup B)^c = A^c \cap B^c, \quad (A \cap B)^c = A^c \cup B^c \quad \text{and} \quad (A^c)^c = A.$$

We write $A \setminus B = A \cap B^c$ and $A \triangle B = (A \setminus B) \cup (B \setminus A)$. ($A \triangle B$ is called the symmetric difference *of A and B.) Write $\mathbb{I}_{A \setminus B}$ and $\mathbb{I}_{A \triangle B}$ in terms of $\mathbb{I}_A$ and $\mathbb{I}_B$. Write down simple expressions for $\mathbb{I}_\Omega$ and $\mathbb{I}_\varnothing$, where, as usual, $\varnothing$ denotes the empty set.*

Using the results of Exercise 2.3.7 we see, for example, that

$$
\begin{aligned}
\mathbb{I}_{A \cup B \cup C} &= 1 - \mathbb{I}_{(A \cup B \cup C)^c} = 1 - \mathbb{I}_{A^c \cap B^c \cap C^c} \\
&= 1 - \mathbb{I}_{A^c} \mathbb{I}_{B^c} \mathbb{I}_{C^c} = 1 - (1 - \mathbb{I}_A)(1 - \mathbb{I}_B)(1 - \mathbb{I}_C) \\
&= \mathbb{I}_A + \mathbb{I}_B + \mathbb{I}_C - \mathbb{I}_A \mathbb{I}_B - \mathbb{I}_B \mathbb{I}_C - \mathbb{I}_C \mathbb{I}_A + \mathbb{I}_A \mathbb{I}_B \mathbb{I}_C \\
&= \mathbb{I}_A + \mathbb{I}_B + \mathbb{I}_C - \mathbb{I}_{A \cap B} - \mathbb{I}_{B \cap C} - \mathbb{I}_{C \cap A} + \mathbb{I}_{A \cap B \cap C}.
\end{aligned}
$$

Exercise 2.3.8 *Suppose that Ω is finite and A, B, C and D are finite subsets of Ω.*

(i) Show that if we write $|D|$ for the number of elements in D we obtain

$$
|D| = \sum_{\omega \in \Omega} \mathbb{I}_D(\omega)
$$

(ii) Use the result in the paragraph preceding this exercise to show that

$$
|A \cup B \cup C| = |A| + |B| + |C| - |A \cap B| - |B \cap C| - |C \cap A| + |A \cap B \cap C|.
$$

(iii) Let A be the set of students at a university who wear glasses, B the set of students of mathematics and C the set of students who can sing in tune. Express the formula of (ii) in words and explain to a non-mathematician why it is correct.

(iv) Explain why any integer n with $2 \le n < 49$ is either a prime or divisible by 2, 3, 5. Use the formula given in (ii) to find the number of primes less than 49.

(v) Extend the result in the paragraph preceding this exercise and the formula in (iv) to the case of four sets A, B, C and D (or A_1, A_2, A_3 and A_4 if you think this is easier notationally).

(vi) Use the result of (v) to find the number of primes less than 100.

If Ω is a probability space with associated probability Pr, the results of Lemmas 2.3.5 and 2.3.2 give

$$
\begin{aligned}
\Pr(A \cup B \cup C) &= \mathbb{E} \mathbb{I}_{A \cup B \cup C} \\
&= \mathbb{E}(\mathbb{I}_A + \mathbb{I}_B + \mathbb{I}_C - \mathbb{I}_{A \cap B} - \mathbb{I}_{B \cap C} - \mathbb{I}_{C \cap A} + \mathbb{I}_{A \cap B \cap C}) \\
&= \mathbb{E} \mathbb{I}_A + \mathbb{E} \mathbb{I}_B + \mathbb{E} \mathbb{I}_C - \mathbb{E} \mathbb{I}_{A \cap B} - \mathbb{E} \mathbb{I}_{B \cap C} - \mathbb{E} \mathbb{I}_{C \cap A} + \mathbb{E} \mathbb{I}_{A \cap B \cap C} \\
&= \Pr(A) + \Pr(B) + \Pr(C) - \Pr(A \cap B) - \Pr(B \cap C) \\
&\quad - \Pr(C \cap A) + \Pr(A \cap B \cap C).
\end{aligned}
$$

We can clearly extend our results to any number of sets provided we can find an appropriate notation.

Lemma 2.3.9 *Let $A_1, A_2, \ldots, A_n$ be subsets of a set Ω.*

(i) $\mathbb{I}_{\bigcup_i A_i} = \sum_i \mathbb{I}_{A_i} - \sum_{i<j} \mathbb{I}_{A_i \cap A_j} + \sum_{i<j<k} \mathbb{I}_{A_i \cap A_j \cap A_k} - \cdots.$

(ii) If Ω is finite, then

$$\left| \bigcup_i A_i \right| = \sum_i |A_i| - \sum_{i<j} |A_i \cap A_j| + \sum_{i<j<k} |A_i \cap A_j \cap A_k| - \cdots.$$

(iii) If Ω is a probability space with associated probability $\Pr$, *then*

$$\Pr\left(\bigcup_i A_i \right) = \sum_i \Pr(A_i) - \sum_{i<j} \Pr(A_i \cap A_j) + \sum_{i<j<k} \Pr(A_i \cap A_j \cap A_k) - \cdots.$$

Here, for example, $\bigcup_i A_i$ is the union of all the sets A_i with $1 \le i \le n$, that is to say, the set of all points lying in at least one of the A_i. The expression $\sum_{i<j<k} \Pr(A_i \cap A_j \cap A_k)$ means the sum of all the terms $\Pr(A_i \cap A_j \cap A_k)$ with $1 \le i < j < k \le n$ and so on. If the reader is happy with such expressions, she should test her familiarity with them by writing out a proof of Lemma 2.3.9. If not, she should make sure that she is more or less happy with the meaning of the lemma and can write it out in full when n is small. Mathematicians refer to all the results of Lemma 2.3.9 as *inclusion-exclusion formulae*. (Exercise 4.6.11, which the reader could do now, gives an interesting example of its use.)

We can now answer the question with which we started this section.

Exercise 2.3.10 *Suppose we have a well-shuffled pack of n cards numbered from 1 to n. The probability space Ω consists of the $n!$ possible deals and we take $\Pr(\omega) = 1/n!$ for each $\omega \in \Omega$. We write A_j for the set of shuffles such that the jth card dealt bears the number j.*

(i) Explain why the number of ways of choosing integers i, j, k with $1 \le i < j < k \le n$ is the same as the number of ways of arranging 3 red cards and $n-3$ blue cards.

(ii) Describe the event $A_1 \cap A_2 \cap A_3$ in words and find $\Pr(A_1 \cap A_2 \cap A_3)$.

(iii) Show that

$$\sum_{i<j<k} \Pr(A_i \cap A_j \cap A_k) = \frac{1}{3!}.$$

(iv) Find and prove a general result along the lines of (iii) and deduce that

$$\Pr\left(\bigcup_i A_i\right) = \frac{1}{1!} - \frac{1}{2!} + \frac{1}{3!} - \cdots + (-1)^{n-1}\frac{1}{n!}.$$

(v) Describe the event $\bigcup_i A_i$ in words and show that the probability that no card is dealt in the same place as the number it bears is

$$1 - \frac{1}{1!} + \frac{1}{2!} - \frac{1}{3!} + \cdots + (-1)^n\frac{1}{n!}.$$

(vi) Use a calculator to evaluate the formula in (v) for $n = 1, 2, 3, 4, 5, 6$. Use Exercise A.10 to show that

$$1 - \frac{1}{1!} + \frac{1}{2!} - \frac{1}{3!} + \cdots + (-1)^n\frac{1}{n!} \approx e^{-1}$$

for large n and use your calculator to compute e^{-1}. (If you know about such things, you may wish to comment on why the approximation is good even for fairly small n.)

(vii) Answer the question raised in Exercise 2.3.1.

(viii) Show that the probability that exactly k cards are dealt in the same place as the number they bear is approximately $e^{-1}/k!$ if k is small compared with n.

Exercise 2.3.11 *(i) Each member of the Tripos Reform Procrastination Committee has probability p of being available for a meeting on a given date, independent of their availability on other dates and of the obligations of the other members. (In other words, we can use a coin tossing model.) There are m members of the committee and there are n dates on which they might meet. Show that the probability $P(p, m, n)$ that they can find a date to meet is*

$$\binom{n}{1}p^m - \binom{n}{2}p^{2m} + \cdots + (-1)^{n-1}\binom{n}{n}p^{nm}.$$

Show that

$$P(1/2, 8, 30) \approx 0.12.$$

(ii) In an attempt to speed up matters, the chairman tells the members that they must each name k days on which they are available. Each member names the minimum number of days and chooses those days at random. Find the probability $Q(k, m, n)$ that they can find a date to meet.

Give a simple argument to show that that there are values of k, m and n such that $P(k/n, m, n) \neq Q(k, m, n)$. Explain why

$$P(1/2, 8, 30) \approx Q(15, 8, 30).$$

2.4 Independence

Very early in the study of probability, it was realised that if there is a probability p that a multicoloured die shows a red face, then the probability that it shows a red face in two successive throws is p^2. More generally, if we have a die with probability p of showing red and a second die with probability of q of showing red then the probability that both show red when thrown is pq.

This remark proved so useful that it was eventually generalised as follows.

Definition 2.4.1 *Let Ω be a probability space with associated probability* Pr. *We say that two events A and B are* independent *if*

$$\Pr(A \cap B) = \Pr(A)\Pr(B).$$

It often happens in elementary probability that two events are 'obviously independent' in the sense that we can apply the following lemma.

Lemma 2.4.2 *Let U and V be probability spaces with associated probabilities* $\Pr_U$ *and* $\Pr_V$.

Suppose $\Omega = U \times V$ (that is to say, Ω consists of the ordered pairs (u, v) with $u \in U$ and $v \in V$). Then we can define a probability Pr *on Ω in such a way that*

$$\Pr(\{(u, v)\}) = \Pr_U(\{u\})\Pr_V(\{v\})$$

for all $(u, v) \in \Omega$.

With this choice of Pr, *it follows that, if $E \subseteq U$ and $F \subseteq V$, the events*

$$E \times V = \{(e, v) : e \in E, \ v \in V\} \text{ and } U \times F = \{(u, f) : u \in U, \ f \in F\}$$

are independent.

Proof Set $p_u = \Pr_U(\{u\})$ and $q_v = \Pr_V(\{v\})$. We observe that $p_u q_v \geq 0$ for all $u \in U$ and $v \in V$ and that

$$\sum_{u \in U, \, v \in V} p_u q_v = \left(\sum_{u \in U} p_u\right)\left(\sum_{v \in V} q_v\right) = 1^2 = 1$$

and so

$$\Pr(A) = \sum_{(u,v) \in A} p_u q_v$$

defines a probability on Ω. We note also that

$$\Pr(E \times V) = \sum_{(u,v) \in E \times V} p_u q_v = \left(\sum_{u \in E} p_u\right)\left(\sum_{v \in V} q_v\right) = \Pr_U(E) \times 1 = \Pr_U(E)$$

and similarly

$$\Pr(U \times F) = \Pr_V(F).$$

Finally, we observe that

$$(E \times V) \cap (U \times F) = \{(e, f) : e \in E, \ f \in F\} = E \times F$$

and so

$$\Pr\big((E \times V) \cap (U \times F)\big) = \Pr(E \times F) = \sum_{u \in U, \, v \in V} p_u q_v$$

$$= \sum_{u \in U} p_u \sum_{v \in U} q_v = \Pr_U(E)\Pr_V(F) = \Pr(E \times V)\Pr(U \times F).$$

∎

However, not all pairs of independent events arise in this way.

Exercise 2.4.3 *A fair coin is thrown 3 times. Let A be the event that the first two throws are the same (that is both heads or both tails) and B the event that the last two throws are the same. Show that A and B are independent.*

Exercise 2.4.4 *(i) Suppose Ω is a probability space with associated probability* Pr. *Show that if A and B are independent events so are A^c and B. Deduce that if A and B are independent events so are A^c and B^c.*

(ii) Suppose Ω is a probability space with associated probability Pr. *Show that, if A and B are independent events with $A \cup B = \Omega$, then at least one of A or B has probability 1.*

(iii) Suppose that $0 < p, q < 1$,

$$\Omega = \{\omega_1, \omega_2, \ldots, \omega_5\},$$

$A = \{\omega_1, \omega_2, \omega_3\}$ and $B = \{\omega_3, \omega_4\}$. Show that there is a probability Pr *on Ω such that A and B are independent, $\Pr(A) = p$, $\Pr(B) = q$ and $\Pr(\{\omega_j\}) \neq 0$ for all j. Explain why $(\Omega, \Pr)$, A and B cannot be constructed starting from some $(U, \Pr_U)$, $(V, \Pr_V)$, E and F as in Lemma 2.4.2.*

We extend the definition of independence to 3 sets as follows.

Definition 2.4.5 *Let Ω be a probability space with associated probability* Pr. *We say that the events A, B and C are independent if*

$$\Pr(A \cap B) = \Pr(A)\Pr(B), \quad \Pr(B \cap C) = \Pr(B)\Pr(C),$$

$$\Pr(C \cap A) = \Pr(C)\Pr(A),$$

$$\Pr(A \cap B \cap C) = \Pr(A)\Pr(B)\Pr(C).$$

When thinking about this definition, you should bear in mind the results of the following exercises.

Exercise 2.4.6 *(i) A fair coin is thrown 3 times. Let A be the event that the first two throws are the same (that is to say, both heads or both tails), B the event that the last two throws are the same and C the event that the first and last throws are the same. Show that*

$$\Pr(A \cap B) = \Pr(A)\Pr(B), \ \Pr(B \cap C) = \Pr(B)\Pr(C),$$
$$\Pr(C \cap A) = \Pr(C)\Pr(A)$$

but $\Pr(A \cap B \cap C) \neq \Pr(A)\Pr(B)\Pr(C)$.
(ii) Let $\Omega = \{\omega_1, \omega_2, \ldots, \omega_7\}$ *and*

$$A = \{\omega_1, \omega_2, \omega_3, \omega_4\}, \ B = \{\omega_2, \omega_4, \omega_5\}, \ C = \{\omega_3, \omega_4, \omega_6\}.$$

Show that we can find a probability $\Pr$ *on* Ω *such that*

$$\Pr(A) = \Pr(B) = \Pr(C) = 1/3,$$
$$\Pr(A \cap B) = \Pr(A)\Pr(B), \ \Pr(C \cap A) = \Pr(C)\Pr(A),$$
$$\Pr(A \cap B \cap C) = \Pr(A)\Pr(B)\Pr(C).$$

Show that $\Pr(B \cap C) \neq \Pr(B)\Pr(C)$.

Exercise 2.4.7 *Let* Ω *be a probability space with associated probability* $\Pr$. *Show that, if the events A, B and C are independent, then so are the events A, B and* $C^c = \Omega \setminus C$.

Exercise 2.4.8 *State and prove a result corresponding to Lemma 2.4.2 for three sets A, B and C.*

The extension of our definition to the case of n events is now simply a matter of notation.

Definition 2.4.9 *Let* Ω *be a probability space with associated probability* $\Pr$. *We say that the events* A_j $[1 \leq j \leq n]$ *are* independent *if, whenever* $1 \leq j(1) < j(2) < \cdots < j(k) \leq n$, *we have*

$$\Pr(A_{j(1)} \cap A_{j(2)} \cap \cdots \cap A_{j(k)}) = \Pr(A_{j(1)})\Pr(A_{j(2)})\cdots\Pr(A_{j(k)}).$$

It is clear that results like Exercise 2.4.7 and Lemma 2.4.2 extend to the cases of n independent events.

In turns out that we shall be more interested in random variables than in events (that is to say, we shall be more interested in our bets than the events we bet on). We therefore need a definition of independent random variables.

The following definition may not be immediately intuitive, but long experience has convinced mathematicians that it is the most appropriate.

Definition 2.4.10 *Let Ω be a probability space with associated probability* Pr. *The n random variables X_1, X_2, ..., X_n (that is to say, the n functions $X_j : \Omega \to \mathbb{R}$) are said to be independent if whenever $t_j \in \mathbb{R}$ the n events*

$$A_j = \{\omega \in \Omega : X_j(\omega) = t_j\}$$

are independent.

The proof of the next lemma illustrates how the definition is used.

Lemma 2.4.11 *Let Ω be a probability space with associated probability* Pr. *If X and Y are independent random variables, then*

$$\mathbb{E}XY = (\mathbb{E}X)(\mathbb{E}Y).$$

(We shall usually use the unbracketed form $\mathbb{E}XY = \mathbb{E}X\mathbb{E}Y$ rather than $\mathbb{E}XY = (\mathbb{E}X)(\mathbb{E}Y)$.)

Proof Since Ω is finite, X can only take a finite number of values t_1, t_2, ..., t_M. If we write

$$A_m = \{\omega \in \Omega : X(\omega) = t_m\}$$

for $1 \le m \le M$, then

$$X = t_1\mathbb{I}_{A_1} + t_2\mathbb{I}_{A_2} + \cdots + t_M\mathbb{I}_{A_M}.$$

In the same way, we can write

$$Y = s_1\mathbb{I}_{B_1} + s_2\mathbb{I}_{B_2} + \cdots + s_N\mathbb{I}_{B_N},$$

where $s_1, s_2, \ldots, s_N$ are distinct real numbers and

$$B_n = \{\omega \in \Omega : Y(\omega) = s_n\}$$

for $1 \le n \le N$.

By the definition of independent random variables, A_m and B_n are independent events and so

$$\mathbb{E}\mathbb{I}_{A_m}\mathbb{I}_{B_n} = \mathbb{E}\mathbb{I}_{A_m \cap B_n} = \Pr(A_m \cap B_n)$$
$$= \Pr(A_m)\Pr(B_n) = \mathbb{E}\mathbb{I}_{A_m}\mathbb{E}\mathbb{I}_{B_n}.$$

It follows that

$$\mathbb{E}XY = \mathbb{E}\left(\sum_{m=1}^{M} t_m \mathbb{I}_{A_m}\right)\left(\sum_{n=1}^{N} s_n \mathbb{I}_{B_n}\right) = \mathbb{E}\left(\sum_{m=1}^{M}\sum_{n=1}^{N} t_m s_n \mathbb{I}_{A_m}\mathbb{I}_{B_n}\right)$$

$$= \sum_{m=1}^{M}\sum_{n=1}^{N} t_m s_n \mathbb{E}(\mathbb{I}_{A_m}\mathbb{I}_{B_n}) = \sum_{m=1}^{M}\sum_{n=1}^{N} t_m s_n \mathbb{E}\mathbb{I}_{A_m}\mathbb{E}\mathbb{I}_{B_n}$$

$$= \left(\sum_{m=1}^{M} t_m \mathbb{E}\mathbb{I}_{A_m}\right)\left(\sum_{n=1}^{N} s_n \mathbb{E}\mathbb{I}_{B_n}\right) = \mathbb{E}X\mathbb{E}Y,$$

as required. ∎

Exercise 2.4.12 *Many readers will have seen another proof of Lemma 2.4.11. If you are one of those readers, convince yourself that the other proof and the one given here are essentially the same.*

The next exercise shows that Lemma 2.4.11 makes sense when we consider our standard example of horse racing.

Exercise 2.4.13 *We can make an 'accumulator bet' on two races as follows. In the first race, the jth horse has probability p_j of winning and if we pay c we obtain a promise to pay $a_j c$ if it wins $[1 \le j \le J]$. In the second race, the kth horse has probability q_k of winning and if we pay d we obtain a promise to pay $b_k d$ if it wins $[1 \le k \le K]$. We decide to place c_j on the jth horse in in the first race where*

$$c_1 + c_2 + \cdots + c_J = 1.$$

We take our winnings x on the first race and place $x d_k$ on the kth horse in the second race where

$$d_1 + d_2 + \cdots + d_K = 1.$$

Convince yourself that the following is a reasonable model for our bet. Consider a probability space Ω containing the JK points $\omega_{(j,k)}$ (corresponding to the jth horse winning the first race and the kth horse winning the second race) with

$$\Pr(\{\omega_{(j,k)}\}) = p_j q_k$$

for all $1 \le j \le J$ and $1 \le k \le K$. Check that $(\Omega, \Pr)$ is indeed a probability space.

Let $X(\omega_{(j,k)}) = a_j c_j$ and $Y(\omega_{(j,k)}) = b_k d_k$. Show that X and Y are independent random variables and deduce that $\mathbb{E}X\mathbb{E}Y = \mathbb{E}XY$

Interpret $\mathbb{E}X$ *and* $\mathbb{E}Y$ *in terms of the value of bets on the first and second races. Interpret* $\mathbb{E}XY$ *in terms of the value of an accumulator bet on the first and second races.*

Although it is always true that $\mathbb{E}(X + Y) = \mathbb{E}X + \mathbb{E}Y$, it may not be true that $\mathbb{E}(XY) = \mathbb{E}X\mathbb{E}Y$ unless X and Y are independent. (Expectations always add and *independent* expectations multiply.)

Exercise 2.4.14 *Let* $\Omega = \{\omega_1, \omega_2\}$ *and* $\mathrm{Pr}(\{\omega_1\}) = \mathrm{Pr}(\{\omega_2\}) = 1/2$ *(corresponding to throwing a fair coin once). Show that, if we set*

$$X(\omega_1) = 1, \ Y(\omega_1) = 0; \ X(\omega_2) = 0, \ Y(\omega_2) = 1,$$

then $\mathbb{E}XY \neq \mathbb{E}X\mathbb{E}Y$ *and* $\mathbb{E}X^2 \neq (\mathbb{E}X)^2$.

Exercise 2.4.15 *Let* $\Omega = \{\omega_1, \omega_2, \ldots, \omega_5\}$ *and* $\mathrm{Pr}(\{\omega_j\}) = 1/5 \ [1 \leq j \leq 5]$. *If we set*

$$X(\omega_1) = 1, \ Y(\omega_1) = 1; \ X(\omega_2) = 1, \ Y(\omega_2) = -1;$$
$$X(\omega_3) = -1, \ Y(\omega_3) = 1; \ X(\omega_4) = -1, \ Y(\omega_4) = -1;$$
$$X(\omega_5) = 0, \ Y(\omega_5) = 0$$

show that $\mathbb{E}XY = \mathbb{E}X\mathbb{E}Y$ *but that* X *and* Y *are not independent.*

Exercise 2.4.16 [A St Petersburg proposition][10] *I have a well-shuffled pack of 8 cards numbered from 1 to 8. Quickly guess which of the following outcomes is more probable. Is either outcome much more probable than the other?*

(i) If I deal 3 cards the set will either *contain three of the four smallest cards* or *contain three of the four largest cards.*

(ii) If I deal 6 cards the set will both *contain three of the four smallest cards* and *contain three of the four largest cards.*

(iii) Now compute the correct probabilities.
[*As a schoolgirl, a Russian mathematician of my acquaintance observed a gambling game based on a more complicated version of these ideas on the streets of St Petersburg. The gentlemen running the game strongly objected to her attempts to explain probability theory to their customers.*]

Exercise 2.4.17 *In this exercise we prove a couple of results which the reader may consider as obvious.*

---

[10] 'Many citizens prefer betting on propositions to anything you can think of, because they figure a proposition gives them a chance to out-smart somebody, and in fact I know citizens who will sit up all night making up propositions to offer other citizens the next day.' [Damon Runyon, *The Idyll of Miss Sarah Brown*]

(i) Let Ω be a probability space with associated probability Pr. *Show that, if* X, $Y : \Omega \to \mathbb{R}$ *are independent random variables and U and V are subsets of $\mathbb{R}$ then the events*

$$\{\omega \in \Omega : X(\omega) \in U\} \quad and \quad \{\omega \in \Omega : Y(\omega) \in V\}$$

are independent.

(ii) If X and Y are independent random variables and f, $g : \mathbb{R} \to \mathbb{R}$ are functions, show that $f(X)$ and $g(Y)$ are independent random variables.

(iii) State the results corresponding to (i) and (ii) for n random variables and, if you feel it necessary, prove them.

We conclude this section with a few remarks on random variables in general. If you think that they merely labour the obvious,[11] then you may ignore them (apart from Definition 2.4.20) and move on to the next section.

We first introduce a little notation (which certainly labours the obvious).

Definition 2.4.18 *Let (Ω, Pr) be a probability space and $X : \Omega \to \mathbb{R}$ a random variable. If U is a subset of $\mathbb{R}$, then we write*

$$\text{Pr}(X \in U) = \text{Pr}\{\omega \in \Omega : X(\omega) \in U\}.$$

If $U = \{a\}$ then we write

$$\text{Pr}(X = a) = \text{Pr}(X \in U).$$

We can now prove a simple lemma.

Lemma 2.4.19 *Let (Ω, Pr) be a probability space and $X : \Omega \to \mathbb{R}$ be a random variable. Write $X(\Omega)$ for the set of values taken by X.*

If $f : \mathbb{R} \to \mathbb{R}$ is a function then, writing $f(X)(\omega) = f(X(\omega))$, it follows that $f(X)$ is a random variable and

$$\text{Pr}(f(X) \in U) = \sum_{f(x) \in U} \text{Pr}(X = x)$$

where U is subset of $\mathbb{R}$ and we sum over all x with $f(x) \in U$.

We have

$$\mathbb{E}f(X) = \sum_{y \in f(X)(\Omega)} y \, \text{Pr}(f(X) = y)$$

where $f(X)(\Omega)$ is the set of possible values of $f(X(\omega))$.

If U is a subset of $\mathbb{R}$ and we consider the indicator function $\mathbb{I}_U$ given by

$$\mathbb{I}_U(x) = 1 \qquad\qquad for \; all \; x \in U,$$
$$\mathbb{I}_U(x) = 0 \qquad\qquad otherwise,$$

---

[11] And you may well be right.

then

$$\mathbb{E}\mathbb{I}_U(X) = \Pr(X \in U).$$

Proof The first result is easier to prove than to state. We observe that $f(X)$ is a function from Ω to $\mathbb{R}$ so, automatically, a random variable. Changing the way in which we add things up, we get

$$\begin{aligned}
\Pr(f(X) \in U) &= \Pr\{\omega \in \Omega : f(X(\omega)) \in U\} \\
&= \sum_{f(X(\omega)) \in U} \Pr\{\omega\} \\
&= \sum_{f(x) \in U} \sum_{X(\omega)=x} \Pr\{\omega\} \\
&= \sum_{f(x) \in U} \Pr(X = x)
\end{aligned}$$

as claimed.

The other two results are just as trivial. ∎

Halmos gave the result

$$\mathbb{E}f(X) = \sum_{y \in f(X)(\Omega)} y \Pr(f(x) = y)$$

of Lemma 2.4.19 the name 'the law of the unconscious statistician'.[12] The statistician need only observe $f(X(\omega))$ and so need only look at $\Pr(f(X) = y)$ while remaining *unconscious* of the finer structure involving $\Pr(\omega)$. Notice that, if the the statistician can only observe $X(\omega)$, she *must* remain *unconscious* of the finer structure involving $\Pr(\omega)$. For example, consider a die with $3n$ faces numbered 1 to $3n$. Suppose we define the random variable X to be k if the face shown has number $3r + k$ with $0 \le r \le n - 1$ and $1 \le k \le 3$. Then a statistician, who can only observe X, will never be able to tell the value of n.

The following important definition belongs to the same set of ideas.

Definition 2.4.20 *Let $(\Omega, \Pr)$ be a probability space and $X, Y : \Omega \to \mathbb{R}$ be random variables. We say that X and Y are* identically distributed *if*

$$\Pr(X = a) = \Pr(Y = a)$$

for all $a \in \mathbb{R}$.

Exercise 2.4.21 *A fair coin is thrown 3 times. Let $X = 1$, $Y = 2$ if the first throw is heads and let $X = -1$, $Y = -2$ if the first throw is tails. Let $Z = 1$ if the second throw is heads and let $Z = -1$ if the second throw is tails. Let*

[12] This rather startling nomenclature is now standard.

$W = 1$ *if the first and second throws are both heads or both tails and let* $W = -1$ *if they are different. Let* $V = 1$ *if all three throws are heads or all three are tails and let* $V = -1$ *otherwise.*

Show that X, Z and W are identically distributed. Show that no pair of X, Y and V are identically distributed.

The notion of a random variable is so attractive that it seems to take on a life of its own without reference to any underlying probability space. It is perfectly possible to operate at the elementary level of this book with a very vague idea of what a random variable might be. However, mathematicians[13] have two good reasons for insisting that a real-valued[14] random variable is a function

$$X : \Omega \to \mathbb{R}.$$

The first is purely pragmatic. The best and most exciting theorems in modern probability are proved by mathematicians like Kolmogorov for whom a random variable is a function $X : \Omega \to \mathbb{R}$.

The second reason is that ideas from probability theory are now used throughout pure mathematics. Since pure mathematicians require exact definitions, they can only use the concept of a random variable if it is precisely defined and the definition of a random variable as a function $X : \Omega \to \mathbb{R}$ is the best we have.[15]

Exercise 2.4.22 *(i) Compute the probability of throwing a total of r using two ordinary six-sided unbiased dice for* $2 \le r \le 12$.

(ii) Expand $(x + x^2 + x^3 + x^4 + x^5 + x^6)^2$.

(iii) A non-standard pair of dice is a pair of six-sided unbiased dice whose faces are numbered with strictly positive integers in a non-standard way. Find the probability of throwing a total of r with non-standard dice numbered $(2, 2, 2, 3, 5, 7)$ *and* $(1, 1, 5, 6, 7, 8)$ *for* $3 \le r \le 15$.

(iv) Expand $(3x^2 + x^3 + x^5 + x^7)(2x + x^5 + x^6 + x^7 + x^8)$.

(v) Show that there exists a non-standard pair of dice A and B such that, when thrown,

$$\mathrm{Pr}(\textit{total shown by A and B is r}) = \mathrm{Pr}(\textit{total shown by ordinary dice is r}).$$

[13] The needs of non-mathematicians and practical statisticians may be very different. Moroney's book [44] is an excellent introduction to practical statistics which does not use the words 'random variable' anywhere.

[14] A complex-valued random variable is a function $Z : \Omega \to \mathbb{C}$, a vector-valued random variable is a function $\mathbf{X} : \Omega \to \mathbb{R}^n$ and so on.

[15] Of course this means that, just as the Holy Roman Empire in 1790 was neither holy nor Roman nor an empire so, at least for mathematicians, a random variable is neither random nor a variable.

[*Hint:* $(x + x^2 + x^3 + x^4 + x^5 + x^6) = x(1 + x)(1 + x^2 + x^4)$
$$= x(1 + x + x^2)(1 + x^3).]$$

(vi) (Think about this. It may require more mathematical tools than you have.) Show that there is, essentially, only one non-standard pair of dice satisfying the conditions of (v).

Exercise 2.4.23 *Suppose I have a six-sided die numbered in the usual manner such that the probability of throwing r is p_r for $1 \leq r \leq 6$. Is it possible to choose the p_r in such a way that the probability that the sum of two independent throws takes the value k is $1/11$ for each $2 \leq k \leq 12$?*

Random variables are much easier to study under the assumption of independence. Sometimes, however, it may be more profitable to study random variables without this assumption.

Exercise 2.4.24 *(i) In the game of Simplejack a large pack of cards containing equal numbers of cards marked 2, 3 and 4 is shuffled. One card is dealt face up to the banker and one to the player. The player may (but need not) ask for a second card. After looking at both cards the player may (but need not) ask for a third card. When the player has finished, the banker must obey the following rules. If her card is a 2 or a 3, she must ask for a further card; if it is a 4, she must not. If both the player and the banker have a total score of 7 or more, the player pays one unit to the banker. If one of them has a total score of 7 or more and the other does not, then the higher scorer pays one unit to the lower. Otherwise, if the scores are unequal, the lower scorer pays one unit to the higher and, if the scores are equal, no money changes hands.*

Suppose first that the pack is so large that you may assume that the probability of any particular type of card is $1/3$ independent of the other cards dealt. Find the correct action for the player if she has r and the banker has s after the first cards are dealt. Show that the game favours the banker.

(ii) Suppose that, instead of adding the used cards to the pack and reshuffling, the player and banker continue playing a series of games without *replacing previously used cards. At some point, the player realises that most of the 3s and 4s have been dealt so the pack now consists of a large number of 2s and a few other cards. Show that, with the appropriate strategy, the game now favours the player.*

[The reader will probably find part (i) tedious rather than difficult. She will also be prepared to believe that, as we increase the number of different types of cards, the player's strategy will become more complicated.]

Simplejack is a very simplified version of 'Twenty-One' (also called Black-jack). In order to speed up the game, casinos play Blackjack in the manner described in (ii), only gathering the used cards and reshuffling after most of the pack has been used.

In 1960, *The Proceedings of the National Academy of Sciences of the United States of America* carried a paper with the title *A Favourable Strategy For Twenty-One* which began

> It has long been an open question whether those of the standard gambling games which are not repeated independent trials admit strategies favourable to the player.... In this note we settle the issue by showing that there is a markedly favourable mathematical strategy for one of the most widely played games, twenty-one or blackjack. [62]

A young professor of mathematics, called Edward Thorp, realised that, since casinos did not reshuffle their cards at the end of each deal, situations would arise in which the pack was poor in certain cards and the advantage shifted from the Blackjack banker to the Blackjack player. Using a computer to do the calculations for a large number of carefully chosen situations, he was able to produce a fairly simple method of 'card counting' which enabled players to recognise such situations. By betting heavily when the pack was favourable and lightly otherwise, it became possible to beat the casino. After a successful trial in Las Vegas he published his result in the paper just quoted and then in book entitled *Beat the Dealer* [63].

It is possible for a casino to counter Thorp's strategy by refusing to play with obvious 'card counters', by shuffling more frequently and by using larger packs of cards. Although 700 000 copies of the book were sold, it did not produce 700 000 millionaires and casinos continue to offer Blackjack to their patrons. Professor Thorp switched his attentions to the financial markets where the rewards are better and the participants do not resort to physical violence. He is reported to be a millionaire many times over and attributes some of his success to the use of the Kelly criterion which we shall talk about in Section 2.6.

Ill-disposed individuals sometimes ask mathematicians 'If you are so clever why are you not rich?'. Edward Thorp is both extremely clever and extremely rich. His exploits are described with gusto in Poundstone's *Fortune's Formula* [54].

Exercise 2.4.25 *In the popular TV quiz game* It's Your Choice *contestants are presented with n questions Q_1, Q_2, ... which they may choose to answer in any order. If they answer question Q_j correctly they receive a_j and proceed to the next question. If they get the answer wrong the game ends, but they keep all the money they have gained so far. Suppose that you know that you have probability p_j of answering question Q_j correctly. This exercise seeks the correct order to answer the questions if you wish to maximise your expected winnings. We assume $a_j > 0$ for all j.*

(i) What should you do if $p_j = 1$? What should you do if $p_j = 0$? From now on we suppose $0 < p_j < 1$ for all j.

(ii) Suppose you decide to ask the questions in order $j(1), j(2), \ldots, j(n)$. Show that your expected winnings are

$$a_{j(1)} p_{j(1)} (1 - p_{j(2)}) + (a_{j(1)} + a_{j(2)}) p_{j(1)} p_{j(2)} (1 - p_{j(3)})$$
$$+ (a_{j(1)} + a_{j(2)} + a_{j(3)}) p_{j(1)} p_{j(2)} p_{j(3)} (1 - p_{j(3)}) + \cdots$$
$$= p_{j(1)} a_{j(1)} + p_{j(1)} p_{j(2)} a_{j(2)} + p_{j(1)} p_{j(2)} p_{j(3)} a_{j(3)} + \cdots .$$

(iii) Suppose that you consider two possible orders of asking questions Plan A and Plan B. The two plans are identical except that in Plan A you ask $Q(i)$ as your rth question and $Q(j)$ as your $r + 1$th question while in you ask $Q(j)$ as your rth question and $Q(i)$ as your $r + 1$th question. If your expected winnings under plan A are e_A and your expected winnings under plan B are e_B show that

$$e_A - e_B = q\big(p_i(1 - p_j)a_i - p_j(1 - p_i)a_j\big)$$

where q is the probability that you answer the first $r - 1$ questions of Plans A and B correctly.

Show that you should prefer plan A to plan B if

$$\frac{a_i p_i}{1 - p_i} > \frac{a_j p_j}{1 - p_j}$$

and plan B to plan A if the inequality is reversed. If we have equality the plans are equally advantageous.

(iv) Show that you should ask the questions in order $j(1), j(2), \ldots, j(n)$ where

$$\frac{a_{j(r)} p_{j(r)}}{1 - p_{j(r)}} \geq \frac{a_{j(r+1)} p_{j(r+1)}}{1 - p_{j(r+1)}}$$

for all $1 \leq r \leq n - 1$.

(v) What does the strategy reduce to if $p_1 = p_2 = \cdots = p_n$? Why is this reasonable? What does the strategy reduce to if $a_1 = a_2 = \cdots = a_n$? Why is this reasonable? What does the strategy reduce to if p_j is very small? Why is this reasonable?

2.5 A law of large numbers

It is always enjoyable and sometimes useful to be able to carry out calculations along the lines of the next two exercises.

Exercise 2.5.1 *An ordinary pack of cards contains* 52 *cards of which* 4 *are aces. A bridge hand contains* 13 *cards. If the cards are well-shuffled what is the probability that a particular hand contains exactly* 3 *aces?*

Exercise 2.5.2 *According to de Moivre,*[16] *some players at the Royal Oak lottery, at the beginning of the eighteenth century,*

> ... who lost considerably by it, had their losses chiefly occasioned by an argument of which they could not perceive the fallacy. The odds against any particular point of the ball [in effect, one face of a 32-sided die] were one and thirty to one, which entitled the adventurers, in case they were winners, to have thirty two stakes returned, including their own; instead of which they having but eight and twenty, it was very plain that on the single account of the disadvantage of the play, they lost one eighth part of all the money they played for. But the Master of the Ball maintained they had no reason to complain; since he would undertake that any particular point of the ball would come up in two and twenty throws; of this he would offer to lay a wager,[17] and actually laid it when required. The seeming contradiction between the odds of one and thirty to one, and twenty two throws for any chance to come up, so perplexed the adventurers, that they began to think that the advantage was on their side; for which reason they played on, and continued to lose. [16]

Why did de Moivre say that the players 'lost one eighth part of all the money they played for'? Why was the Master of the Ball prepared to make the wager described?

You may be interested in the general case which is treated in Exercise C.2.

However, an endless collection of such calculations does not constitute a deep mathematical theory. In this section we use the ideas of a random variable and expectation to obtain the kind of general results which a proper mathematical theory should reveal.

We start with a series of inequalities associated with the great Russian mathematician Tchebychev.[18]

---

[16]   At this juncture I may mention
       That this erudition sham
       Is but classical pretension,
       The result of steady cram.
       Periphrastic methods spurning,
       To this audience discerning
       I admit this show of learning
       Is the fruit of steady cram!

                    [*Gilbert and Sullivan*, The Grand Duke]

[17] That is, he would offer to pay a given sum if a named point did not appear in the first 22 throws, provided that the bettor paid him the same sum if it did.

[18] This is the old fashioned transliteration of his name. Modern authors tend to write Chebychev or, in the case of some advanced spirits, Chebychov. Mathematical fame consists in having your name mispronounced and misspelt by generations yet to come.

Lemma 2.5.3 *Let* $(\Omega, \Pr)$ *be a probability space and* X *a random variable. Suppose* $f : \mathbb{R} \to \mathbb{R}$ *is a function such that* $f(x) = f(-x) \geq 0$ *for all* $x \in \mathbb{R}$ *and* $f(x)$ *is increasing as* x *runs from* 0 *to* ∞. *Then, if* $a \geq 0$ *and* $f(a) > 0$,

$$\Pr(|X| \geq a) \leq \frac{\mathbb{E}f(X)}{f(a)}.$$

Proof The conditions on f imply that

$$f(x) \geq 0 \qquad\qquad \text{for all } |x| < a,$$
$$f(x) \geq f(a) \qquad\qquad \text{for all } |x| \geq a.$$

In other words,

$$f(x) \geq f(a)\mathbb{I}_{(-a,a)^c}(x)$$

for all x, where

$$(-a, a)^c = \{x \in \mathbb{R} : |x| \geq a\},$$

and so

$$\mathbb{I}_{(-a,a)^c}(x) = 0 \qquad\qquad \text{for all } |x| < a,$$
$$\mathbb{I}_{(-a,a)^c}(x) = 1 \qquad\qquad \text{for all } |x| \geq a.$$

We thus have

$$f(X(\omega)) \geq f(a)\mathbb{I}_{(-a,a)^c}(X(\omega))$$

for all $\omega \in \Omega$ and, taking expectations, we obtain

$$\begin{aligned}
\mathbb{E}f(X) &\geq \mathbb{E}f(a)\mathbb{I}_{(-a,a)^c}(X)\\
&= f(a)\mathbb{E}\mathbb{I}_{(-a,a)^c}(X)\\
&= f(a)\Pr\{\omega \in \Omega : X(\omega) \in (-a, a)^c\}\\
&= f(a)\Pr(|X| \geq a).
\end{aligned}$$

The result follows.   ∎

Exercise 2.5.4 *Let* $(\Omega, \Pr)$ *be a probability space and* X *a random variable. Suppose that* $f : \mathbb{R} \to \mathbb{R}$ *is an increasing function with* $f(x) > 0$ *for all* x. *Then*

$$\Pr(X \geq a) \leq \frac{\mathbb{E}f(X)}{f(a)}.$$

In general, we are more interested in how X differs from $\mathbb{E}X$ than in how X differs from 0.

Lemma 2.5.5 *Let* (Ω, Pr) *be a probability space and* Y *a random variable. Suppose that* $f : \mathbb{R} \to \mathbb{R}$ *is a function such that* $f(x) = f(-x) \geq 0$ *for all* $x \in \mathbb{R}$ *and* $f(x)$ *is increasing as* x *runs from 0 to* ∞. *Then, if* $a \geq 0$ *and* $f(a) > 0$,

$$\mathrm{Pr}(|Y - \mathbb{E}Y| \geq a) \leq \frac{\mathbb{E}f(Y - \mathbb{E}Y)}{f(a)}.$$

Proof Set $X = Y - \mathbb{E}Y$ and apply Lemma 2.5.3. ∎

In order to make use of Lemma 2.5.5, we need to choose an f which is easy to work with. Long experience has shown that $f(x) = x^2$ gives such a function.

Definition 2.5.6 *Let* (Ω, Pr) *be a probability space and* X *be a random variable. We define the* variance *of* X *to be the number* var X *given by*

$$\mathrm{var}\, X = \mathbb{E}(X - \mathbb{E}X)^2.$$

In some sense, var X measures the 'amount of scatter about the mean $\mathbb{E}X$', but the reader should remember that there are many competing 'measures of scatter' and we use the variance for mathematical convenience. If we make this choice, then Lemma 2.5.5 takes the following form.

Lemma 2.5.7 *Let* (Ω, Pr) *be a probability space and* Y *a random variable. Then, if* $a > 0$,

$$\mathrm{Pr}(|Y - \mathbb{E}Y| \geq a) \leq \frac{\mathrm{var}\, Y}{a^2}.$$

Exercise 2.5.8 *Suppose* $1 \geq p \geq 0$ *and* $a > 0$. *Let* $\Omega = \{\omega_1, \omega_2, \omega_3\}$,

$$\mathrm{Pr}(\omega_1) = 1 - p, \ \mathrm{Pr}(\omega_2) = \mathrm{Pr}(\omega_3) = p/2$$

and define $X : \Omega \to \mathbb{R}$ *by*

$$X(\omega_1) = 0, \ X(\omega_2) = a, \ X(\omega_3) = -a.$$

Show that (Ω, Pr) *is a probability space and* X *is random variable such that*

$$\mathrm{Pr}(|X - \mathbb{E}X| \geq a) = p = \frac{\mathrm{var}\, X}{a^2}.$$

Thus Tchebychev's inequality given in Lemma 2.5.7 cannot be improved without further conditions on X.

The following useful results are easy to prove.

Lemma 2.5.9 *Let* $(\Omega, \Pr)$ *be a probability space and* X, Y *be random variables.*

(i) $\operatorname{var}(X + a) = \operatorname{var} X$ *for all real* a.
(ii) $\operatorname{var}(aX) = a^2 \operatorname{var} X$ *for all real* a.
(iii) $\operatorname{var}(X) = \mathbb{E}X^2 - (\mathbb{E}X)^2$.
(iv) If X *and* Y *are independent then* $\operatorname{var}(X + Y) = \operatorname{var} X + \operatorname{var} Y$.

Proof (i) Observe that $\mathbb{E}(X + a) = (\mathbb{E}X) + a$, so

$$\operatorname{var}(X + a) = \mathbb{E}\big((X + a) - \mathbb{E}(X + a)\big)^2 = \mathbb{E}(X - \mathbb{E}X)^2 = \operatorname{var} X.$$

(ii) We have

$$\operatorname{var}(bX) = \mathbb{E}(bX - \mathbb{E}bX)^2 = \mathbb{E}(bX - b\mathbb{E}X)^2$$
$$= \mathbb{E}b^2(X - \mathbb{E}X)^2 = b^2\mathbb{E}(X - \mathbb{E}X)^2 = b^2 \operatorname{var} X.$$

(iii) We have

$$\operatorname{var} X = \mathbb{E}(X - \mathbb{E}X)^2 = \mathbb{E}\big(X^2 - (2\mathbb{E}X)X + (\mathbb{E}X)^2\big)$$
$$= \mathbb{E}X^2 - (2\mathbb{E}X)\mathbb{E}X + (\mathbb{E}X)^2 = \mathbb{E}X^2 - 2(\mathbb{E}X)^2 + (\mathbb{E}X)^2$$
$$= \mathbb{E}X^2 - (\mathbb{E}X)^2.$$

(iv) Let $U = X - \mathbb{E}X$ and $V = Y - \mathbb{E}Y$. Then $\mathbb{E}U = \mathbb{E}V = 0$, so

$$\mathbb{E}(U + V) = \mathbb{E}U + \mathbb{E}V = 0.$$

By part (i), $\operatorname{var} X = \operatorname{var} U$, $\operatorname{var} Y = \operatorname{var} V$ and $\operatorname{var}(X + Y) = \operatorname{var}(U + V)$. By independence

$$\mathbb{E}UV = \mathbb{E}U\mathbb{E}V = 0$$

and

$$\operatorname{var}(U + V) = \mathbb{E}(U + V)^2 = \mathbb{E}(U^2 + 2UV + V^2)$$
$$= \mathbb{E}U^2 + 2\mathbb{E}UV + \mathbb{E}V^2 = \mathbb{E}U^2 + \mathbb{E}V^2 = \operatorname{var} U + \operatorname{var} V.$$

Thus $\operatorname{var}(X + Y) = \operatorname{var} X + \operatorname{var} Y$ as stated. ∎

Exercise 2.5.10 *(i) Show that, if* X *is a random variable,*

$$\mathbb{E}(X - a)^2 = \operatorname{var} X + (\mathbb{E}X - a)^2.$$

(You may recognise this as the parallel axis theorem for computing moments of inertia in mechanics.) What value of a *minimises* $\mathbb{E}(X - a)^2$?
(ii) If X *is a random variable and* $Y = -X$, *show that*

$$\operatorname{var} X + \operatorname{var} Y = 2 \operatorname{var} X, \quad \text{but} \quad \operatorname{var}(X + Y) = 0.$$

Why does this not contradict part (iv) of Lemma 2.5.9?

(iii) Prove part (iv) of Lemma 2.5.9 by direct computation of $\mathrm{var}(X + Y)$.

Exercise 2.5.11 *Investigate how much of Lemma 2.5.9 carries over (with appropriate changes) if we replace* var X *by* $\mathbb{E}|X - \mathbb{E}X|$ *and if we replace* var X *by* $\mathbb{E}(X - \mathbb{E}X)^4$.

Exercise 2.5.12 *Let* (Ω, Pr) *be a probability space and* $X_1, X_2, \ldots, X_n$ *be independent random variables. Show that*

$$\mathrm{var}(X_1 + X_2 + \cdots + X_n) = \mathrm{var}\, X_1 + \mathrm{var}\, X_2 + \cdots + \mathrm{var}\, X_n.$$

It is traditional to write $\mathbb{E}X = \mu$ (small Greek mu) and to denote the positive square root $(\mathrm{var}\, X)^{1/2}$ by σ (small Greek sigma) so that var $X = \sigma^2$. We introduce the square for dimensional reasons[19] (observe that, if var $X = \sigma^2$, then var $aX = (a\sigma)^2$). We return briefly to this point on page 290

We can now prove the *weak law of large numbers*. (The word 'weak' is technical.[20] As we shall see, this is a very powerful result.)

Jacob Bernoulli introduced the first version of this law with the proud words 'Both its novelty and its great utility combined with its equally great difficulty can add to the weight and and value of all the other chapters of this theory' [6] and the result together with its many variations has remained central to probability and statistics for 300 years.[21]

Theorem 2.5.13 *Let* (Ω, Pr) *be a probability space. Suppose that* $X_1, X_2, \ldots,$ X_n *are independent identically distributed random variables each with mean* μ *and variance* σ^2. *Then, if* $a > 0$,

$$\mathrm{Pr}\left(\left|\frac{X_1 + X_2 + \cdots + X_n}{n} - \mu\right| \geq a\right) \leq \frac{\sigma^2}{na^2}.$$

Proof By Exercise 2.5.12,

$$\mathrm{var}(X_1 + X_2 + \cdots + X_n) = \mathrm{var}\, X_1 + \mathrm{var}\, X_2 + \cdots + \mathrm{var}\, X_n = n\sigma^2,$$

---

[19] We call σ the standard deviation of X and think of it as a length.

[20] In informal discussions, I will follow common practice and simply refer to 'the law of large numbers'.

[21] We shall meet several Bernoullis during the course of the book. Jacob Bernoulli wrote *Ars Conjectandi* [6] (The Art of Conjecturing) which begins by developing the ideas of Huygens (see Appendix C) and concludes with the first 'law of large numbers'. His brother Johann Bernoulli, another fine mathematician had three sons all of whom became mathematicians. One of these was Daniel Bernoulli whose work on smallpox will discussed in Section 3.3. He wrote on the St Petersburg paradox (see Exercise 9.3.4) first proposed by his cousin Nicholas Bernoulli. Nicholas Bernoulli gave the first solution to the problem considered in Exercise 2.3.10.

so, by Lemma 2.5.9 (ii),

$$\text{var}\left(\frac{X_1 + X_2 + \cdots + X_n}{n}\right) = n^{-2} n \sigma^2 = \frac{\sigma^2}{n}.$$

The parallel results on expectation show that

$$\begin{aligned}
\mathbb{E}\left(\frac{X_1 + X_2 + \cdots + X_n}{n}\right) &= \frac{1}{n}\mathbb{E}(X_1 + X_2 + \cdots + X_n) \\
&= \frac{1}{n}(\mathbb{E}X_1 + \mathbb{E}X_2 + \cdots + \mathbb{E}X_n) \\
&= \frac{1}{n}(n\mu) = \mu.
\end{aligned}$$

The theorem now follows from Tchebychev's inequality in the form of Lemma 2.5.7 applied to

$$Y = \frac{X_1 + X_2 + \cdots + X_n}{n}.$$

∎

Exercise 2.5.14 *Let* (Ω, Pr) *be a probability space. Suppose that* $Y_1, Y_2, \ldots,$ Y_n *are independent random variables such that* $\mathbb{E}Y_j = \mu_j$ *and* $\text{var}\, Y_j \le \sigma^2$ *for all* $1 \le j \le n$. *Show that, if* $a > 0$,

$$\text{Pr}\left(|(Y_1 + Y_2 + \cdots + Y_n) - (\mu_1 + \mu_2 + \cdots + \mu_n)| \ge n^{1/2}a\right) \le \frac{\sigma^2}{a^2}.$$

Suppose that $Y_1, Y_2, \ldots, Y_n$ represent the results of a long series of bets on different horse races. We know that μ_j is the expected value of the jth bet and that the actual outcome Y_j may be very different. Even if we make a long series of bets we certainly do not expect our actual gain $Y_1 + Y_2 + \cdots + Y_n$ to equal our expected gain $\mu_1 + \mu_2 + \cdots + \mu_n$. However, if we choose a so that σ^2/a^2 is small, Exercise 2.5.14 tells us that (under the hypotheses of the exercise) with high probability, our actual gain will not differ from our expected gain by more than some fixed multiple of $n^{1/2}$. Suppose, in particular, that our bets are all sufficiently favourable that $\mu_j > \mu$ where $\mu > 0$. Then $\mu_1 + \mu_2 + \cdots + \mu_n > n\mu$ and our expected gain increases at least linearly with n while the range by which we expect our gain to vary from its expected value only increases as the square root of n.

Exercise 2.5.15 *Let* (Ω, Pr) *be a probability space. Suppose that* $Y_1, Y_2, \ldots,$ Y_n *are independent random variables such that* $\mathbb{E}Y_j \ge \mu$ *and* $\text{var}\, Y_j \le \sigma^2$ *for* *all* $1 \leq j \leq n$. *Suppose that c and b are fixed with* $1 > b$, $c > 0$. *Find an* N *(depending on* μ, σ^2, *b and c) such that*

$$\Pr\left(Y_1 + Y_2 + \cdots + Y_n > (1-c)n\mu\right) \geq 1 - b$$

whenever $n \geq N$. *(You are not asked to find the best possible* N. *If we knew a little more about the random variables and a lot more about probability theory we could do much better.)*

Explain the meaning of your result to an intelligent but non-mathematical gambler.

Results like Exercise 2.5.14 strengthen our feeling that the mathematically defined 'expected value' of a bet is a good indication of the necessarily undefined 'value to the sensible bettor'.

However, it is important to keep in mind examples like the following.

Example 2.5.16 *Let* $1 > \epsilon, \delta > 0.$[22] *Suppose that* X_1, X_2, ..., X_n *are independent random variables such that*

$$\Pr(X_j = 0) = 1 - 2^{-j} \text{ and } \Pr(X_j = 2^j) = 2^{-j},$$

so that, in particular, $\mathbb{E}X_j = 1$. *Then, if* $n > N$, *where* N *is some integer, depending on* ϵ *and* δ, *to be determined, we have*

$$\Pr\left(\left|\frac{X_1 + X_2 + \cdots + X_n}{n} - 1\right| \leq \delta\right) \geq 1 - \epsilon,$$

and so, in particular,

$$\Pr\left(\left|\frac{X_1 + X_2 + \cdots + X_n}{n} - 1\right| \geq 1 - \delta\right) \geq 1 - \epsilon.$$

Proof Choose an integer $k \geq 1$ with $2^{-k} < \epsilon$. Now choose N so that $2^{k+1}/N < \delta$. From now on, we suppose that $n \geq N$.

Let A_j be the event that $X_j \neq 0$. We have

$$\Pr\left(\bigcup_{j=k+1}^{N} A_j\right) \leq \sum_{j=k+1}^{N} \Pr(A_j) = \sum_{j=k+1}^{N} 2^{-j} < 2^{-k} < \epsilon.$$

---

[22] Very few of my readers will fail to recognise δ (small Greek delta) and ϵ (small Greek epsilon). Traditionally these are used to represent small strictly positive quantities.

Thus $X_{k+1} = X_{k+2} = \cdots = X_n = 0$ and

$$\frac{X_1 + X_2 + \cdots + X_n}{n} = \frac{X_1 + X_2 + \cdots + X_k}{n} \leq \frac{2^1 + 2^2 + \cdots + 2^k}{n}$$

$$< \frac{2^{k+1}}{n} \leq \frac{2^{k+1}}{N} < \delta$$

with probability at least $1 - \epsilon$ and we are done.   ∎

Example 2.5.16 is a warning that both our theorems and our intuition may fail when we consider a sequence of bets which increase in a 'wild manner'.

Exercise 2.5.17 *In Example 2.5.16, we considered a sequence of random variables without giving an associated probability space* (Ω, Pr). *Fill this gap.*

From time to time, I shall give a sequence of random variables without giving an associated probability space. When this occurs, you should convince yourself that it is an easy matter to provide an appropriate (Ω, Pr).

We conclude this section with Exercise 2.5.18, which illustrates the kind of estimates we can make using the weak law of large numbers, and Exercise 2.5.20, whose second part gives a technique for finding the variance of sums of random variables which are not independent.

Exercise 2.5.18 *Consider a public opinion survey. We use the following model. Each person interviewed has probability* p *of answering yes to a particular question and the answers of the various respondents are independent. In effect, if we conduct n interviews, this is equivalent to making n throws of a coin which has probability* p *of coming down heads.*

Write $X_j = 1$ *if the jth throw is heads (that is, if the jth respondent says yes) and* $X_j = 0$ *otherwise.*

(i) Show that

$$\mathbb{E}X_j = p \text{ and } \operatorname{var} X_j = p(1 - p).$$

(ii) Show that $p(1 - p) \leq 1/4$ *for all* $0 \leq p \leq 1$.

(iii) Use the weak law of large numbers to show that, if we take $a > 0$ *and write*

$$\bar{X} = \frac{X_1 + X_2 + \cdots + X_n}{n},$$

we have

$$\mathrm{Pr}(|\bar{X} - p| \geq a) \leq \frac{1}{4na^2}.$$

(iv) Deduce that if we do a survey with 1000 *people, the probability that* $\bar{X}$ *will differ from p by more than* $1/10$ *is less than* $1/40$.

(v) How large will n have to be for the argument above to show that the probability that $\bar{X}$ will differ from p by more than $1/20$ is less than $1/40$?

In Section 11.3 we shall see that the estimates of this question can be improved but it will remain true that an increase in accuracy by a factor of K (with the same risk of error) demands an increase in the number of people surveyed by a factor of K^2. This 'iron law'[23] reappears throughout probability and statistics.

However, few users of opinion polls require high accuracy. Newspapers use polls to entertain us and pressure groups to try to influence us and neither purpose demands a high standard of evidence. If you are a manufacturer, it is reassuring to know that most people have heard of your product and disturbing if very few have, but the exact percentage hardly matters. Moreover, we know that people's answers to questions depend on who asks them and how the questions are asked.[24] Human behaviour renders chimeric any attempt to use increasing sample size to reduce uncertainty below a certain level.

Exercise 2.5.19 (i) If $p_1 + p_2 + \cdots + p_m = 1$ show, by considering

$$(p_1 - m^{-1})^2 + (p_2 - m^{-1})^2 + \cdots + (p_m - m^{-1})^2,$$

or otherwise, that

$$p_1^2 + p_2^2 + \cdots + p_m^2 \geq \frac{1}{m}$$

with equality if and only if

$$p_1 = p_2 = \cdots = p_m = \frac{1}{m}.$$

(ii) You and n other hermits (call them $A_1, A_2, \ldots, A_n$) can choose to retire to m secluded grottos. Each of you chooses independently of the others and there is a probability p_j that any hermit will choose the jth grotto. Let $Y_j = 1$ if you choose the jth grotto and $Y_j = 0$ if you do not. Let $X_{jk} = 1$ if A_k chooses the jth grottos and $X_{jk} = 0$ if A_k does not. Show that the number Z of hermits (excluding yourself) who choose the same grotto as you do is given by

$$Z = \sum_{k=1}^{n} Y X_{1k} + \sum_{k=1}^{n} Y X_{2k} + \cdots + \sum_{k=1}^{n} Y X_{mk}.$$

---

[23] Which like all 'iron laws' only operates under certain circumstances.

[24] 'A referendum', de Gaulle is alleged to have said, 'will certainly give you an answer, but not necessarily to the question asked'. A couple of hours spent with [30] will prove at least as useful and amusing as a longer time spent with its numerous successors.

Deduce that

$$\mathbb{E}Z = n(p_1^2 + p_2^2 + \cdots + p_m^2)$$

and conclude that the expected number of other hermits who choose the same grotto as you is strictly greater than the average number of other hermits per grotto unless all grottos are equally likely to be chosen.

Show that the expected number of hermits (including yourself) who choose your grotto is always strictly greater than the average number of hermits per grotto. Thus 'on average the grottos you choose will be more crowded than the average grotto'. Explain to a non-mathematician why we should expect this.

Results of this kind are called 'inspection paradoxes' but pessimists will consider this exercise as a confirmation of Murphy's law.[25]

Even if $p_1 = p_2 = \cdots = p_m = \frac{1}{m}$, we shall see in Exercise 10.5.10 (viii) that m has to be rather large compared with n for this method of choosing grottos to give satisfactory results.

Exercise 2.5.20 *(i) The $2n$ children at* Miss Prism's Academy for the Offspring of Gentlefolk *sit at n double desks. Each day they sit themselves at random, but keep the same seat throughout the day. A child without the sniffles will certainly catch them if their neighbour has the sniffles, but will not catch them otherwise.*

Suppose that, at the start of the day, the probability that a child has the sniffles is p, independent of what is true for the other children. Let $X_j = 1$ if one of the children at the jth desk catches sniffles during the day and $X_j = 0$ otherwise. If the number of children who catch the sniffles during the day is X, use the fact that

$$X = X_1 + X_2 + \cdots + X_n$$

to find $\mu_{p,n} = \mathbb{E}X$ and $\sigma_{p,n}^2 = \operatorname{var} X$.

Show that $\mu_{p,n}$ is largest when $p = 1/2$.

(ii) The conditions of the first paragraph of (i) continue to apply, but now we know that k children have the sniffles at the beginning of the day. Let $Y_j = 1$ if one of the children at the jth desk catches sniffles during the day and $Y_j = 0$ otherwise. Compute $\mathbb{E}Y_1$, $\mathbb{E}Y_1^2$ and $\mathbb{E}Y_1 Y_2$.

If the number of children who catch the sniffles during the day is Y, use the fact that

$$Y = Y_1 + Y_2 + \cdots + Y_n$$

---

[25] If something can go wrong it will.

to find $\tilde{\mu}_{k,n} = \mathbb{E}Y$ and $\tilde{\sigma}^2_{k,n} = \mathrm{var}\, Y$. *(There is no need to seek the simplest form form $\tilde{\sigma}^2_{k,n}$.)*

(iii) If $2n - 1 \geq k \geq 1$, show that $\tilde{\mu}_{k,n} \neq \mu_{k/2n,n}$.

If $1 > p > 0$ and k_n is a sequence of integers with $0 \leq k_n \leq 2n$ such that $k_n/2n \to p$, show that

$$\frac{\tilde{\mu}_{k_n,n}}{\mu_{p,n}} \to 1$$

as $n \to \infty$.

2.6 A long day at the races

In Section 1.8 we identified a class of *knowledgeable bettors* who, from time to time, could identify a race and a horse such that up (the product of the payout ratio u and the probability p) of the horse winning was greater than 1. Such bettors can buy a bet whose expected value kup exceeds its cost k.

It might be thought that knowledgeable bettors could make an easy living out of betting, but things are not quite as simple as that.

Exercise 2.6.1 *A fair coin is tossed n times in succession. If I pay k ahead of a particular throw, a very rich (and not very bright) individual is prepared to return $\frac{11}{5}k$ if the throw is heads but nothing if the throw is tails. Show that the expected return on a bet of k is $\frac{11}{10}k$.*

I start with 1 and decide to stake everything I have on each throw. Show that my expected winnings are $(11/10)^n$ and that, at the end of the game, I will have the sum of $(11/5)^n$ with probability 2^{-n} and 0 with probability $1 - 2^{-n}$. Evaluate the expressions in the previous sentence, either using a calculator or appropriate approximations, when $n = 5$, $n = 10$, $n = 20$ and $n = 30$.

I hope the reader will agree that the strategy set out in the previous exercise is not an attractive one when n is large. I have a vanishingly small probability of gaining an astronomical sum but a near certainty of ending up bankrupt. Surely it would be better to keep something in reserve so that a single throw will not bankrupt me. So much is obvious – but how much should I keep in reserve?

Theorem 2.6.2 *Let $1 > p > 0$, $u > 0$ and $1 > t \geq 0$. Suppose that a coin is tossed n times in succession. The probability that it comes down heads is p and, if I pay k ahead of a particular throw, then I get back uk if the throw is heads but nothing if the throw is tails. I start with a fortune of $X_0 = 1$. If I have* *a fortune X_j after the jth throw, then I bet tX_j, retaining $(1 - t)X_j$. Thus my fortune X_{j+1} after the $j + 1$th throw is given by*

$$X_{j+1} = \begin{cases} X_j(tu + (1 - t)) & \text{if the } j\text{th throw is heads,} \\ X_j(1 - t) & \text{if the } j\text{th throw is tails.} \end{cases}$$

We set

$$Y_{j+1} = \frac{X_{j+1}}{X_j} = \begin{cases} tu + (1 - t) & \text{if the } j\text{th throw is heads,} \\ 1 - t & \text{if the } j\text{th throw is tails.} \end{cases}$$

(i) $Y_1, Y_2, \ldots Y_n$ is a sequence of independent identically distributed random variables.

(ii) $\log Y_1$, $\log Y_2$, $\ldots$, $\log Y_n$ is a sequence of independent identically distributed random variables.

(iii) Set $\mathbb{E} \log Y_1 = \tilde{\mu}$ and $\operatorname{var} \log Y_1 = \tilde{\sigma}^2$. Then, if $a > 0$,

$$\Pr\left(\left| \frac{\log Y_1 + \log Y_2 + \cdots + \log Y_n}{n} - \tilde{\mu} \right| \geq a \right) \leq \frac{\tilde{\sigma}^2}{na^2}.$$

(iv) If $a > 0$,

$$\Pr\left(\left| \frac{\log X_n}{n} - \tilde{\mu} \right| \geq a \right) \leq \frac{\tilde{\sigma}^2}{na^2}.$$

(v) Given ϵ, $\delta > 0$, we can find an N such that

$$\Pr\left(\left| \frac{\log X_n}{n} - \tilde{\mu} \right| \geq \delta \right) \leq \epsilon$$

for all $n \geq N$.

(vi) Write $L = \exp \tilde{\mu}$. Given $\epsilon > 0$ and $k > 1$, we can find an N such that

$$\Pr\left((k^{-1}L)^n < X_n < (kL)^n \right) > 1 - \epsilon$$

for all $n \geq N$.

Exercise 2.6.3 *Check that you can prove each part of Theorem 2.6.2. In part (vi) you may want to take $\delta = \log k$.*

Lemma 2.6.4 *Consider a single toss of a coin with probability p of coming down heads with $p < 1$. Suppose that a bet on heads of k has payout ratio u. Suppose that we have 1 unit and we bet t on heads retaining $1 - t$ units $[0 \leq t < 1]$. If Y is the expected value of our fortune after the throw, then*

$$\mathbb{E} \log Y = p \log \left(1 + (u - 1)t \right) + (1 - p) \log(1 - t).$$

The value of $\mathbb{E} \log Y$ *is maximised by taking* $t = 0$ *if* $up \le 1$ *and by setting*

$$t = \frac{up - 1}{u - 1}$$

if $up > 1$.

Proof If $u \le 1$ then we should take $t = 0$ (since we never get back more than we spent on a bet). From now on we suppose $u > 1$.

If the coin comes down heads, Y takes the value $(1 - t) + ut = 1 + (u - 1)t$ and, if it comes down tails, Y takes the value $1 - t$. Thus $\mathbb{E} \log Y = f(t)$ with

$$f(t) = p \log \big(1 + (u - 1)t\big) + (1 - p) \log(1 - t)$$

as required. Now

$$\begin{aligned}
f'(t) &= \frac{p(u - 1)}{1 + (u - 1)t} - \frac{1 - p}{1 - t} = \frac{p(u - 1)(1 - t) - (1 - p)(1 + (u - 1)t)}{(1 + (u - 1)t)(1 - t)} \\
&= \frac{\big(p(u - 1) - (1 - p)\big) - \big(p(u - 1) + (1 - p)(u - 1)\big)t}{(1 + (u - 1)t)(1 - t)} \\
&= \frac{(up - 1) - (u - 1)t}{(1 + (u - 1)t)(1 - t)}.
\end{aligned}$$

Thus, if $up \le 1$, $f'(t) < 0$ for all $0 \le t < 1$ so f is decreasing and the maximum occurs when $t = 0$.

If $up > 1$, then $f'(t) > 0$ when $(up - 1) > (u - 1)t$ and $f'(t) < 0$ when $(up - 1) < (u - 1)t$. Thus f attains its maximum when

$$t = \frac{up - 1}{u - 1}$$

as stated. ∎

Exercise 2.6.5 *Consider the situation described in Theorem 2.6.2 and suppose* $up > 1$. *Three players A, B and C decide to follow the strategy described with A betting a proportion* t_A *of her fortune on each throw, B betting a proportion* t_B *and C a proportion* t_C. *Suppose that*

$$t_B = \frac{up - 1}{u - 1}$$

and $0 < t_A < t_B < t_C$. *Let* X_n, Y_n *and* Z_n *be their fortune at the end of the game (all starting with a fortune of 1).*

(i) *If* $\epsilon > 0$, *show that there exists an N (depending on* ϵ, u, p, t_A *and* t_C) *such that*

$$\Pr(X_n, Z_n < Y_n) > 1 - \epsilon,$$

whenever $n \geq N$.

(ii) If $\epsilon > 0$, *show that there exists an* N' *(depending on* $\epsilon > 0$, u, p *and* t_A*) such that*

$$\Pr(X_n > 1) > 1 - \epsilon,$$

whenever $n \geq N'$.

(iii) Show that there exists a t_D *such that* $1 > t_D > t_B$ *with the following property. If* $t_B < t_C < t_D$ *and* $\epsilon > 0$, *then there exists an* N'' *(depending on* ϵ, u, p *and* t_C*) such that*

$$\Pr(Z_n > 1) > 1 - \epsilon,$$

whenever $n \geq N''$. *However, if* $t_C > t_D$ *and* $\epsilon > 0$, *then there exists an* N''' *(depending on* ϵ, u, p *and* t_C*) such that*

$$\Pr(Z_n < 1) > 1 - \epsilon$$

whenever $n \geq N'''$.

The ideas in this section and the next entered mathematics in a paper of Kelly [33] dealing with sending signals over noisy channels. For this reason, the advice to take

$$t = \frac{up - 1}{u - 1}$$

is known as Kelly's rule. A bettor who seeks to maximise the logarithm of her fortune is known as a Kelly bettor.

Exercise 2.6.6 *Show that, if we use Kelly's rule, then, when we bet on an event with probability* p, *we will never bet more than a proportion* p *of our fortune however good the odds offered.*

Very informally, Exercise 2.6.5 shows that, if we bet a proportion less than Kelly recommends, our fortune will increase (in the long run, with high probability) but more slowly than it could. If we bet a bit more of a proportion than Kelly recommends, our fortune will again increase (in the long run, with high probability) but more slowly than it could. If we bet a much higher proportion than Kelly recommends, then our fortune will decrease (in the long run, with high probability).

Any reader who hopes to use these ideas to get rich quickly should consider their application to the situation described in Exercise 2.6.1. Here, Kelly recommends $t = 1/10$. In Section 10.4 we shall look at similar problems and come to a similar conclusion. *Unless a bet is extremely favourable, you should never risk more than a small proportion of your fortune on a single throw.*

Because we only risk a small proportion of our fortune, our wealth can only change slowly. If we win one toss and lose the next or vice versa our fortune will only increase by a factor of

$$\frac{9}{10} \times \left(\frac{9}{10} + \frac{1}{10} \times \frac{11}{5} \right) = \frac{504}{500}$$

over those two goes. If we throw exactly 87 heads and 87 tails in some order (making a total of 174 throws) we will double our money.

But things are rather more uncomfortable that this. If we throw 84 heads and 90 tails our fortune will be practically unchanged from the start and if we throw 81 heads and 93 tails we will have halved our fortune. (An optimist would turn these figures round and remark that if we throw 90 heads and 84 tails we will quadruple initial fortune and if throw 81 heads and 93 tails we will have multiplied our fortune by 8.) At some stage the reader may wish to try Exercise C.5 (vi).

As we say repeatedly throughout the book, mathematical theorems are like legal contracts. They say exactly what they say and not what we believe they say. The weak law of large numbers tells us that, after a large number of independent trials, the average value of the results obtained is likely to be close to the mean. In the form just stated, it does not tell us how large the 'large number' has to be. If we do the calculations, we find that, for the game described in Exercise 2.6.1, $\tilde{\sigma}$ (defined in Theorem 2.6.2) is very big compared to $\tilde{\mu}$ and this means that we must take n very large before we can say that is highly likely that we will more than double our fortune. (Of course, if we take n this large, we are quite likely to multiply our fortune many times over.)

Exercise 2.6.7 *(i) Check that Kelly recommends* $t = 1/10$ *and use a calculator to compute the associated* $\exp \tilde{\mu}$, *and* $\tilde{\sigma}^2$ *for Exercise 2.6.1. (We use the notation of Theorem 2.6.2.) Roughly how many throws does it take to have a reasonable probability that we will at least double our stake? Is there a useful exact answer to this question?*

(ii) (This parts require more sophistication than most of our exercises.) We know (see Appendix A) that, if x is small,

$$\log(1 + x) \approx x - x^2/2 \text{ and } \exp x \approx 1 + x.$$

As before, let

$$t = \frac{up - 1}{u - 1}$$

and (mainly to simplify the calculation) choose $p = 1/2$ *If t is small and* $t > 0$, *estimate* $\tilde{\mu}, \tilde{\sigma}^2$. *Explain why your answers mean that if t is small it is going to take a very long time indeed before you can be reasonably confident of doubling your stake.*[26]

Exercise 2.6.8 *Suppose that I play a game in which I have probability 9/10 of getting back double my stake and probability 1/10 of getting back nothing. What proportion of my fortune does Kelly recommend me to gamble? Suppose that I play 20 such games. Draw up a table showing the probability that I win n of the games (and lose the others) and my associated fortune for $20 \geq n \geq 14$ if I use the following strategies.*

(i) I stake 1/2 of my fortune each go.
(ii) I follow the Kelly recommendation.
(iii) I stake my entire fortune at each go.

Suppose that I know that the coin in Theorem 2.6.2 is tossed m times a year. I have a sum K which I can either invest at compound interest, so that after M years I have Kl^M, or use to bet in the manner advised by Kelly to obtain a sum close to KL^{Mm} (with high probability when Mm is large). If l is substantially greater than L^m, I should leave my money in the bank (unless I love gambling for its own sake) and if l is substantially smaller than L^m, I should use it to gamble (unless I have other reasons for disliking gambling).

Observe that the decision whether to try to increase my capital by gambling depends not only on the advantage L that I have on the bet, but also on the number of times m per year that I can make it. As we observed earlier, even the knowledgeable bettor will only occasionally see a horse and a race for which she knows that $up > 1$. If such occasions are too rare, betting will cease to be worthwhile even for knowledgeable bettors.

In real life, we are not presented with a series of identical bets but, just as Theorem 2.5.13 can be extended to Exercise 2.5.14, so, under quite weak hypotheses, Kelly's law can be extended to show that, if we have a long series of independent bets, the way to obtain the largest final sum which can be attained with high probability is to maximise $\mathbb{E} \log Y_r$, where Y_r is the ratio between our fortune after the rth bet and that before. As we have seen, if the rth bet is on a coin falling heads with probability p_r and the payout ratio is u_r, then we should bet nothing if $u_r p_r \leq 1$ and a proportion

$$\frac{u_r p_r - 1}{u_r - 1}$$

---

[26] This is not a fault of Kelly's criterion. Schemes which seek to get rich quick by making bets at slim favourable odds must involve 'picking up pennies in front of a steamroller'.

of our fortune otherwise. However, Example 2.5.16 shows that these conclusions may fail if the proportion of capital that we bet varies wildly. (We need to worry if we bet almost all our fortune at any time.)

Exercise 2.6.9 *If you are interested, state and prove theorems that make precise the statements of the previous paragraph.*

In practice, we are unlikely to know $u_r p_r$ exactly. Exercise 2.6.5 suggests that, in such circumstances, it is better to veer on the side of caution and bet a smaller proportion of your fortune than Kelly's formula suggests.

Kelly's rule is inapplicable to the great majority of bettors. It tells us how to bet if we start off with a certain amount of betting capital and never replenish it (except from our winnings) or spend it. Most bettors do not act like this. Since they bet for recreation, they stake a certain fixed amount each week. If they win they spend their winnings. If they lose they hope for better luck next week.[27]

Exercise 2.6.10 *Let $0 < p < 1$, $u > 0$ and $1 \geq t \geq 0$. Suppose that a coin is tossed n times in succession. The probability that it comes down heads is p and, if I pay k ahead of a particular throw, then I get back uk if the throw is heads but nothing if the throw is tails. I bet a fixed amount t retaining $(1 - t)$. Thus my fortune $X_{j+1}(t)$ after the $j + 1$th throw is given by*

$$X_{j+1}(t) = \begin{cases} X_j(t) + (tu + (1-t)) & \text{if the jth throw is heads,} \\ X_j(t) + (1-t) & \text{if the jth throw is tails.} \end{cases}$$

(i) If $up < 1$, $\epsilon > 0$ and $0 < t \leq 1$, show that there exists an N (depending on ϵ and p) such that

$$\Pr\left(X_n(t) < X_n(0)\right) > 1 - \epsilon,$$

whenever $n \geq N$.

(ii) If $up > 1$, $\epsilon > 0$ and $0 \leq t < 1$, show that there exists an N' (depending on ϵ, u, p and t) such that

$$\Pr\left(X_n(t) < X_n(1)\right) > 1 - \epsilon,$$

whenever $n \geq N'$.

In other words, anyone who has a fixed sum to bet each time should bet the full sum if $up > 1$ and nothing if $up < 1$.

---

[27] To this group we may add those gamblers who are in the process of making a small fortune by the simple expedient of starting with a large one. They believe that they are keeping their gambling capital separate, but cannot resist replenishing it from time to time.

The answer to 'how to bet', even in favourable circumstances, is not unique but depends on 'why we bet'.

2.7 The two-horse race

In the previous section, we derived Kelly's rule for the proportion of our capital we should risk on a simple bet such as heads or tails. In this section, we discuss the more complicated case of a race in which we can bet on several horses simultaneously.

Some key points emerge when we discuss the rather artificial situation of a two-horse race. To fix notation, assume that the first horse has probability p_1 of winning and has a payout ratio u_1 and the second has probability $p_2 = 1 - p_1$ of winning and a payout ratio u_2. We assume $u_1 p_1 \geq u_2 p_2$.

So far we have considered three kinds of bettors. The first was discussed in Section 1.1. She dislikes betting and leaving things to chance. She will only bet if she is certain of winning whatever happens. When she does bet, her concern is to maximise her winnings, whichever horse wins. We saw, in Lemma 1.1.5, that she will only bet if

$$\frac{1}{u_1} + \frac{1}{u_2} < 1$$

and will then divide a unit bet so that she places

$$\frac{u_2}{u_1 + u_2} \text{ on the first and } \frac{u_1}{u_1 + u_2} \text{ on the second horse.}$$

The second type, whom we discussed at the end of the previous section, is the occasional bettor who bets a fixed sum. She wishes to maximise $\mathbb{E}X$ where X is the result of betting one unit. As we have seen, she should place all her money on the first horse if $p_1 u_1 > 1$ and not bet at all if $1 > p_1 u_1$.

The third type is the Kelly bettor. She keeps a permanent betting capital and decides not only which bets to make but how much of her capital to risk. If X is the result of a unit bet, she wishes to maximise $\mathbb{E} \log X$. Suppose that

$$\frac{1}{u_1} + \frac{1}{u_2} < 1.$$

Under these circumstances, we know that we can construct a bet which leaves us better off whatever the result of the race. It would be a mistake for the Kelly bettor to keep back some her capital, since she could use it on a 'sure thing bet'. She should therefore commit her whole capital. If she divides a unit bet by placing t on the first horse and $1 - t$ on the second, she then has $\mathbb{E} \log X = f(t)$ with

$$f(t) = p_1 \log(u_1 t) + p_2 \log(u_2(1 - t))$$
$$= p_1 \log t + p_2 \log(1 - t) + p_1 \log u_1 + p_2 \log u_2.$$

Since

$$f'(t) = \frac{p_1}{t} - \frac{p_2}{1 - t} = \frac{p_1(1 - t) - p_2 t}{t(1 - t)} = \frac{p_1 - t}{t(1 - t)},$$

$f(t)$ is maximised by taking $t = p_1$. The Kelly bettor should divide her bet in the ratio of the horses' probability of winning. We have supposed that $u_1^{-1} + u_2^{-1} > 1$ but it is not hard to see that she should follow the same plan if $u_1^{-1} + u_2^{-1} = 1$.

Exercise 2.7.1 *(i) We use the notation of this section. Suppose that*

$$\frac{1}{u_1} + \frac{1}{u_2} = \frac{1}{K}$$

with $K > 1$.

(i)We have just seen that the 'sure thing' bettor will divide a unit bet so that she places $u_2/(u_1 + u_2)$ on the first horse and $u_1/(u_1 + u_2)$ on the second. Show that she will always get K back.

(ii) We have just seen that, if X is the result of a unit bet, the Kelly bettor wishes to maximise $\mathbb{E} \log X$ and will do so by placing a bet of p_1 on the first horse and p_2 on the second. Show that, with this choice, $\mathbb{E} \log X = F(u_1, u_2)$ where

$$F(u_1, u_2) = p_1 \log p_2 + p_2 \log p_2 + p_1 \log u_1 + p_2 \log u_2.$$

Find the values of u_1 and u_2 which minimise $F(u_1, u_2)$ and deduce that, for our Kelly bettor,

$$\mathbb{E} \log X \geq \log K$$

with equality only if u_1 and u_2 take specified values.[28]

(ii) We have just seen that, if X is the result of a unit bet, the maximum expectation bettor wishes to maximise $\mathbb{E}X$ and will do so by placing all their money on the horse or horses with largest $u_j p_j$. Show that, for this bettor,

$$\mathbb{E}X \geq K$$

with equality only if u_1 and u_2 take specified values.

---

[28] Since the bettor always has the option of making a 'sure thing' bet, the inequality is obvious. However, the cases of equality are not.

[*Note that, unlike the 'sure thing' bettor, both the Kelly and the maximum expectation bettors have to estimate p_j. If their estimates are seriously wrong they may do rather badly.*]

The case which is more likely to occur in practice, when

$$\frac{1}{u_1} + \frac{1}{u_2} > 1,$$

is harder and we reserve it for the final exercise of this section (look at cases (A) and (C) of Exercise 2.7.2). In the case which we have dealt with, the reader should observe that both the 'sure thing bettor' and the Kelly bettor may bet on a horse with $p_r u_r < 1$ (so that the expected value of a simple bet on the rth horse is *less* than its cost). In the case of the two horse race, they do this because, if one horse fails to win, the other must. In any race, the fact that one horse loses tells us that one of the others must win.

We observe, once more, that rational bettors with different goals will adopt different betting strategies. We note that the 'highest expectation' bettor's strategy depends only on the values of $p_r u_r$, the 'sure thing' bettor's strategy depends only on the u_r and (if $u_1^{-1} + u_2^{-1} \leq 1$) the Kelly bettor's strategy depends only on the p_r.

Exercise 2.7.2 *In this exercise we extend the Kelly criterion to horse races with many horses. The details are fairly complicated and we will make no use of the results elsewhere. Readers should only do this exercise if they really wish to see how things work out. The exercise follows the pattern of Exercise 1.7.8 and you should reread that exercise and the associated discussion.*

Consider an n-horse race in which the jth horse has probability p_j of winning and payout ratio u_j and $p_1 u_1 \geq p_2 u_2 \geq \cdots \geq p_n u_n$. We make a bet of t_j on the jth horse and keep t_{n+1} back in such a way that

$$t_1 + t_2 + \cdots + t_{n+1} = 1, \quad t_j \geq 0 \text{ for } 1 \leq j \leq n+1.$$

Thus the value of our fortune after the race is given by

$$X = u_j t_j + t_{n+1}$$

if the jth horse wins. We thus wish to maximise $\mathbb{E}\log X = f(t_1, t_2, \ldots, t_n, t_{n+1})$ where

$$f(t_1, t_2, \ldots, t_n, t_{n+1})$$
$$= p_1 \log(u_1 t_1 + t_{n+1}) + p_2 \log(u_2 t_2 + t_{n+1}) + \cdots + p_n \log(u_n t_n + t_{n+1}).$$

Suppose that the maximum occurs when $t_j = s_j$ $[1 \leq j \leq n+1]$.

(i) Let $1 \leq i < j \leq n$. By considering

$$g(x) = f(s_1, \ldots, s_{i-1}, s_i + x, s_{i+1}, \ldots, s_{j-1}, s_j - x, s_{j+1}, \ldots, s_n, s_{n+1})$$

for permissible values of x, show that either

$$\frac{p_i u_i}{u_i s_i + s_{n+1}} = \frac{p_j u_j}{u_j s_j + s_{n+1}},$$

or

$$s_i \neq 0, \ s_j = 0 \ s_{n+1} \neq 0 \quad and \quad \frac{p_i u_i}{u_i s_i + s_{n+1}} > \frac{p_j u_j}{s_{n+1}},$$

or

$$s_i = s_j = 0, \ s_{n+1} \neq 0.$$

(ii) By considering what happens if we vary t_1 and t_{n+1} and what happens when we vary t_n and t_{n+1}, show that either

$$p_1 u_1 \leq 1 \quad and \quad s_{n+1} = 1, \tag{A}$$

or

$$\frac{1}{u_1} + \frac{1}{u_2} + \cdots + \frac{1}{u_n} \leq 1 \ and \ s_{n+1} = 0, \tag{B}$$

or

$$\frac{p_1}{u_1 s_1 + s_{n+1}} + \frac{p_2}{u_2 s_2 + s_{n+1}} + \cdots + \frac{p_n}{u_n s_n + s_{n+1}} = \frac{p_1 u_1}{u_1 s_1 + s_{n+1}}. \tag{C}$$

(iii) Show that, in the case (A), when $p_1 u_1 \leq 1$, the Kelly bettor does not bet. In case (B), when

$$\frac{1}{u_1} + \frac{1}{u_2} + \cdots + \frac{1}{u_n} \leq 1$$

show that $s_j = p_j$ for $1 \leq j \leq n$ and she bets her entire capital in the ratio of the probabilities.

(iv) In case (C), show that there is an m with $1 \leq m < n$ and a $k > 0$ such that

$$\frac{p_j u_j}{u_j s_j + s_{n+1}} = k \ for \ 1 \leq j \leq m,$$

$$s_j = 0, \ \frac{p_j u_j}{s_{n+1}} \leq k \ for \ m + 1 \leq j \leq n$$

and

$$\frac{p_1}{u_1 s_1 + s_{n+1}} + \frac{p_2}{u_2 s_2 + s_{n+1}} + \cdots + \frac{p_n}{u_n s_n + s_{n+1}} = k.$$

Show that, in fact, $k = 1$,

$$s_j = \begin{cases} p_j - \frac{s_{n+1}}{u_j} & \text{for } 1 \le j \le m, \\ 0 & \text{for } m+1 \le j \le n \end{cases}$$

and

$$s_{n+1} = \frac{1-p}{1-T} \text{ where } p = p_1 + p_2 + \ldots + p_m \text{ and } T = \frac{1}{u_1} + \frac{1}{u_2} + \ldots + \frac{1}{u_m}.$$

3

The vice of gambling and the virtue of insurance

3.1 Bernard Shaw

The title for this chapter is taken from an essay in which Shaw explains the principle of insurance and its relation to the welfare state. Shaw begins, as we have done, on the race-track and continues as follows.

[A] bookmaker must never gamble though he lives by gambling. There are practically always enough variable factors in the game to tax the bookmaker's financial ability to the utmost. He must budget to come out at worst still solvent. A bookmaker who gambles will ruin himself as certainly as a ... publican who drinks, or a picture dealer who cannot bear to part with a good picture.

The question at once arises, how is it possible to budget for solvency when dealing with matters of chance? The answer is that when dealt with in sufficient numbers matters of chance become matters of certainty, which is one of the reasons why a million people organised as a State can do things that cannot be dared by private individuals. The discovery of this fact nevertheless was made in the course of private business.

In ancient days, when travelling was dangerous, and people before starting a journey overseas solemnly made their wills and said their prayers as if they were going to die, trade with foreign countries was a risky business, especially when the merchant, instead of staying at home and consigning his goods to a foreign firm, had to accompany them to their destination and sell them there. To do this he had to make a bargain with a ship owner or a ship captain.

Now ship captains, who live on the sea, are not subject to the terrors that it inspires in a landsman. To them the sea is safer than the land; for shipwrecks are less frequent than diseases or disasters on shore. And ship captains make money by carrying passengers as well as cargo. Imagine then a business talk between a merchant greedy for foreign trade but desperately afraid of being shipwrecked or eaten by savages, and a skipper greedy for cargo and passengers. The captain assures the merchant that his goods will be perfectly safe, and himself equally so if he accompanies them. But the merchant, with his head full of the adventures of

Jonah, St Paul, Odysseus and Robinson Crusoe, dares not venture. The conversation will be like this:

CAPTAIN. Come! I bet you umpteen pounds that if you sail with me you will be alive and well this day year.

MERCHANT. But if I take the bet I shall be betting you that sum that I shall die within the year.

CAPTAIN. Why not if you lose the bet as you certainly will?

MERCHANT. But if I am drowned you will be drowned too; and then what becomes of our bet?

CAPTAIN. True. But I will find you a landsman who will make the bet with your wife and family.

MERCHANT. That alters the case of course; but what about my cargo?

CAPTAIN. Pooh! The bet can be on the cargo as well. Or two bets: one on your life, the other on the cargo. Both will be safe, I assure you. Nothing will happen; and you will see all the wonders which are to be seen abroad.

MERCHANT. But if I and my goods get through safely I shall have to pay you the value of my life and of the goods into the bargain. If I am not drowned I shall be ruined.

CAPTAIN That also is very true. But there is not so much for me in it as you think. If you are drowned I shall be drowned first; for I must be the last man to leave the sinking ship. Still let me persuade you to venture. I will make the bet ten to one. Will that tempt you?

MERCHANT. Oh in that case —

The captain has discovered insurance, just as the goldsmiths discovered banking.

It is a lucrative business; and if the insurer's judgement and information are sound a safe one. But it is not so simple as bookmaking on the turf, because in a race, as all the horses but one must lose and the bookmaker gain, in a shipwreck all the passengers may win and the insurer be ruined. Apparently he must therefore own, not one ship only, but several, so that, as many more ships come to port than sink, he will win on half a dozen ships and lose on one only. But in fact the marine insurer need no more own ships than the bookmaker need own horses. He can insure the cargoes and lives in a thousand ships owned by other people without his having owned or even seen as much as a canoe. The more ships he insures the safer are his profits; for half a dozen ships may perish in the same typhoon or be swallowed by the same tidal wave; but of a thousand ships most by far will survive.

Shaw goes on to explain why, because of what we have called the law of large numbers,

[An] insurance company, sanely directed, and making scores of thousands of bets, is not gambling at all; it knows with sufficient accuracy at what age its clients will die, how many of their houses will be burnt every year, how often their houses will be broken into by burglars, to what extent their money will be embezzled by their cashiers, how much compensation they will have to pay to persons injured in their employment, how many accidents will occur to their motor cars and themselves, how much they will suffer from illness or unemployment, and what births and deaths will cost them: in short, what will happen to every thousand or ten thousand or even a million people even when the company cannot tell what will happen to any individual among them.

He concludes that

> ... it is clear that nobody who does not understand insurance and comprehend to some degree its enormous possibilities is qualified to meddle in national business. And nobody can get that far without an acquaintance with the mathematics of probability, not to the extent of making its calculations and filling examination papers with typical equations, but enough to know when they can be trusted, and when they are cooked. For when their imaginary numbers correspond to exact quantities of hard coins unalterably stamped heads and tails they are safe to within certain limits but [in other cases] guesswork, personal bias, and pecuniary interests, come in so strongly that those who began by ignorantly imagining that statistics cannot lie end by imagining, equally ignorantly, that they never do anything else. [59]

Shaw's discussion suggests two reasons why people or firms may take out insurance. The first is that the two parties involved disagree about the probability of the event insured. A major British electrical goods retailer enjoyed substantial profits for some years, not from selling its goods, but by persuading its customers to take out substantial insurance against the breakdown of the goods it sold!

The second reason does not depend on differing views of risk but on the law of large numbers. It is unlikely that my factory will burn down but, if it does, I shall be ruined unless I am insured. The company that insures me insures so many other factory owners that the probability that so many factories will burn down as to ruin it is vanishingly small.

To these reasons we may add a third which is that, although the two parties may agree on the probability of certain events, they may differ on the value they place on the outcome.

Exercise 3.1.1 *Consider A and B who await the outcome of a certain random event as a result of which A will have a fortune X and B a fortune Y. A wishes to maximise $\mathbb{E}X$ and B wishes to maximise $\mathbb{E}\log Y$.*

(i) Suppose that, as a result of the event, A and B will both have a fortune $1/2$ (we call this the unlucky outcome) with probability $1/2$ or will both have a fortune 2 (we call this the lucky outcome) with probability $1/2$. Compute $\mathbb{E}X$ and $\mathbb{E}\log Y$.

(ii) Now suppose that before the event A agrees to pay B the sum of $1/2$ in the case of the unlucky outcome (so that A then has 0 and B has 1) and B agrees to pay A the sum of $3/4$ in the case of the lucky outcome (so that A then has $11/4$ and B has $5/4$). Compute $\mathbb{E}X$ and $\mathbb{E}\log Y$. Observe that this arrangement is better for both A and B! (We return to this point in Exercise 3.5.4.)

3.2 Annuities

Although Shaw's account of the birth of insurance is correct in spirit, insurance could not become widespread until it was possible to calculate its cost correctly. This problem was particularly clear in the case of annuities.

In its simplest form, an annuity is a promise by A to pay B a certain sum every year for the rest of B's life.[1] Most pensions are, in effect, annuities. Annuities go back to Roman times, but they first became important in seventeenth-century Europe when states like France and the Netherlands used them to raise money for war. The cost of an annuity was established by guesswork but mistakes could be very expensive.

Appendix C describes how the the first book on probability theory was published by Huygens. This dealt entirely with games of chance, but the ideas were soon used by de Witt[2] and others to try and find the correct price of an annuity. In this section we discuss the annuity problem starting with very simple models.

Exercise 3.2.1 *Show that*

$$(1 - x)(1 + x + x^2 + \cdots + x^m) = 1 - x^{m+1}$$

and hence deduce the well-known formula

$$1 + x + x^2 + \cdots + x^m = \frac{1 - x^{m+1}}{1 - x}$$

for $x \neq 1$.

Suppose that A promises to pay B an amount 1 unit each year until B's death or until N years have passed, whichever is the sooner, and suppose that the probability that B dies in a given year is $1 - p$. We assume that the first payment is made at once. If X_j is the amount paid at the beginning of the jth year we see that

$$\Pr(X_j = 1) = \Pr(B \text{ survives } j \text{ years}) = p^j$$

so

$$\mathbb{E}X_j = (1 - p^j) \times 0 + p^j \times 1$$

---

[1] Jorge Guinle who ran through an enormous fortune in a most enjoyable manner (he spent his money on Hollywood starlets, high living and jazz) put the case for annuities very clearly. 'The secret to living well is to die without a cent in your pocket. But I seem to have miscalculated, the money ran out before it was supposed to'.

[2] A great Dutch statesman who met his death at the hands of a Dutch mob.

Thus the sum X that A has to pay out has expected value

$$\begin{aligned}
\mathbb{E}X &= \mathbb{E}(X_0 + X_1 + \cdots + X_N) \\
&= \mathbb{E}X_0 + \mathbb{E}X_1 + \cdots + \mathbb{E}X_N \\
&= 1 + p + \cdots + p^N = \frac{1 - p^{N+1}}{1 - p}.
\end{aligned}$$

If N is very large, we see that $\mathbb{E}X \approx (1 - p)^{-1}$ and B should be prepared to pay A at least $(1 - p)^{-1}$ in exchange for the guarantee of 1 unit a year for life.

There are many oversimplifications in the above account but perhaps the most serious is that we ignore the fact that money earns interest.[3] Suppose that, if we place 1 unit in the bank at the beginning of a year then the bank pays back k units at its end.

Exercise 3.2.2 *Show that B can obtain an income of 1 unit a year forever by placing*

$$1 + \frac{1}{k - 1} = \frac{k}{k - 1}$$

in the bank and withdrawing 1 unit a year. (We assume that the first withdrawal is made at once.)

If the annuity provider A places k^{-j} in the bank now, she will have 1 unit after j years. Thus the cost to A of a promise to pay 1 in j years time is k^{-j} and the expected cost to A of a promise to pay B 1 unit after j years, provided that B is still alive, is $p^j k^{-j}$. The expected value of the sum Y which A would need to bank now to pay B (that is to say, its *expected present value*) is

$$\begin{aligned}
\mathbb{E}Y &= \mathbb{E}(X_0 + k^{-1}X_1 + \cdots + k^{-N}X_N) \\
&= \mathbb{E}X_0 + k^{-1}\mathbb{E}X_1 + \cdots + k^{-N}\mathbb{E}X_N \\
&= 1 + k^{-1}p + \cdots + (k^{-1}p)^N = \frac{1 - (k^{-1}p)^{N+1}}{1 - (k^{-1}p)}.
\end{aligned}$$

If N is large, we see that $\mathbb{E}X \approx (1 - k^{-1}p)^{-1}$ and B should be prepared to pay A at least

$$\frac{1}{1 - k^{-1}p} = \frac{k}{k - p}$$

in exchange for the guarantee of 1 unit a year for life.

---

[3] We ignore inflation, or, if the reader prefers, work in inflation adjusted terms.

Exercise 3.2.3 *Suppose that N is large and the expected present cost to the guarantor is l. Find p in terms of l and k.*

In the seventeenth century an annuity typically cost something like '12 years' purchase' (that is to say, took $l = 12$)[4] and interest rates might give $k = 1.04$. Estimate p.

If we write $k = 1 + a$ and $p = 1 - b$, we see that you would have to place $P = (1 + a)/a$ in a bank to obtain an income of 1 unit a year but that, if someone was prepared to sell you an annuity at its expected present cost, you would only have to pay $Q = (1 + a)/(a + b)$. The difference is accounted for by the fact that if you followed the first course of action you would leave P in the bank after your death.

Exercise 3.2.4 *We have*

$$\frac{P}{Q} = \frac{a + b}{a}.$$

Explain to a non-mathematician what this ratio means and why it is large when b is large compared to a and close to 1 when a is large compared to b. If someone retires at 70 today, what sort of choices of a and b would be appropriate?

When a husband and wife buy an annuity, they frequently want a fixed sum to be paid annually until they are both dead. How can we find the value of this more complicated annuity? Recall that, by the inclusion-exclusion formula,

$$\Pr(U \cup V) = \Pr(U) + \Pr(V) - \Pr(U \cap V)$$

and that, if the events U and V are independent, then $\Pr(U \cap V) = \Pr(U)\Pr(V)$. If we take U to be the event that the husband is alive at the beginning of the jth year and V to be the event that the wife is alive (and we assume independence) we see that the probability that at least one is alive at the beginning of the jth year is $2p^j - p^{2j}$.

Consider an annuity which pays out 1 at the beginning of each year. Let us take $Z_j = 1$ if at least one partner is alive at the beginning of the jth year, and

---

[4] So someone whose life 'is not worth an hour's purchase' is in great danger.

$Z_j = 0$ otherwise. If W is the value of our payout (adjusted for bank interest as before),

$$
\begin{aligned}
\mathbb{E}W &= \mathbb{E}(Z_0 + k^{-1}Z_1 + \cdots + k^{-N}Z_N) \\
&= \mathbb{E}Z_0 + k^{-1}\mathbb{E}Z_1 + \cdots + k^{-N}\mathbb{E}Z_N \\
&= 1 + k^{-1}(2p - p^2) + \cdots + k^{-N}(2p^N - p^{2N}) \\
&= 2(1 + k^{-1}p + (k^{-1}p)^2 + \cdots + (k^{-1}p)^N) \\
&\quad - (1 + k^{-1}p^2 + (k^{-1}p^2)^2 + \cdots + (k^{-1}p^2)^N) \\
&= 2\frac{1 - (k^{-1}p)^{N+1}}{1 - (k^{-1}p)} - \frac{1 - (k^{-1}p^2)^{N+1}}{1 - (k^{-1}p^2)}.
\end{aligned}
$$

If N is large, we see that

$$
\mathbb{E}W \approx \frac{2}{1 - k^{-1}p} - \frac{1}{1 - k^{-1}p^2} = \frac{2k}{k - p} - \frac{k}{k - p^2}.
$$

The simple model above gives considerable insight into the nature of annuities, but it has the peculiar feature that it assigns the same value to the annuity whatever their initial age of the person paid the annuity. This is not unreasonable if the person involved is quite young at the start of the annuity, but it is clearly unrealistic to assign the same value to an annuity for a 90 year old as to that for a 20 year old.

To deal with this, we consider another model in which the probability of the annuitant surviving to the beginning of the jth year of the contract is $(N - j)/N$ for $0 \le j \le N$.

Exercise 3.2.5 *Suppose that we ignore bank interest (this is reasonable if N is small and the rate of interest is low). Show that the expected value of an annuity which pays 1 unit at the beginning of each year (including the one in which the annuity starts) is $(N + 1)/2$.*

Thus if I have a fortune of $(N + 1)/2$ at the start of the first year, I can either buy an annuity which pays 1 unit a year (ignoring additional costs), or I can live off my savings at the rate of $(N + 1)/2N$ a year and probably die with some of my savings unspent, or I can live off my savings at the rate greater than $(N + 1)/2N$ a year and risk living so long that all my savings are gone. Looking at this example, we see that the invention of the annuity was a step forward in civilisation.

Now let us look at the effect of bank interest. Suppose, as before, that, if we place 1 unit in the bank at the beginning of a year then the bank pays back

k units at the end. If Y is the value of the annuity, then our earlier arguments show that

$$\mathbb{E}Y = \mathbb{E}(Z_0 + k^{-1}Z_1 + \cdots + k^{-N}Z_N)$$

$$= \mathbb{E}Z_0 + k^{-1}\mathbb{E}Z_1 + \cdots + k^{-N}\mathbb{E}Z_N$$

$$= 1 + \left(1 - \frac{1}{N}\right)k^{-1} + \left(1 - \frac{2}{N}\right)k^{-2} + \cdots + \left(1 - \frac{N-1}{N}\right)k^{-N+1}$$

$$= \left(1 + k^{-1} + \cdots + k^{-N+1}\right) - \frac{1}{N}\left(k^{-1} + 2k^{-2}\cdots + (N-1)k^{-N+1}\right).$$

We can simplify this expression by the neat trick of differentiating both sides of the equation

$$1 + x + x^2 + \cdots + x^m = \frac{1 - x^{m+1}}{1 - x}$$

to obtain

$$1 + 2x + \cdots + mx^{m-1} = \frac{d}{dx}\frac{1 - x^{m+1}}{1 - x} = \frac{1 - x^{m+1}}{(1 - x)^2} - \frac{(m+1)x^m}{1 - x}$$

for $x \neq 1$. Thus

$$\mathbb{E}Y = \frac{1 - k^{-N}}{1 - k^{-1}} - \frac{k^{-1}}{N}\left(1 + k^{-1} + 2k^{-2} + \cdots + (N-1)k^{-N+2}\right)$$

$$= \frac{1 - k^{-N}}{1 - k^{-1}} - \frac{k^{-1}}{N}\left(\frac{1 - k^{-N}}{(1 - k^{-1})^2} - \frac{Nk^{1-N}}{1 - k^{-1}}\right).$$

Exercise 3.2.6 *(i) Find a general, reasonably compact, formula for*

$$(1 \times 2) + (2 \times 3)x + (3 \times 4)x^2 + \cdots + m(m-1)x^{m-2}$$

when $x \neq 1$.

(ii) Use (i) and the observation that $r^2 = r(r-1) + r$, to find a general, reasonably compact, formula for

$$1^2 + 2^2x + 3^2x^2 + \cdots + m^2x^{m-2}$$

when $x \neq 1$.

(iii) Consider the model in which the probability of the annuitant surviving to the beginning of the jth year of the contract is $(N - j)/N$ for $0 \leq j \leq N$ and a sum of 1 deposited at the beginning of the year is worth k at the end. If $k = 1$, find a reasonably compact formula for the expected present value of a joint annuity which pays 1 until both annuitants are dead. Indicate how to deal with the problem when $k > 1$ but do not too much algebra. (We assume that the deaths are independent and the model applies to both.)

Because neither of the two models we have considered is very realistic, sellers of annuities who used such models would have to make very conservative estimates of the quantities involved. (That is to say, they would have to use low estimates of k and high estimates of N and p.) This would make their annuities less attractive to buyers than they could be.

The obvious way forward is to find better estimates of p_j, the probability that someone of age j survives to age $j + 1$. In the UK and similar countries today, $p_{30} \approx 0.999$, p_j decreases as j increases (for $j \geq 30$) but, even at very high ages,[5] it appears that p_j does not fall below $1/2$. However, for mathematical simplicity, our models will assume the existence of an N such that $p_j > 0$ for $0 \leq j \leq N - 1$ and $p_N = 0$. We take $p_0 = 1$.

Suppose, as before, that, if we place 1 unit in the bank at the beginning of a year, then the bank pays back k units at the end. If B is of age r and wishes to buy an annuity which pays 1, at the beginning of each year we then know that the probability that B reaches age $r + s$ is

$$q_{r,r+s} = p_r p_{r+1} \cdots p_{r+s-1}$$

and the expected present value of the annuity is

$$A_r = 1 + k^{-1}q_{r,r+1} + k^{-2}q_{r,r+2} + \cdots + k^{r-N}q_{r,N}.$$

Exercise 3.2.7 *(i) Show that*

$$A_r = 1 + p_r k^{-1} A_{r+1}.$$

Explain this equation to a non-mathematician.

(ii) We obtained the values of $q_{r,s}$ with $N \geq s \geq r$ from the values of p_j with $N \geq j \geq r$. Show that we can obtain the values of p_j with $N \geq j \geq r + 1$ from the values of $q_{r,s}$ with $s \geq r$. In particular we can obtain the values p_j with $N \geq j \geq 0$ from the values of $q_{0,s}$ with $N \geq s \geq 0$. Explain the meaning of N and $q_{0,s}$ to a non-mathematician.

Exercise 3.2.8 *Under the assumptions just discussed, show how to calculate the expected present value of a 'joint annuity', which pays out 1 unit a year until two named individuals are both dead, if one of them has age r when the annuity starts and the other has age t.*

The first table of $q_{0,s}$ (or 'life table' as such things were called) which did not involve a substantial amount of guess work was drawn up by Halley[6] using the exceptionally well kept record of births and deaths of Breslau (now

---

[5] See [32].

[6] The Halley of Halley's comet. He made many contributions to human knowledge but his greatest was to persuade Newton to write the *Principia*. Halley then paid the costs of printing and saw the book through the press.

Wrocław). Advanced states like Britain and France already collected statistics in an attempt to guide policy. The evident utility of life tables reinforced and accelerated the process.

The buyer of an annuity buys freedom from risk. Her income from the annuity is guaranteed however long she lives. She is prepared to pay more than the expected value of the annuity in order to enjoy this freedom from risk. The extra sum she is prepared to pay allows the seller of the annuity to place a bet (involving the length of the buyer's life) whose expected total value to the seller (the cost of the annuity minus its expected value to the buyer) is strictly positive.

A bet, even with large expected total value, is still a bet and one which the seller may lose. A ninety year-old Frenchwoman called Madame Calment gave possession of her apartment on her death to her lawyer in exchange for a life annuity of 2500 francs a month. She lived to the age of 122, becoming the world's oldest woman. The lawyer died a year earlier at the age of 77. Altogether, the lawyer's family paid more than 900 000 francs, three times the value of the apartment.[7]

The selling of a single annuity is a bet, and a risky one at that. However, a firm which sells many (suitably priced) annuities can rely on the law of large numbers to ensure that the probability that they fail to make a good profit is very small indeed.

There remains the question of what represents a 'suitable price'. The reader will already have spotted various ways in which simple life tables might prove inadequate. Is $q_{0,s}$ different for men and women? (Yes, and the difference is important.) Are the death dates of husband and wife independent? (No, but the lack of independence is not large enough to make a great difference to the cost of joint annuities.) Life tables are now sufficiently detailed to deal with such questions.[8]

The real risk to a company which issues many annuities lies in the future behaviour of interest rates. In our calculations, we assumed that k was constant. There have been golden periods[9] when k was effectively constant but the twentieth century was not and I suspect the twentyfirst century will not be. It may be possible to state with some accuracy the probability that someone of a certain age will be alive in ten years' time but the interest rate in ten years' time is much more uncertain. For this reason the State often acts directly (through

---

[7] But they do seem to have enjoyed their association with a national celebrity.

[8] Provided you use the right tables. Not surprisingly, those who take out annuities turn out to have a longer life expectancy than those who take out life insurance.

[9] For those who lived on the interest on their capital.

State pensions) or indirectly (through the issue of securities with a fixed rate of interest) as guarantor for annuities.[10]

3.3 Smallpox

Most children in seventeenth-century London could expect to get smallpox. Some would die, some would be blinded and many would be disfigured.

Those who recovered from smallpox were immune for life. Europeans who travelled to the East brought back news of a system of *inoculation* by which children were deliberately infected with smallpox by inserting matter from smallpox pustules under the skin. This usually produced a relatively mild attack of smallpox and, with it, lifetime immunity.

The terror smallpox inspired is amply demonstrated by the fact that people were willing to contemplate such a desperate expedient as inoculation. However, the practice spread among the English upper classes who could afford the required nursing and isolation. (Since the patients had genuine smallpox, they could infect others with the full blown disease.)

Voltaire noted that continental Europeans considered that the English were

> . . . fools and madmen: fools because they gave smallpox to their children to prevent them having smallpox; madmen because they wantonly infect those children with a certain and unpleasant disease to prevent an uncertain evil.
>
> Lettres Philosophiques *[65]*

In [5], Daniel Bernoulli used Halley's life tables to investigate whether the certain risks of smallpox inoculation were outweighed by the uncertain benefits. He makes clear the various assumptions he makes. He assumes that the probability p of contracting smallpox during a year (if the subject is not immune) remains the same at every age as does the probability q of dying from smallpox once contracted. He then proceeds as follows.[11]

Let $r(t)$ be the number of people in some population who survive to age t and let $s(t)$ be the number of people who survive to age t without catching smallpox. We consider what happens between the age of t and $t + \delta t$ where δt is small. We write

$$\delta r = r(t + \delta t) - r(t) \quad \text{and} \quad \delta s = s(t + \delta t) - s(t).$$

Between the ages of t and $t + \delta t$ the probability that someone, who has survived to age t without having smallpox, will catch smallpox is about $p\delta t$, the

---

[10] As I write, some British universities are taking on a similar role. The possibility arises because people believe that they are unlikely to disappear or to evade their debts.

[11] I have changed the notation but the underlying argument is unaltered.

probability that they will die of smallpox is $pq\delta t$ and the probability that they will die of some other disease is about $u(t)\delta t$ for some unknown $u(t)$. Thus

$$\delta s \approx -s(t)(p\delta t + u(t)\delta t) = -s(t)(p + u(t))\delta t.$$

To calculate δr, we need to consider the number of deaths from other diseases, amongst both the immune and the non-immune, and the number of those who die from smallpox during the period. We see that

$$\delta r \approx -r(t)u(t)\delta t - s(t)pq\delta t = -(r(t)u(t) + s(t)pq)\delta t.$$

We thus have

$$\frac{\delta s}{\delta t} \approx -s(t)(p + u(t)) \quad \text{and} \quad \frac{\delta r}{\delta t} \approx -r(t)u(t) - s(t)pq,$$

so, assuming that the approximation behaves well as we let $\delta t \to 0$, we have

$$s'(t) = -s(t)(p + u(t)) \quad \text{and} \quad r'(t) = -r(t)u(t) - s(t)pq.$$

Thus

$$\frac{s'(t)}{s(t)} = -p - u(t) \quad \text{and} \quad \frac{r'(t)}{r(t)} = -u(t) - pq\frac{s(t)}{r(t)}$$

and we may eliminate the unknown function $u(t)$ to obtain

$$\frac{r'(t)}{r(t)} - \frac{s'(t)}{s(t)} = p - pq\frac{s(t)}{r(t)}.$$

At this point, Bernoulli remarks 'It is very remarkable that our differential equation admits of a solution, ... a situation which is rare in problems which investigate a state of Nature and which differ greatly from abstract problems'. He considers the natural substitution $f(t) = r(t)/s(t)$ with the results set out in the next exercise.

Exercise 3.3.1 *Show that*

$$\frac{f'(t)}{f(t)} = \frac{r'(t)}{r(t)} - \frac{s'(t)}{s(t)}$$

and deduce that

$$\frac{f'(t)}{f(t)} = p - \frac{pq}{f(t)}.$$

Hence, show that

$$\frac{f'(t)}{f(t) - q} = p,$$

and, by showing that

$$\frac{d}{dt}\left(\log(f(t)-q)-pt\right)=0,$$

or otherwise, deduce that

$$\log(f(t)-q)=pt+A$$

for some constant A.

Explain why $f(0)=1$ and deduce that $A=\log(1-q)$. Now show that

$$f(t)=q+(1-q)e^{pt}$$

and

$$s(t)=\frac{r(t)}{q+(1-q)e^{pt}}.$$

Bernoulli's formula contains two constants p and q. He notes that the death rate varies between epidemics but (in agreement with others who had studied the available statistics) takes $q=1/8$ as a reasonable estimate. To estimate p, he observes that there were very few smallpox cases among those older than 23. He interprets this as meaning that almost everybody alive at the age of 23 will have had smallpox, so $s(23)\approx 0$ and e^{23p} must be large. He decides that $p=1/8$ gives the best fit to the figures he has.

He now uses Halley's life table to construct Table 3.1. The second column labelled 'Survivors according to Halley' shows the number of children of a given age who have survived from an initial group of 1300 and is based on Halley's data from births and deaths registered in Breslau.[12] The remaining columns are calculated by Bernoulli using the formula just derived. Since there were no records showing both the age of death and the cause of death,[13] the column labelled 'Dying of smallpox each year' could not then be checked against real data. However, extensive London records (the 'bills of mortality') showed that, in London, about one in fourteen deaths was attributed to smallpox and estimates based on other data which Bernoulli relied on suggested a figure of about one in thirteen. Since very few people died of smallpox beyond the age of 24, Bernoulli chose p and q so that the entry for 'Total smallpox deaths' at age 24 was roughly one thirteenth of the initial size of the group.

---

[12] For comparison, if we took a group of 100 000 newborns in the UK in 2003, we could expect about 1345 deaths before the end of their first year, about 24 deaths per year between the ages of 1 and 4, and about 11 per year between the ages of 5 and 9. Figures from UK National Statistics Series DH3 no.36.

[13] The Breslau records showed the age but not the cause and the London records showed the cause but not the age.

Table 3.1. *Estimated smallpox deaths at each age*

Age in years	Survivors according to Halley	Not having had smallpox	Having had smallpox	Catching smallpox each year	Dying of smallpox each year	Total smallpox deaths	Deaths from other diseases each year
0	1300	1300	0				
1	1000	895	104	137	17.1	17.1	283
2	855	685	170	99	12.4	29.5	133
3	798	571	227	78	9.7	39.2	47
4	760	485	275	66	8.3	47.5	30
5	732	416	316	56	7.0	54.5	21
6	710	359	351	48	6.0	60.5	16
7	692	311	381	42	5.2	65.7	12.8
8	680	272	408	36	4.5	70.2	7.5
9	670	237	433	32	4.0	74.2	6
10	661	208	453	28	3.5	77.7	5.5
11	653	182	471	24.4	3.0	80.7	5
12	646	160	486	21.4	2.7	83.4	4.3
13	640	140	500	18.7	2.3	85.7	3.7
14	634	123	511	16.6	2.1	87.8	3.9
15	628	108	520	14.4	1.8	89.6	4.2
16	622	94	528	12.6	1.6	91.2	4.4
17	616	83	533	11.0	1.4	92.6	4.6
18	610	72	538	9.7	1.2	93.8	4.8
19	604	63	541	8.4	1.0	94.8	5
20	598	56	542	7.4	0.9	95.7	5.1
21	592	48.5	543	6.5	0.8	95.7	5.2
22	586	42.5	543	5.6	0.7	97.2	5.3
23	579	37	542	5.0	0.6	97.8	6.4
24	572	32.4	540	4.4	0.5	98.3	6.5

Exercise 3.3.2 *A certain population suffers from diseases A and B. Everyone first catches disease A and one half die. All those who remain alive then catch B and one half die. All survivors are now immune from A and B and go on to die from other causes. What is the probability that someone with disease A will die from it? What is the probability that someone with disease B will die from it? What is the probability (at birth) that someone will die from disease A? What is the probability (at birth) that someone will die from disease B?*

Explain to a non-mathematician why Bernoulli can say both that the probability of dying from an attack of smallpox, is $1/8$ and that the probability of dying from smallpox is $1/13$. Explain why the probability of someone dying from smallpox cannot exceed the probability of dying from an attack of smallpox. Use Table 3.1 to explain why the probability of someone dying from smallpox is so much less than the probability of dying from an attack of smallpox.

Exercise 3.3.3 *Table 3.1 is what we might now call a 'spread sheet'. Construct a spread sheet to verify Bernoulli's calculations. (Note that the number $a(n)$ corresponding to age n in the column labelled 'Catching smallpox each year' is obtained by setting*

$$a(n) = \frac{1}{8} \frac{b(n) + b(n-1)}{2}$$

where $b(n)$ is the number corresponding to age n in the column labelled 'Not having had smallpox'.) Investigate the effect of changing p and q. You should recall the constraints that almost everybody should have had smallpox by the age of 23 and that about 1/13th of the population dies of smallpox.

We can now see what the life table would look like if there were no smallpox deaths. Let $A(n)$ be the number alive after n years in the absence of small-pox, $B(n)$ the actual number alive (corresponding to the figure in the column labelled 'Survivors according to Halley') and $C(n)$ the estimated number of deaths from other causes between the ages of $n - 1$ and n (the last column of Table 3.1). It seems reasonable to suppose that $C(n)/B(n-1)$ is roughly equal to the probability of dying from some other cause between the ages of $n - 1$ and n, in the absence of smallpox, and that

$$A(n) \approx A(n-1) \left(1 - \frac{C(n)}{B(n-1)} \right).$$

Replacing $\approx$ by $=$, Bernoulli can now compute Table 3.2.

Exercise 3.3.4 *Construct a spread sheet to verify Bernoulli's calculations.*

Exercise 3.3.5 *Why do the numbers in the last column increase and then start to decrease?*

Exercise 3.3.6 *Bernoulli gives 'a pretty theorem' which provides an alternative method of estimating the numbers in the column labelled 'State without smallpox'. We go back to the argument which culminated in the final formula of Exercise 3.3.1. Let $R(t)$ be the number of people whom one would expect to be alive at time t in the absence of smallpox and let $r(t)$, $s(t)$, $u(t)$, p and q have the meanings that we gave them earlier. Show, by imitating our earlier arguments, that*

$$\frac{R'(t)}{R(t)} = -u(t),$$

and eliminate $u(t)$, as before, to get

$$\frac{R'(t)}{R(t)} - \frac{r'(t)}{r(t)} = pq \frac{s(t)}{r(t)}.$$

Table 3.2. *Deaths with and without smallpox*

Age in years	Natural state with smallpox	State without smallpox	Gain
0	1300	1300	0
1	1000	1017.1	17.1
2	855	881.8	26.8
3	798	833.3	35.3
4	760	802.0	42.0
5	732	77.98	47.8
6	710	762.8	52.8
7	692	749.1	57.2
8	680	740.9	60.9
9	670	734.4	64.4
10	661	728.4	67.4
11	653	722.9	69.9
12	646	718.2	72.2
13	640	741.1	74.1
14	634	709.7	75.7
15	628	705.0	77.0
16	622	700.1	78.1
17	616	695.0	79.0
18	610	689.6	79.6
19	604	684.0	80.0
20	598	678.2	80.2
21	592	672.3	80.3
22	586	666.3	80.3
23	579	659.0	80.0
24	572	644.3	79.3

Conclude that

$$\frac{R'(t)}{R(t)} - \frac{r'(t)}{r(t)} = \frac{pq}{q + (1-q)e^{pt}}.$$

Observe that

$$\frac{pq}{q + (1-q)e^{pt}} = p - \frac{p(1-q)e^{pt}}{q + (1-q)e^{pt}}$$

and, by showing that

$$\frac{d}{dt}\left(\log R(t) - \log r(t) - pt + \log(q + (1-q)e^{pt})\right) = 0,$$

or otherwise, conclude that

$$\log R(t) - \log r(t) - pt + \log(q + (1 - q)e^{pt}) = C$$

where C is a constant.

Find C by considering what happens when t = 0 and show that

$$\frac{R(t)}{r(t)} = \frac{e^{pt}}{q + (1 - q)e^{pt}}.$$

What happens to the ratio $R(t)/r(t)$ as t becomes large? Should we expect this?

Bernoulli states that, if we put $p = q = 1/8$, as before, his new formula 'does not differ appreciably from those shown in the table, particularly towards the end'.

3.4 Should we inoculate?

Table 3.2 shows what we would expect to happen if there was no smallpox or, equivalently, if everyone was inoculated at birth and inoculation carried no risk. Inoculation was not risk-free,[14] but Bernoulli's tables enable us to take this into account.

Exercise 3.4.1 *Let D(n) be the number of children who would have been killed by inoculation but in fact survive to the age of n and let B(n) the total number of children alive at age n (corresponding to the figure in the column labelled 'Survivors according to Halley' in Table 3.1). Explain why we should expect*

$$D(n) \approx D(n - 1) \times \frac{B(n)}{B(n - 1)}$$

and why subtracting D(n) from the corresponding number in the column labelled 'Gain' in Table 3.2 gives the total extra number of children we would expect to be alive at age n if every child was inoculated at birth.

Bernoulli suggests that the probability v of dying as result of inoculation is less than 1/200. Draw up the appropriate table showing the gain in the case of universal inoculation at birth if $v = 1/200$.

---

[14] When the Empress of Russia, Catherine the Great, was inoculated by an English doctor, she arranged for relays of post-horses to be ready to carry him from St Petersburg to the frontier and give him a chance of escape if she died.

As the century progressed, doctors became more skilled at inoculation and Bernoulli was probably right in supposing $v < 1/200$ when he wrote. However, when inoculation was introduced, the choice $v = 1/50$ would be more appropriate. Draw up the appropriate table.

From now on (except in Exercises 3.4.2 and 3.4.3) we shall assume that the risk of dying from the inoculation is $1/200$ and that the choice is between inoculating a child at birth and not inoculating at all. Bernoulli writes

> A great deal of trouble has been taken to evaluate the gain which could be hoped from inoculation if it were generally introduced, and the advantage to each individual who was inoculated. It is, in general, clear that this profit and this advantage could not fail to be considerable and infinitely precious, but what sort of units could we use to measure it? By the average life which could be expected after inoculation? Are all the years of life equally valuable?

To make clear the distinction between the advantage to the individual and the 'advantage to the Prince' (we might say 'advantage to the Economy') he points out that, even if inoculation killed as many children as were killed by smallpox before, it would increase the wealth of the State, since it would reduce the cost of supporting 'non-productive' children who would not reach 'productive' adulthood.

But what is the advantage to the individual? Surely we should start by looking at the expected length of life with and without inoculation. Suppose that, under given conditions, there are $U(n)$ members of our group alive at age n. Our first estimates of the average length of life might be

$$\frac{U(0) + U(1) + U(2) + U(3) + \cdots}{1300},$$

if we assume that deaths take place just before birthdays, and

$$\frac{U(1) + U(2) + U(3) + \cdots}{1300},$$

if we assume that deaths take place just after birthdays. Taking the average of the optimistic and the pessimistic view, we estimate the average length of life as

$$\frac{U(0)/2 + U(1) + U(2) + U(3) + \cdots}{1300}.$$

Bernoulli's tables give us the appropriate $U(n)$ for $n \leq 25$. Since there are very few deaths from smallpox above the age of 25, the remaining values of $U(n)$ are essentially the same whether we inoculate or not and may be derived from Halley's original table. Bernoulli calculates that the expected length of life L_1 of a child at present is 26 years 7 months and that the average length of life L_2 if there was no smallpox would be 29 years 9 months.

If we inoculate each child at birth, then the expected length of life of each surviving child will be L_2, but $1/200$ of the children will not survive the inoculation so the expected length of life of a child at birth will be

$$L_3 = \frac{199}{200} L_2.$$

Bernoulli calculates that $L_2 - L_3$ is about 1 month and 20 days and 'notwithstanding this risk, the gain is still 3 years on the average life for the natural state'.

Exercise 3.4.2 *Let us say that someone who has had smallpox or been inoculated is immune. Consider a simple model in which all deaths occur at the end of year. Suppose that someone who is alive and immune at the end of year n has probability a of being alive at the end of year n + 1. Suppose that someone who is alive at the end of year n and is not immune has probability b of being alive and not having had smallpox at the end of year n + 1 and a probability c of being alive and having had smallpox at the end of year n + 1. (To complete the model we may suppose that everyone alive at the end of year N − 1 dies at the end of year N, but we take N so large that this may be ignored and we can replace sums by their limits as N → ∞.) Show that, if someone is alive and immune at the end of year k, their expected lifetime after year k is*

$$\frac{1}{1-a}.$$

(Thus their total expected life time is $k + (1 - a)^{-1}$.)

Show that if someone who refuses to be inoculated is alive and not immune at year k their expected lifetime after year k is

$$\frac{1}{1-b} \left(1 + \frac{c}{1-a} \right).$$

Suppose that the probability of dying as a result of inoculation is r. Show that someone, who is not immune at age k, will increase their expected lifetime by being inoculated at once if and only if

$$\frac{1-r}{1-a} > \frac{1}{1-b} \left(1 + \frac{c}{1-a} \right).$$

Observe that the increase in expected lifetime is the same whatever the age of the individual inoculated. Explain to a non-mathematician why, in spite of this, the increase in expected lifetime for a child at birth is greatest if all children are inoculated immediately after birth.

Exercise 3.4.3 *Explain how you would use Bernoulli's and Halley's table to evaluate the expected additional lifetime from inoculation obtained by inoculating a child of age 4 who has not yet had smallpox.*

Bernoulli views his work as a first step rather than a final answer. He calls for doctors to keep a record of both the cause and age of death of each of their patients and points out that, in the case of smallpox, it would be particularly useful to have figures for the early years of life. Even so, experts seem to agree that his estimates give a pretty good picture of what was going on.

Although he accepted that inoculation was beneficial, d'Alembert attacked Bernoulli's use of 'expected additional lifetime' as a measure of benefit. His chief argument is summed up in the following thought experiment.

Exercise 3.4.4 *Suppose that some benevolent and truth telling being offers you a potion which will kill you instantly and painlessly with probability p but will otherwise guarantee N further years of happy life.*
  (i) What will you choose if $p = 1/2$ and $N = 1000$?
  (ii) What will you choose if $p = 10^{-7}$ and $N = 100$?
  (iii) If $p = 9/10$ is there any N which will cause you to drink the potion?
  (iv) If $N = 100$ what is the largest value of p for which you will choose to drink?

Exercise 3.4.4 convinces most people that 'expected additional length of life' should not be the sole criterion in such decision making. Similar arguments will convince most people that there can be no single 'figure of merit' which allows us to compare different patterns of mortality.

Bernoulli's tables avoid this problem, since they allow us to look at two different patterns of mortality and choose between them.

D'Alembert also argues, in effect, that we live according to the moral rule 'it is worse to cause harm by action than by inaction'[15] and the prudential rule 'do not sacrifice a certain present good in the hope of a larger uncertain future good'[16] and that Bernoulli's analysis fails to take account of this.

He raises a third objection closely related to the previous two that we cannot assign the same value to each year of life. Perhaps it is foolish to risk one year in the prime of life for the sake of a few years of old age.[17]

---

[15] Above all, do no harm.

[16] A bird in the hand is worth two in the bush.

[17] This point of view may seem less attractive as the reader grows older. But why, for example, should thirty-year olds forgo the pleasures of a holiday or a new car in order to provide pensions for their potential older selves whom they do not know and probably would dislike if they did?

The arguments of intellectuals in general and mathematicians in particular may have profound long term consequences but rarely have much immediate effect. The death of Louis XV from smallpox and the consequent conversion of the French royal family to inoculation probably had more effect on French public opinion than all the efforts of Bernoulli and his fellow thinkers.

Throughout the eighteenth century, doctors improved their methods and inoculation became widespread. Unfortunately, although it was possible to inoculate all the inhabitants of a village simultaneously, this could not be done in towns. Whilst the wealthy could be nursed in isolation, the poor could not and this meant that the newly inoculated could spread smallpox amongst the uninoculated. Although inoculation was good for individual it remained unclear whether it was good for the community.

Fortunately, a country doctor named Jenner noted that the inoculation 'did not take' amongst those of his patients who had had cowpox (a mild disease caught from cows). This suggested trying inoculation with cowpox rather than smallpox. He found that this prevented smallpox inoculation from 'taking' indicating that cowpox inoculation (which caused no illness) protected against smallpox as, indeed, proved to be the case.[18]

Jenner's vaccination was almost risk-free for the patient, caused little discomfort and avoided the risk that others might catch smallpox from the patient. By means of Jenner's discovery, smallpox has, at least for the moment, vanished from the earth. 'If you seek his monument, look around you'.[19]

Exercise 3.4.5 *Estimate the number of deaths in your country which would be expected each year from smallpox if it were still endemic.*

3.5 Utility and Jensen's inequality

In the previous section I gave some of d'Alembert's arguments against the assumption that people wish to maximise their expected lifetime. Do similar objections apply to the assumption of the first two chapters that people wish to maximise their expected fortune?

---

[18] For the full story see [24]. I have relied heavily on [24] and [57] throughout the last two sections.

[19] But not in Trafalgar Square. His statue was exiled from there to Kensington Gardens by nineteenth-century Britain as unworthy to stand with military heroes at the centre of a 'city of empire' and remains exiled by twentyfirst-century Britain as unworthy to stand at the centre of a 'city of culture'.

For most people money is not an end in itself. They wish to be happy and view money as a means to that end. However, it is a common observation that having twice as much money does not make people twice as happy.

A simple-minded mathematician might seek to make sense of this observation as follows. Everyone has an associated function f (which will vary from person to person) such that, if she has a fortune x, she has 'happiness' (or, to use less loaded words, 'satisfaction' or 'utility') $f(x)$. What can we say about the *utility function* f? Having more money increases satisfaction, so f is increasing and $f'(x) \geq 0$ for all x. It is plausible that the rate of increase of satisfaction declines as we have more money. (An extra euro gives more satisfaction to a beggar than to a millionaire.) Thus f' is decreasing and $f''(x) \leq 0$ for all x.

Definition 3.5.1 *We say that a function f is a smooth[20] concave[21] function if it is well-behaved and $f''(x) \leq 0$.*

Lemma 3.5.2 *If f is a smooth concave function, then*

$$f(x) - f(y) \leq f'(y)(x - y)$$

for all x and y.

Proof If we fix y and set $g(x) = f'(y)(x - y) - \big(f(x) - f(y)\big)$, then

$$g'(x) = f'(y) - f'(x) \text{ and } g''(x) = -f''(x) \geq 0$$

for all x. Thus g is increasing. Since $g'(y) = 0$, it follows that

$$g'(x) \begin{cases} \leq 0 & \text{for } x \leq y, \\ \geq 0 & \text{for } x \geq y, \end{cases}$$

so g attains a minimum at y. This means that

$$g(x) \geq g(y) = 0$$

for all x and this is what we wished to prove.   ∎

We can now prove a version of Jensen's inequality.

----

[20] In more advanced work, mathematicians use a more general definition of concave which does not require the function to be differentiable. However, we shall assume that all the functions we consider are twice continuously differentiable.

[21] Unfortunately, half the world calls this a convex function. The half which agrees with our definition, remembers that the graph of the concave function $1 - x^2$ 'looks like a cave'. Everyone is agreed that f is convex if $-f$ is concave.

Theorem 3.5.3 *If X is a random variable and f is a smooth concave function, then*

$$\mathbb{E}f(X) \le f(\mathbb{E}X).$$

Proof If we set $x = X$ and $y = \mathbb{E}X$ in Lemma 3.5.2, we obtain

$$f(X) - f(\mathbb{E}X) \le f'(\mathbb{E}X)(X - \mathbb{E}X).$$

Applying the rules governing expectation, this gives

$$\mathbb{E}f(X) - f(\mathbb{E}X) = \mathbb{E}\big(f(X) - f(\mathbb{E}X)\big) \le \mathbb{E}\big(f'(\mathbb{E}X)(X - \mathbb{E}X)\big)$$
$$= f'(\mathbb{E}X)(\mathbb{E}X - \mathbb{E}X) = 0$$

which is the required result. ∎

Jensen's inequality has an immediate interpretation in terms of the utility function. If, as suggested above, my utility function f is a smooth concave function, then, if my fortune is some random variable X with mean μ, my expected utility $\mathbb{E}f(X)$ will be greatest when

$$\Pr(X = \mu) = 1.$$

In other words, I am at least as happy with the certain sum μ as I would be with any random sum X which had mean μ. Economists would say that I am risk averse.

Risk averse people will tend to buy insurance, but the amount they buy will depend on its cost and the exact shape of their utility function. Let us return to the situation discussed in Exercise 3.1.1.

Exercise 3.5.4 *If $f(t) = \log t$, show that f is a smooth concave function.*

Let f be my utility function. Suppose that my fortune tomorrow is a random variable X such that $\Pr(X = a/2) = \Pr(X = 2a) = 1/2$ for some $a > 0$. For the price of $a\upsilon$ paid now, an insurance company promises to pay me $ka\upsilon$ if $X = a/2$. What are the possible values of my fortune Y tomorrow if I pay $a\upsilon$ for insurance and what are their probabilities?

I wish to maximise my expected utility $\mathbb{E}f(Y)$ but I must take $\upsilon \ge 0$. For what values of k will I buy insurance (that is to say, take $\upsilon > 0$) and what will be my choice of υ if I do? What will be the expected gain of the insurance company (that is to say, the difference between what I pay them and the expected value of what they pay out)? What value of k should the insurance company take?

The argument that I gave, to show that utility functions ought to be concave, may appeal to middle aged professors, but it is less likely to appeal to river boat gamblers, captains of industry and all those who adopt Cesare Borgia's motto 'Aut Caesar, aut nihil'.[22] We note the following converse to Jensen's inequality.

Lemma 3.5.5 *If f is a well-behaved function such that*

$$\mathbb{E} f(X) \leq f(\mathbb{E}(X))$$

for all random variables X, then f is a smooth concave function.

We give the proof in the form of an exercise.

Exercise 3.5.6 *Suppose that f satisfies the hypotheses of Lemma 3.5.5.*
(i) By considering an appropriate X, show that

$$t f(x) + (1 - t) f(y) \leq f\big(t x + (1 - t) y \big)$$

for all t with $0 \leq t \leq 1$ and all x and y.
(ii) If $x < v < y$, show, by writing $v = t x + (1 - t) y$, or otherwise, that

$$\frac{f(v) - f(x)}{v - x} \geq \frac{f(y) - f(v)}{y - v}.$$

Draw a diagram to illustrate your result.
(iii) If $a < b < c < d$, show, by using (ii), or otherwise that

$$\frac{f(b) - f(a)}{b - a} \geq \frac{f(d) - f(c)}{d - c}.$$

Draw a diagram to illustrate your result.
(iv) Suppose that $x < y$. Explain why

$$\frac{f(x + k) - f(x)}{k} \geq \frac{f(y + k) - f(y)}{k}$$

when k is small and deduce that $f'(x) \geq f'(y)$. Conclude that f is a smooth concave function.

If I go for dinner in a restaurant at which the cheapest dish costs €2 and discover that I have left my wallet at home, then finding a one euro coin in my coat lining will not make me much happier but finding a two euro coin will greatly increase my happiness. In this situation my utility function is certainly not concave.

Exercise 3.5.7 *In golf the player with the* lowest *score wins. I must choose one member of my team to play golf against a member of an opposing team who*

---

[22] Either Caesar or nothing.

consistently scores 72. *I know that the score of the j th member of my team will be a random variable* X_j *with*

$$\Pr(X_1 = 73) = \Pr(X_1 = 75) = 1/2,$$
$$\Pr(X_2 = 80) = \Pr(X_2 = 70) = 1/2,$$
$$\Pr(X_3 = 90) = \Pr(X_3 = 80) = \Pr(X_3 = 70) = 1/3.$$

Find the mean and variance of X_j *for each* j. *Which player should I choose?*

The notion of a utility function is an attractive one to mathematicians and economists but much of this attractiveness vanishes when we move from the abstract to the concrete.

One problem is that, even after considerable thought, I cannot graph my own utility function f and I suspect my readers cannot graph their own.[23]

Exercise 3.5.8 *Explain why, if* $a > 0$, *a person with utility function* f *will make the same choices as someone with utility function* af. *Thus utility functions are only determined up to multiplication by a constant. We shall ignore this and continue to talk about 'the utility function'.*

In theory, I could ask myself a series of questions like 'Would I prefer a lottery which gave me one chance in a million of winning €1 000 000 and nothing otherwise or one which gave me one chance in twenty of winning €200?' and work out my utility function. In practice, my answers would be inconsistent and, in any case, my feelings about imaginary choices may not represent my feelings about real choices.[24] (Two and a half centuries later, I have no difficulty in accepting Bernoulli's arguments for inoculation, but, if faced with a one in a hundred chance of an unpleasant death within a month from inoculation, I might not find it so easy to appreciate the balancing benefit of an 'average' increase in lifetime.)

Even our crude division of people into risk-avoiders and risk-seekers has to be modified after the first time that we see someone take out house insurance and then buy a ticket for a lottery.

---

[23] There are other problems. Happiness seems partly to depend on comparisons with other people. A millionaire may be happy in a society in which she is the richest person but unhappy in a society of billionaires. There is also a curious 'hysteresis effect' in that, if somebody gains something and then loses it, they may end up more unhappy than if they had never had it.

[24] It can be argued that it is not necessary for individuals to know their utility function. According to this view, people *cannot fail* to maximise their expected utility. They do not decide their actions by looking at their utility function, but their utility function can be deduced from their actions. 'How do I know what I think until I hear what I say?' In this book we assume that decision makers know what they want before they act rather than afterwards.

How can we deal with a function which we do not know? One way out is to observe that reasonable functions satisfy the condition

$$f(x + \delta x) \approx f(x) + f'(x)\delta x$$

when δx is small. Thus, *provided we only consider small decisions*, we may suppose that our utility function is linear, that is to say,

$$f(x) = ax + b.$$

We can also argue that, although individuals may have non-linear utility functions, institutions ought to have linear utility functions. (A chain of stores should be exactly as anxious to gain an extra dollar as a single store.) From now on, we shall implicitly assume that we are dealing with linear utility functions.[25] We shall return to the discussion of utility briefly on page 194.

I end this section with some exercises on Jensen's inequality. They are not required for the rest of the book and are intended for readers with an interest in mathematics for its own sake.

Exercise 3.5.9 *Jensen showed that many well-known inequalities were special cases of the inequality named after him.*[26]

(i) Suppose that f is a smooth concave function. Show, by applying Theorem 3.5.3, that, if

$$t_1, t_2 \ldots, t_n \geq 0 \text{ and } t_1 + t_2 + \cdots + t_n = 1,$$

then

$$t_1 f(x_1) + t_2 f(x_2) + \cdots + t_n f(x_n) \leq f(t_1 x_1 + t_2 x_2 + \cdots + t_n x_n).$$

for all $x_1, x_2, \ldots, x_n$.

(ii) By taking $f(x) = \log x$ and $t_j = 1/n$ in (i), prove Cauchy's arithmetic-geometric inequality which states that

$$(x_1 x_2 \cdots x_n)^{1/n} \leq \frac{x_1 + x_2 + \cdots + x_n}{n}$$

whenever $x_1, x_2, \ldots, x_n > 0$.

(iii) Suppose that $p > 1$ (we do not assume that p is an integer) and let $g(x) = (1 + x^{1/p})^p$. Show that g is a smooth concave function.

---

[25] This is a book about mathematics and not about human behaviour. Luce and Raiffa [39] discuss in a sensible and humane way how far abstract notions like utility can be linked to real life.

[26] In fact, Jensen has the rare distinction of having given his name to *two* powerful inequalities.

Suppose that $a_1, a_2, \ldots, a_n > 0$, $a_1^p + a_2^p + \cdots + a_n^p = 1$ and $b_1, b_2, \ldots, b_n > 0$. By applying Jensen's inequality with $x_k = b_k^p/a_k^p$ and t_k chosen appropriately, show that

$$\left((a_1+b_1)^p + (a_2+b_2)^p + \cdots + (a_n+b_n)^p\right)^{1/p} \leq 1 + \left(b_1^p + b_2^p + \cdots + b_n^p\right)^{1/p}.$$

(iv) Suppose that $p > 1$ and $t_1, t_2, \ldots, t_n > 0$ and $s_1, s_2, \ldots, s_n > 0$. By applying the result of (iii), show that

$$\left((t_1 + s_1)^p + (t_2 + s_2)^p + \cdots + (t_n + s_n)^p\right)^{1/p}$$
$$\leq \left(t_1^p + t_2^p + \cdots + t_n^p\right)^{1/p} + \left(s_1^p + s_2^p + \cdots + s_n^p\right)^{1/p}.$$

Suppose that u_j and v_j are real. Show that

$$\left(|u_1 + v_1|^p + |u_2 + v_2|^p + \cdots + |u_n + v_n|^p\right)^{1/p}$$
$$\leq \left(|u_1|^p + |u_2|^p + \cdots + |u_n|^p\right)^{1/p} + \left(|v_1|^p + |v_2|^p + \cdots + |v_n^p|\right)^{1/p}.$$

This is the famous Minkowski inequality. If $n = 2$ and $p = 2$, show that it corresponds to the statement that the distance[27] from the point U with coordinates (u_1, u_2) to the point V with coordinates (v_1, v_2) is less than or equal to the sum of the distance from the origin to U and the distance from the origin to V (the sum of the lengths of two sides of a triangle is never less than the length of the third side). Give a similar interpretation when $n = 3$ and $p = 2$.

(v) Let $h(x) = -x \log x$ for $x > 0$. Show that h is smooth concave function and deduce that

$$(x + y) \log \frac{x+y}{a+b} \leq x \log \frac{x}{a} + y \log \frac{y}{b}$$

for all $a, b, x, y > 0$.

Exercise 3.5.10 *Suppose $p \geq 1$. Find the smallest value of c_p such that*

$$(|a| + |b|)^p \leq c_p |a + b|^p$$

for all values of a and b and show that your answer is correct.

Exercise 3.5.11 *(If you are put off by the notation used, just ignore this question. Even those familiar with the notation may find the question a bit of a brain-teaser.) Suppose that $\pi_{ij} > 0$ for $1 \leq i \leq n$, $1 \leq j \leq m$ and*

$$\sum_{i=1}^{n} \sum_{j=1}^{m} \pi_{ij} = 1.$$

---

[27] Euclidean distance, if you know enough to worry about such things.

We write $p_i = \sum_{j=1}^{n} \pi_{ij}$, $q_j = \sum_{i=1}^{m} \pi_{ij}$ *and*

$$I = \sum_{i=1}^{n} \sum_{j=1}^{m} \pi_{ij} \log \frac{\pi_{ij}}{p_i q_j}.$$

By considering the function $h(x) = -x \log x$ *of Exercise 3.5.9, or otherwise, show that* $I \leq 0$. *When is* $I = 0$?

Can you give an interpretation of this result in terms of random variables?

4

Passing the time

4.1 The three towers

Mr Claus, mandarin of the College of Li-Sou-Stan[1] reported that, during his travels,

> ... he saw in the great temple at Benares, beneath the dome which marks the centre of the world, a brass plate in which are fixed three diamond needles, each a cubit high and as thick as the body of a bee. On one of these needles, God placed, at the Creation, sixty-four discs of pure gold, the largest disc resting on the brass plate, the others getting smaller and smaller up to the top. This is the sacred tower of Brahma. Night and day, teams of priests follow each other on the steps of the altar transferring the discs from one diamond needle to another according to the fixed and immutable laws of Brahma [which require that the priests on duty must not move more than one disc at a time, and that no disc may be placed on a needle which already holds a smaller disc]. When the sixty-four discs shall have been thus transferred from the needle on which, at the creation, God placed them to one of the other needles, then towers and priests alike will vanish and the universe will end. ([38], page 57, Volume 3.)

Exercise 4.1.1 *Quickly guess the time to the end of the world.*

In order to find how much time remains before the end of the world, we consider the more general case in which the priests have n discs on one needle A and must transfer them to the second needle B making use of the third needle C. In order to move the largest disc to B they will have had to move the remaining $n - 1$ discs to form a tower on C (that is to say, they must perform the required task with $n - 1$ discs). They now move the largest disc to B and must

---

[1] Édouarde Lucas, professor at the Lycée Saint Louis, the inventor, in 1883, of the game described. The toy is usually known as the 'Tower of Hanoi'. Lucas was inspired by the theory of Chinese Rings, a game which is genuinely old (see page 351) and seems indeed to have originated in China.

now move the tower of $n - 1$ smaller discs from C to rest above the largest disc on B (that is they must perform the required task with $n - 1$ discs).

Thus, if it takes them k_{n-1} moves to perform the task with $n - 1$ discs in the most efficient manner it will take them

$$k_n = k_{n-1} + 1 + k_{n-1} = 2k_{n-1} + 1$$

moves to perform the task with n discs in the most efficient manner. If $n = 1$ there is only one disc, so we need only one move to transfer from one needle to the other. Thus $k_1 = 1$.

Exercise 4.1.2 *(i) Compute k_2, k_3 and k_4.*
(ii) Quickly guess the time to the end of the world.
(iii) It is not hard to guess that $k_n = 2^n - 1$. Prove this by induction.
(iv) Quickly guess the time to the end of the world.

We now recall the useful approximation

$$2^{10} = 1024 \approx 10^3$$

and obtain

$$2^{64} - 1 \approx 2^{64} = 2^4 \times (2^{10})^6 \approx 16 \times (10^3)^6 = 16 \times 10^{18}.$$

Those with access to appropriate calculators can obtain the exact result

$$2^{64} - 1 = 18\,446\,744\,073\,709\,551\,615$$

but will be little wiser.

Exercise 4.1.3 *(i) It must surely take more than a second to transfer a disc. Using a calculator or working by hand, show that there are fewer than* $31\,600\,000$ *seconds in the average year. Conclude that the universe will last more than* $580\,000$ *million years from its date of creation.*

(ii) Suppose that the priests receive divine sanction to replace the actual towers by a computer simulation.[2] Making what assumptions you please, and assuming that they have access to the most powerful machine you know, estimate how long the universe will now last.

(iii) In a dream, the chief priestess is allowed to choose between the old fashioned system of real towers with 64 discs and computer simulation using the most powerful machines available as they come onto market but with four times as many simulated discs (that is to say, 256 discs). Modern physics suggests that no operation can take less than about 5×10^{-44} *seconds (the Planck*

---

[2] For an elegant variation on the theme, see Arthur C. Clarke *The Nine Billion Names of God.*

time). The devout priestess wishes to bring about the end of the universe as quickly as possible. Advise her.

As a matter of interest, some modern estimates make the universe roughly 13 000 million years old.

Exercise 4.1.4 *Let us label the discs 1 to n in order of size, starting with the smallest. Show, by induction, or otherwise, that the following rules for the kth move [$1 \leq k \leq 2^n - 1$] will give a most efficient way of moving a tower of n from one peg to another.*

(a) Suppose that k is divisible by 2^{r-1} but is not divisible by 2^r. Then move the rth disc.

(b) If r is odd, always move the rth disc clockwise (so that you go from peg A to peg B, peg B to peg C and from peg C to peg A, say) when you move it. If r is even, move it anti-clockwise (from peg C to peg B, from peg B to peg A and from peg A to peg C, say).

An *algorithm* is a method for carrying out a mathematical task. In order to qualify as an algorithm, we must be able to describe the method in sufficient detail that it can be carried out by a junior priest (or computer) by simply following the given rules (or program) without exercising any initiative.

A computer program must be very detailed and may take into account the particular properties of a computer (for example it may take a long time to fetch information from parts of the machine memory). Mathematicians think more like junior priests and are satisfied by an algorithm if they are convinced that it could, in principle, be converted into a computer program.

In this section we have found an algorithm for solving the 'Tower of Hanoi' and shown that no better algorithm exists (in the sense that no algorithm can use fewer moves). We have also discovered that we can specify very simple tasks which are easy to perform in theory, but which cannot be completed in practice because they take too long.

4.2 Euclid's algorithm

Mathematicians distinguish between *non-constructive proofs*[3] which show that something exists but do not tell us how to find it and *constructive proofs* which show that something exists by giving us a method for finding it.

We illustrate the idea by giving non-constructive proofs for Lemmas 4.2.1 and 4.2.3 which follow. Recall that we say that an integer u divides an integer v if we can find an integer k such that $v = uk$.

---

[3] Sometimes called *existence proofs.*

Lemma 4.2.1 *If a and b are non-zero integers, then there exists a unique integer d with the following properties.*

(i) d divides a and b.

(ii) If the integer e divides a and b, then $e \leq d$.

Proof Consider the set F of positive integers which divide a and b. We observe that $1 \in F$, so F is non-empty. On the other hand, if $u \in F$, then $|u| \leq |a|$, so F is finite. Every finite non-empty set has a largest member, so F has a largest member d. ∎

We call d the *greatest common divisor* (or greatest common factor) of a and b.

Exercise 4.2.2 *(i) Explain briefly why a and b have the same greatest common divisor as a and $-b$.*

(ii) Check that you agree that the highest common divisor of 182 and 140 is 14.

(iii) Choose two three-figure numbers at random and find their highest common divisor. Repeat the exercise with two four-figure numbers (or admit defeat). Repeat the exercise with two five-figure numbers (or admit defeat). Repeat with 815 055 and 208 427 (or admit defeat).[4]

The reader may consider that the proof of Lemma 4.2.1 is little more than the restatement of the obvious. However, the same ideas give our next result which is much less obvious. (If you disagree, try and find your own proof.)

Lemma 4.2.3 [Bézout's identity][5] *(i) If a and b are non-zero integers with greatest common divisor d, then we can find integers m and n such that*

$$d = ma + nb.$$

(ii) If a and b are non-zero integers with greatest common divisor d, then we can find integers M and N such that $y = Ma + Nb$ if and only if y is an integer multiple of d.

Proof (i) Let E be the set of integers of the form $ua + vb$ with u and v integers and $ua + vb \geq 1$. Since

$$aa + bb = a^2 + b^2 \geq 1,$$

---

[4] These numbers were not chosen at random.

[5] The first European statement of this result is due to Bachet, but Bézout proved the extension to polynomials given in Exercise 4.2.5. The great Indian mathematician Brahmagupta understood the result a thousand years earlier.

we know that E is non-empty. Since any non-empty set of positive integers has a least member, E has a least member

$$d_0 = u_0 a + v_0 b,$$

with u_0 and v_0 integers.

Since $|a| = (|a|/a)a + 0b \in E$, we know that $1 \leq d_0 \leq |a|$ and so $a = kd_0 + r$ where k and r are integers and $d_0 > r \geq 0$. If $r \geq 1$, then

$$r = a - kd_0 = (1 - ku_0)a + (-kv_0)b \in E$$

contradicting the definition of d_0 as the least element of E. Thus $r = 0$, so $a = kd_0$ and d_0 divides a. Similarly d_0 divides b. It follows that $d \geq d_0$.

We also know that $d_0 = u_0 a + v_0 b$ and d divides both a and b, so d divides d_0 and $d_0 \geq d$. Thus $d = d_0$ and the result follows on setting $m = u_0$ and $n = v_0$.

(ii) If $y = Ma + Nb$ then, since d divides a and b, it follows that d divides y. Conversely if d divides y, then $y = kd$ for some integer k. Set $M = km$ and $N = kn$, where m and n are the integers obtained in (i). ∎

Exercise 4.2.4 *Use Lemma 4.2.3 to show that, if a and b are non-zero integers, any integer which divides both a and b also divides their greatest common divisor. Spend a little time trying to find a rigorous proof that does not use Lemma 4.2.3.*

Exercise 4.2.5 *This exercise should be omitted if you are not happy with the ideas it deals with.*

We consider polynomials with real coefficients. We say that a non-zero polynomial Q divides a polynomial P if we can find a polynomial S such that $P(x) = Q(x)S(x)$. For example, $3x + 1$ divides $\frac{2}{5}x^2 + \frac{11}{15}x + \frac{1}{5}$ since

$$\frac{2}{5}x^2 + \frac{11}{15}x + \frac{1}{5} = (3x + 1)\left(\frac{2}{15}x + \frac{1}{5}\right).$$

We say that a polynomial is monic if the coefficient of its highest term is 1.

(i) Explain why, if P is a non-zero polynomial, we can find a unique monic polynomial Q and a unique constant a such that $P = aQ$.

(ii) Explain why, if Q is a polynomial of degree $n \geq 0$ and P is any polynomial, we can find polynomials K and R such that

$$P(x) = K(x)Q(x) + R(x)$$

where R is a polynomial of degree strictly less than n. (By convention, we take the zero polynomial to have degree -1.) You are not asked to supply a *rigorous proof, but, if you wish to provide one, you could use induction on the degree of P.*

(iii) If P and Q are non-zero polynomials, show that there is a monic polynomial S of highest degree dividing both P and Q.

(iv) Continuing with the notation of (iii), show that we can find polynomials U and V such that

$$S(x) = U(x)P(x) + V(x)Q(x).$$

Explain why S is unique.

(v) State and prove a result for polynomials which corresponds to the result of Lemma 4.2.3 (ii).

Exercise 4.2.6 *(i) Find integers u and v such that $14 = 182u + 140v$.*

(ii) In Exercise 4.2.2 (iii) you were invited to choose a and b of a particular size and to find their greatest common divisor d. Repeat the exercise but this time find integers m and n such that

$$d = ma + nb.$$

Exercise 4.2.6 should convince you that there is great difference between knowing that something exists and being able to find it.

Fortunately, there is a constructive approach to the problems considered in this section. Euclid's algorithm is a very clever way of finding greatest common divisors which goes back at least as far as Euclid.

Lemma 4.2.7 *(i) If (a, b) is a pair of non-zero integers with $|a| \geq |b|$, then, either $|b|$ divides $|a|$ and $|b|$ is the greatest common divisor of a and b, or we can find integers k and c such that*

$$a = kb + c \quad and \quad |c| \leq |b|/2.$$

(ii) Suppose (a, b) is a pair of non-zero integers with $|a| \geq |b|$ and $|b|$ does not divide $|a|$. Then, if c is defined as in (i), (a, b) and (b, c) have the same highest common divisor.

Proof (i) We have $a/b = r + x$ where r is an integer and $0 \leq x < 1$. If $x = 0$, then $|b|$ divides $|a|$. If $0 < x \leq 1/2$, set $k = r$ and $c = xb$. If $1/2 < x < 1$, set $k = r + 1$ and $c = (1 - x)b$.

(ii) Let d be the highest common divisor of a and b and let D be the highest common divisor of b and c. Since d divides a and b, it follows that d divides $a - kb = c$. We already know that d divides b, so, since d divides b and c and we know that D is the largest integer with that property, $d \leq D$.

In the same way, since D divides b and c, it follows that D divides $a = kb + c$. Thus D divides a and b and so $D \leq d$. Combining the results of the two paragraphs, we obtain $d = D$. ∎

Let us call (b, c) the pair derived from (a, b). We can now describe Euclid's algorithm. Suppose that we want to find the greatest common divisor of two integers a and b with $|a| \geq |b| > 0$. We set $a_1 = a$, $b_1 = b$ and proceed according to the following rule.

Rule If at the rth step we are given a pair (a_r, b_r) with $|a_r| \geq |b_r| > 0$, then either $|b_r|$ divides $|a_r|$, in which case we write $E = |b_r|$ and stop the process, or not, in which case we let (a_{r+1}, b_{r+1}) be the the pair derived from (a_r, b_r) and proceed to the $r + 1$th step.

Lemma 4.2.8 *With the notation just introduced, $|b_{r+1}| \leq |b_r|/2$ for all $r \geq 1$ and so $|b_r| \leq 2^{-r+1}|b_1|$. In particular, if $|b| < 2^N$, Euclid's algorithm will stop after at most N steps.*

Proof Left to the reader. Observe that, if b is an integer with $|b| < 2$ and a a non-zero integer, then $|b| = 1$ so b divides $|a|$. ∎

Lemma 4.2.8 tells us that Euclid's algorithm will always stop and Lemma 4.2.7 tells us that the integer E it produces will be the highest common divisor of a and b.

Exactly the same calculations that told us that it will take an interminable time to complete the Tower of Hanoi, even when the number of discs is quite small, tell us that Euclid's algorithm will work like greased lightning, even when the numbers involved are quite large.

Exercise 4.2.9 *Show that, if $|a|$, $|b| \leq 10^{300}$, then Euclid's algorithm will require fewer than 1000 steps.*[6]

Euclid's algorithm also gives us a constructive approach to Lemma 4.2.3.

Lemma 4.2.10 *(i) Suppose that (a, b) is a pair of non-zero integers with $|a| \geq |b|$ and suppose that $|b|$ does not divide $|a|$. Let d be the highest common divisor of a and b and let (b, c) be the pair derived from (a, b). If we can find integers u and v such that*

$$d = ub + vc,$$

then we can find integers U and V such that

$$d = Ua + Vb.$$

---

[6] Of course, it takes a certain amount of work to program a computer to handle such numbers but it is not really hard.

(ii) If a and b are non-zero integers with greatest common divisor d, then we can find integers m and n such that

$$d = ma + nb.$$

(iii) If a and b are non-zero integers with greatest common divisor d, then we can find integers M and N such that $y = Ma + Nb$ if and only if y is an integer multiple of d.

Proof (i) We know that $a = kb + c$ for some integer k and so

$$d = ub + vc = ub + v(a - kb) = va + (u - kv)b.$$

Set $U = v$ and $V = u - kv$.

(ii) Without loss of generality, suppose that $|a| \geq |b|$. If $|b|$ divides $|a|$, then $d = |b|$ and the result follows on taking $m = 0$ and $n = |b|/b$. If not, we consider the pairs produced by Euclid's algorithm and use part (i).

(iii) Do this as a revision exercise or consult part (ii) of Lemma 4.2.3. ∎

We illustrate Euclid's algorithm by applying it to $815\,055$ and $208\,427$. Using a calculator we see that

$$815\,055 \approx 3.91 \times 208\,427$$

and so $815\,055 = 4 \times 208\,427 + r$ with $|r| \leq 208\,427/2$. Proceeding in this way, Euclid's algorithm gives us the successive steps

$$815\,055 = 4 \times 208\,427 - 18\,653$$
$$208\,427 = (-11) \times (-18\,653) + 3244$$
$$18\,653 = 6 \times 3244 - 811$$
$$3244 = (-4) \times (-811).$$

Thus we start with the pair $(815\,055, 208\,427)$ and obtain the successive pairs $(208\,427, -18\,653)$, $(-18\,653, 3244)$ and $(3244, -811)$ stopping at this pair since 811 divides 3244.

We can reverse our calculations, in the manner of Lemma 4.2.10, to obtain

$$811 = 6 \times 3244 - 18\,653 = 6 \times (208\,427 - 11 \times 18\,653) - 18\,653$$
$$= 6 \times 208\,427 - 67 \times 18\,653$$
$$= 6 \times 208\,427 - 67 \times (4 \times 208\,427 - 815\,055)$$
$$= -262 \times 208\,427 + 67 \times 815\,055.$$

Exercise 4.2.11 *(i) Choose a pair of six-figure numbers a and b. Find their highest common divisor d and find integers u and v such that*

$$d = ua + vb.$$

(ii) Repeat the process with different pairs of six-figure numbers until you are satisfied that you understand exactly how to solve the problem. Now write out the algorithm explicitly (as a program, flow chart, instructions to a junior priest or whatever you prefer).

Exercise 4.2.12 *The five-man crew of the HMS Bézout are shipwrecked on a tropical island together with the cabin boy's pet monkey. In order to pass the time they collect an enormous pile of coconuts which they then divide as follows.*

The Captain gives one coconut to the monkey and takes exactly one fifth of the remaining coconuts. Then the First Mate gives one coconut to the monkey and takes exactly one fifth of the remaining coconuts. The Second Mate does the same, followed by the Bosun and finally the cabin boy. The number of coconuts remaining is exactly divisible by 5 so they take equal shares of the remaining coconuts without giving any to the monkey. What is the smallest number of coconuts they could have started with? (For more on this puzzle consult [23]. Although you may need a calculator to solve the problem the result can be verified without one.)

Exercise 4.2.13 *Suppose that you are given two non-zero integers a and b together with positive non-zero integers u and v such that d divides a and b and $d = ua + vb$. Explain why you know, without further calculation, that $|d|$ is the greatest common divisor of a and b.*

We shall also need the concept of a lowest common multiple.

Exercise 4.2.14 *Let a and b be non-zero integers.*

(i) Give a non-constructive proof along the lines of Lemma 4.2.1 that there exists a unique smallest integer e with $e \geq 1$ such that a and b divide e. We call e the lowest common multiple of a and b.

(ii) Show, by considering integers of the form $n + ve$ or otherwise, that, if a and b divide n, then e divides n.

(iii) Suppose $a, b \geq 1$ and that a and b have highest common divisor d. Show that a and b divide ab/d and deduce that $ab/d \geq e$. Show that ab/e is an integer which divides a and b and deduce that $ab/e \leq d$. Conclude that

$$ab = ed.$$

If we drop the condition $a, b \geq 1$, show that we obtain $|ab| = ed$.

(iv) Find the lowest common multiple of $22\,015$ and 5291.

Exercise 4.2.15 *If you have done Exercise 4.2.5, show how to adapt Euclid's algorithm to find the unique monic polynomial S of highest degree dividing both P and Q and to find polynomials U and V such that*

$$S(x) = U(x)P(x) + V(x)Q(x).$$

4.3 Arithmetic modulo n

Sometime between AD270 and AD480 there appeared in China[7] a book *Sun Zi Suanjing* (Sun Zi's Mathematical Manual) containing the following problem together with an ingenious solution.

Exercise 4.3.1 *We have a number of things, but we do not know exactly how many. If we count them by threes we have two left over. If we count them by fives we have three left over. If we count them by sevens we have two left over. How many things are there?*

The object of this section is to produce a method for solving such problems. Before proceeding, the reader should put a little effort into solving Exercise 4.3.1. If she succeeds, she will find the discussion very much easier. Whether she succeeds or fails, she should acquire considerable respect for Sun Zi.

Monsieur Jourdain[8] was surprised and delighted to learn that he had been speaking prose all his life. The reader will be equally delighted to learn that she has been doing modular arithmetic all her life.

Exercise 4.3.2 *(i) What month are you in today? What month will you be in when exactly* 279 *months have passed?*

(ii) A mathematician decides that the natural length of a day is 22 *hours and so she will breakfast every* 22 *hours. If she had her first breakfast at* 8pm *today, at what time will she eat her* 200th *breakfast?*

---

[7] See [72], which contains a translation.

[8] MONSIEUR JOURDAIN. There is nothing but prose or verse?
PHILOSOPHY MASTER. No, sir, everything that is not prose is verse, and everything that is not verse is prose.
MONSIEUR JOURDAIN. And when one speaks, what is that then?
PHILOSOPHY MASTER. Prose.
MONSIEUR JOURDAIN. What! When I say, 'Nicole, bring me my slippers, and give me my nightcap,' that's prose?
PHILOSOPHY MASTER. Yes, Sir.
MONSIEUR JOURDAIN. Good heavens! For more than forty years I have been speaking prose without knowing anything about it.
                                                    Moliére, The Bourgeois Gentleman

(iii) If I add an odd and an even integer together, will the result be even or odd?

Definition 4.3.3 *If n is a non-zero integer and a and b are integers, we write*

$$a \equiv b \quad \mathrm{mod}\ n$$

if and only if b − a is divisible by n.

If $a \equiv b$ mod n we say that 'a equals b modulo n'. Sometimes, as in Lemma 4.3.4 below, we abbreviate $a \equiv b$ mod n to $a \equiv b$ when it is clear that we are working modulo n.

Lemma 4.3.4 *Suppose that n is a non-zero integer and a, a′, b, b′ and c are integers. Then*

(i) $a \equiv a$ mod n.
(ii) If $a \equiv b$, then $b \equiv a$ mod n.
(iii) If $a \equiv b$ and $b \equiv c$, then $a \equiv c$ mod n.
(iv) If $a \equiv a′$ and $b \equiv b′$, then $a + a′ \equiv b + b′$ mod n.
(v) If $a \equiv a′$ and $b \equiv b′$, then $ab \equiv a′b′$ mod n.
However,
(vi) If $a \equiv a′$ and $b \equiv b′$ mod n and $a′ > b′$, it does not follow that $a > b$.

Proof We shall prove parts (iii), (v) and (vi), leaving the rest to the reader.

(iii) If $a \equiv b$ and $b \equiv c$ mod n, then we can find integers k and l such that $b - a = kn$ and $c - b = ln$. It follows that

$$c - a = (c - b) + (b - a) = kn + ln = (k + l)n,$$

so $a \equiv c$, as stated.

(v) If $a \equiv a′$ and $b \equiv b′$ mod n, then we can find integers k and l such that $a′ - a = kn$ and $b′ - b = ln$. It follows that

$$a′b′ - ab = (a + kn)(b + ln) - ab = aln + bkn + kln^2 = (al + bk + kln)n,$$

so $ab \equiv a′b′$, as stated.

(vi) Take $n = 3, a = -1, b = b′ = 1$ and $a′ = 2$ to obtain a counterexample. ■

Exercise 4.3.5 *(i) If a is an integer and n is a positive integer, show that $10^n a \equiv a$ mod 9. If $a_0, a_1, \ldots, a_n$ are integers, show that*

$$a_n 10^n + a_{n-1} 10^{n-1} + \cdots + a_0 \equiv a_n + a_{n-1} + \cdots + a_0 \quad \mathrm{mod}\ 9.$$

If $b_0, b_1, \ldots, b_n$ are integers, show that

$$\left(a_n 10^n + a_{n-1} 10^{n-1} + \cdots + a_0\right) \times \left(b_n 10^n + b_{n-1} 10^{n-1} + \cdots + b_0\right)$$
$$\equiv (a_n + a_{n-1} + \cdots + a_0) \times (b_n + b_{n-1} + \cdots + b_0).$$

(ii) The following procedure is called 'casting out nines'. It was used before the days of pocket calculators by people like the present author to check long sums. We illustrate by checking the multiplication

$$897\,457 \times 584\,762 \overset{?}{=} 52\,4798\,750\,234.$$

To 'cast out nines' take each number in the calculation and add the digits. If the new number is bigger than 9, repeat the operation and continue until you have an integer between 0 and 9. If you have a 9 replace it by 0.

In the case given,

$$897\,457 \longrightarrow 40 \longrightarrow 4$$
$$584\,762 \longrightarrow 32 \longrightarrow 5$$
$$524\,798\,750\,234 \longrightarrow 56 \longrightarrow 11 \longrightarrow 2.$$

Carry out the calculation again with the reduced numbers and cast out nines again. In this case

$$4 \times 5 = 20 \longrightarrow 2.$$

If your initial calculation was right, the new answer will agree with the answer obtained from the old answer by casting out nines. Is the converse true?

Explain why this method works. Show that it will also work for addition and subtraction.

(iii) Try out the method of casting out nines on an addition sum.

Exercise 4.3.6 *(i) By considering what happens to the equality*

$$(1 + 1)^n = \binom{n}{0} + \binom{n}{1} + \binom{n}{2} + \cdots + \binom{n}{n}$$

when we work modulo 2, show that there are always an even number of odd binomial coefficients $\binom{n}{r}$ for $n \geq 1$. Why does your result fail when $n = 0$?

(ii) If you know about Pascal's triangle, consider the pattern that results when you replace each even term in it by 0 and each odd term by 1. Give a rule for constructing your new triangle and build it up to, say, the tenth row. The result should look quite pretty.

Exercise 4.3.7 *Find a solution of*

$$x \equiv -1 \quad \text{mod } 6,$$
$$x \equiv -1 \quad \text{mod } 10.$$

What is the smallest positive solution?
What is the smallest solution x with $x \geq 10^6$?

Exercise 4.3.8 *(i) In the following table* $r_1 \equiv r \mod 3$, $0 \leq r_1 \leq 2$ *and* $r_2 \equiv r \mod 4$, $0 \leq r \leq 3$

$r = 0$	1	2	3	4	5	6	7	8	9	10	11
$r_1 = 0$	1	2	0	1	2	0	1	2	0	1	2
$r_2 = 0$	1	2	3	0	1	2	3	0	1	2	3

How many solutions of

$$x \equiv 14 \mod 3,$$
$$x \equiv 21 \mod 4$$

are there with $0 \leq x \leq 35$?

(ii) By writing out the appropriate table, find the number of solutions of

$$x \equiv 2 \mod 4,$$
$$x \equiv 4 \mod 6$$

with $0 \leq x \leq 47$ *and the number of solutions of*

$$x \equiv 3 \mod 4,$$
$$x \equiv 4 \mod 6.$$

We now start our attack on Sun Zi's problem. Naturally enough, the key idea is called the *Chinese remainder theorem*.

Theorem 4.3.9 *Suppose that* a_1 *and* a_2 *are non-zero integers with highest common divisor d and lowest common multiple e. Then we can find an integer x satisfying the conditions*

$$x \equiv u_1 \mod a_1, \qquad \qquad ★$$
$$x \equiv u_2 \mod a_2$$

if and only if d divides $u_2 - u_1$. *If d divides* $u_2 - u_1$ *and* x_0 *is a solution, then all solutions of* ★ *are given by* $x \equiv x_0 \mod e$.

The reader should try and prove Theorem 4.3.9 for herself.

Our proof follows the traditional approach of mathematicians when attacking a hard problem. We try to solve particularly simple versions of our problem and then to build up a full solution from the solution of the simpler problems.

Lemma 4.3.10 *(i) Suppose that* b_1 *and* b_2 *are non-zero integers with highest common divisor* 1. *Then we can find a* y_1 *with*

$$y_1 \equiv 1 \mod b_1,$$
$$y_1 \equiv 0 \mod b_2$$

and a y_2 with

$$y_2 \equiv 0 \quad \text{mod } b_1,$$
$$y_2 \equiv 1 \quad \text{mod } b_2.$$

(ii) We continue with the notation of (i). If we set $y = v_1 y_1 + v_2 y_2$, then

$$y \equiv v_1 \quad \text{mod } b_1,$$
$$y \equiv v_2 \quad \text{mod } b_2.$$

Further

$$y' \equiv v_1 \quad \text{mod } b_1,$$
$$y' \equiv v_2 \quad \text{mod } b_2$$

if and only if $y \equiv y' \mod b_1 b_2$.
(iii) Theorem 4.3.9 is true.

Proof (i) Observe that $y_1 \equiv 0 \mod b_2$ if $y_1 = k_2 b_2$ for some integer k_2. Further, if $y_1 = k_2 b_2$, then $y_1 \equiv 1 \mod b_1$ if

$$k_1 b_1 + k_2 b_2 = 1$$

for some integer k_1.

Since 1 is the highest common divisor of b_1 and b_2, Euclid's algorithm tells us that we can, indeed, find k_1 and k_2 satisfying this last equation. Setting $y_1 = k_2 b_2$, we have the required result.

We can obtain a suitable y_2 in the same way.

(ii) Observe that

$$y \equiv v_1 y_1 + v_2 y_2 \equiv v_1 \times 1 + v_2 \times 0 \equiv v_1 \quad \text{mod } b_1,$$
$$y \equiv v_1 y_1 + v_2 y_2 \equiv v_1 \times 0 + v_2 \times 1 \equiv v_0 \quad \text{mod } b_2$$

as required.

If

$$y' \equiv v_1 \quad \text{mod } b_1,$$
$$y' \equiv v_2 \quad \text{mod } b_2,$$

then

$$y' \equiv y \quad \text{mod } b_1,$$
$$y' \equiv y \quad \text{mod } b_2,$$

so $y' - y$ is divisible by both b_1 and b_2 and so by their lowest common multiple which we know (by Exercise 4.2.14) to be $b_1 b_2$. Thus $y \equiv y' \mod b_1 b_2$.

The converse is immediate.

(iii) First suppose that we have an x satisfying ★. Then we can find integers k_1 and k_2 such that

$$x - u_1 = k_1 a_1,$$
$$x - u_2 = k_2 a_2.$$

Thus

$$u_2 - u_1 = (x - u_1) - (x - u_2) = k_1 a_1 - k_2 a_2$$

and, since a_1 and a_2 are divisible by d, it follows that $u_2 - u_1$ is.

Now suppose that $u_2 - u_1$ is divisible by d. Set $b_1 = a_1/d$, $b_2 = a_2/d$, $v_1 = 0$ and $v_2 = (u_2 - u_1)/d$. Observe that b_1, b_2, v_1 and v_2 are all integers and that the highest common divisor of b_1 and b_2 is 1. We can therefore apply part (i) and find a y with

$$y \equiv v_1 \quad \bmod b_1,$$
$$y \equiv v_2 \quad \bmod b_2.$$

Automatically,

$$dy \equiv dv_1 \equiv 0 \quad \bmod a_1,$$
$$dy \equiv dv_2 \equiv u_2 - u_1 \quad \bmod a_2$$

and so, setting $x = dy + u_1$, we have

$$x \equiv dy + u_1 \equiv 0 + u_1 \quad \bmod a_1,$$
$$x \equiv dy + u_1 \equiv (u_2 - u_1) + u_1 \equiv u_2 \quad \bmod a_2$$

as required.

We leave the last sentence of Theorem 4.3.9 as an exercise for the reader. ∎

Exercise 4.3.11 *Use Euclid's algorithm and the ideas of the proof of Lemma 4.3.10 (i) and (ii) to find the general solution of*

$$x \equiv 3 \quad \bmod 17,$$
$$x \equiv -2 \quad \bmod 19.$$

What is the smallest positive solution?

Exercise 4.3.12 *(We need the following result for the next exercise.) Suppose that a, b and c are non-zero integers such that the highest common divisor of* *a and b is* 1 *and the highest common divisor of a and c is* 1. *Explain why we can find n, m, r and s such that*

$$na + mb = 1 \quad and \quad ra + sc = 1.$$

By considering $(na + mb)(ra + sc)$, *or otherwise, show that there are integers N and M such that*

$$Na + Mbc = 1.$$

Conclude that the highest common divisor of a and bc is 1.

Exercise 4.3.13 *Prove the following partial generalisation of Theorem 4.3.9. Suppose that each of the pairs of non-zero integers* (a_1, a_2), (a_2, a_3) *and* (a_3, a_1) *have highest common divisor* 1. *Show how to find an integer x satisfying the conditions*

$$x \equiv u_1 \quad \mathrm{mod}\ a_1, \qquad\qquad \bigstar\bigstar$$
$$x \equiv u_2 \quad \mathrm{mod}\ a_2,$$
$$x \equiv u_3 \quad \mathrm{mod}\ a_3.$$

Show that, if x_0 *is a solution, then all solutions of* $\bigstar\bigstar$ *are given by*

$$x \equiv x_0 \quad \mathrm{mod}\ a_1 a_2 a_3.$$

Generalise the result to systems of n equations.

Exercise 4.3.14 *Solve Sun Zi's problem (Exercise 4.3.1). It is clear from the context that Sun Zi is asking for the smallest positive solution.*

Exercise 4.3.15 *Someone who thinks that mathematicians enjoy long sums tells you that their age in years is*

$$A = \left(78^2 + (9!) - 2^{14} - 179\,329\right) \times 6 - 1\,039\,470.$$

Without using a calculator, find A modulo 3, modulo 5 and modulo 7. You know that their age is between 20 and 80. Use the method of the previous exercise to find their age.

Computers can only store and manipulate integers of less than a certain size N. One way of calculating exactly with larger integers is to do the calculations modulo $a_1, a_2, \ldots, a_k$ where the $a_j < N$ and the greatest common divisor is each pair a_i and a_j is 1. At the end of the calculation you then use the Chinese remainder theorem to recover the answer. You have to program your machine to convert inputted integers into the required form (more exactly, given m the machine must produce m_j with $a_j > m_j \geq 0$ and $m_j \equiv m \mod a_j$) and to perform the reverse process. Whether this is worthwhile depends on the calculations to be performed and the architecture of the machine used.[9]

Exercise 4.3.16 *Since computers work in binary, it is often convenient to take* $a_j = 2^{n_j} - 1$. *By applying Euclid's algorithm, or otherwise, show that*

$$\gcd(2^n - 1, 2^m - 1) = 2^{\gcd(n,m)} - 1$$

(where $\gcd(a, b)$ *denotes the greatest common divisor of a and b). Thus, for example, if* $N = 2^{32}$, *we could take* $a_1 = 2^{32}-1$, $a_2 = 2^{31}-1$ *and* $a_3 = 2^{29}-1$.

Exercise 4.3.17 *Exercise 4.3.13 is only a partial generalisation of Theorem 4.3.9, though it is all that is needed for practical purposes. In this exercise, which is for enthusiasts only, we obtain the full generalisation.*

(i) Let a_1, a_2 *and* a_3 *be non-zero integers. Show that, if there exists an x such that*

$$x \equiv u_1 \quad \mathrm{mod}\ a_1, \qquad\qquad ★★$$
$$x \equiv u_2 \quad \mathrm{mod}\ a_2,$$
$$x \equiv u_3 \quad \mathrm{mod}\ a_3,$$

then we must have

$$u_i \equiv u_j \quad \mathrm{mod}\ \gcd(a_i, a_j)$$

for all $1 \leq i < j \leq 3$.

Show that if x is a solution of ★★, *then* x' *is a solution of* ★★ *if and only if* $x \equiv x'$ *mod e where e is the least common multiple of the* a_i *(that is to say the smallest strictly positive integer divisible by all the* a_i).

(ii) Suppose that a_1, a_2 *and* a_3 *are non-zero integers and*

$$u_i \equiv u_j \quad \mathrm{mod}\ \gcd(a_i, a_j)$$

for all $1 \leq i < j \leq 3$.

By first finding a solution of

$$x_1 \equiv u_1 \quad \mathrm{mod}\ a_1$$

and using it to find a solution of

$$x_2 \equiv u_1 \quad \mathrm{mod}\ a_1,$$
$$x_2 \equiv u_2 \quad \mathrm{mod}\ a_2$$

and then using this to find a solution of ★★, *show that* ★★ *does indeed have a solution.*

---

[9] Note that we can perform our calculations to various moduli in parallel if the machine allows it.

(iii) Summarise parts (i) and (ii) as a single theorem. Generalise to the case of n equations.

4.4 Arithmetic modulo p

In many ways, arithmetic modulo n runs in parallel with ordinary arithmetic. However, the following simple example brings us up with a start.

Example 4.4.1 *We have $2 \not\equiv 0$ and $3 \not\equiv 0$ but*

$$2 \times 3 \equiv 0 \quad \mathrm{mod}\ 6.$$

Exercise 4.4.2 *Show that if $|n| \geq 2$ and $|n|$ is not a prime,[10] then we can always find r and s such that $r \not\equiv 0$ and $s \not\equiv 0$ but*

$$r \times s \equiv 0 \quad \mathrm{mod}\ n.$$

Provided we restrict ourselves to working modulo a prime, this unpleasant phenomenon does not occur.

Lemma 4.4.3 *Suppose that p is a prime.*
(i) If $r \not\equiv 0$ mod p, then we can find a u such that

$$ru \equiv 1 \quad \mathrm{mod}\ p.$$

(ii) If $r \times s \equiv 0$ and $r \not\equiv 0$, then $s \equiv 0$ mod p.
(iii) Suppose $r \not\equiv 0$. Then, if $rs \equiv rs'$, it follows that $s \equiv s'$ mod p.

Proof (i) Not surprisingly, we use Euclid's algorithm. Since r is not divisible by p and p is a prime (and is thus only divisible by ± 1 and $\pm p$), the highest common divisor of r and p is 1. We can thus find integers u and v such that

$$ru + pv = 1$$

and so $ru \equiv 1$ mod p.
 (ii) Take u as in (i). Then

$$s \equiv 1 \times s \equiv (u \times r) \times s \equiv u \times (r \times s) \equiv u \times 0 \equiv 0 \quad \mathrm{mod}\ p.$$

 (iii) We have

$$r(s - s') \equiv rs - rs' \equiv 0$$

and so, by (ii), $s - s' \equiv 0$ whence $s \equiv s'$ mod p.   ∎

---

[10] If this were a textbook, I would need to establish the properties of primes. Since it is not, I shall simply assume that the reader knows them.

Exercise 4.4.4 *If you look at the inner title page of almost any book published between 1985 and 2005 you will find its International Standard Book Number.[11] The ISBN uses single digits selected from 0, 1, ..., 8, 9 and X representing 10. Each ISBN consists of nine such digits $a_1, a_2, ..., a_9$ followed by a single check digit a_{10} chosen so that*

$$10a_1 + 9a_2 + \cdots + 2a_9 + a_{10} \equiv 0 \mod 11. \tag{*}$$

(i) Find a couple of books and check that (∗) *holds for their ISBNs.*

(ii) Show that (∗) *will fail if you make a mistake in writing down one digit of an ISBN.*

(iii) Show that (∗) *may not fail if you make a mistake in writing down 2 digits.*

(iv) Show that (∗) *will fail if you interchange two adjacent digits.*

(v) Does (iv) remain true if we replace 'adjacent' by 'different'?
(Errors of type (ii) and (iv) are the most common in typing.)

If we work modulo a prime p and $uv \equiv 1 \mod p$, then we could think of multiplication by v as division by u and write $v = u^{-1}$. If we do this, then the parallel between arithmetic modulo p and ordinary arithmetic on the rational, real or complex numbers becomes very strong indeed. However, there are characteristic differences between the systems and part (iii) of the next lemma illustrates one of them.

Lemma 4.4.5 *Let p be a prime.*
(i) If r is an integer with $1 \le r \le p - 1$, then

$$\binom{p}{r} \equiv 0 \mod p.$$

(ii) If k is any integer, then

$$(k + 1)^p \equiv k^p + 1 \mod p.$$

(iii) **[Fermat's little theorem]** *If k is any integer, then*

$$k^p \equiv k \mod p.$$

(iv) If $k \not\equiv 0$, then

$$k^{p-1} \equiv 1 \mod p.$$

Proof (i) Observe that $r!$ is the product of positive integers strictly less than p and so is not divisible by p. Similarly, $(p - r)!$ is not divisible by p. However, $p!$ is divisible by p. We know that the binomial coefficient

---

[11] After this date a new system was phased in.

$$\binom{p}{r} = \frac{p!}{r!(p-r)!}$$

is an integer and the previous two sentences show that it must be divisible by p.

(ii) Using (i) and the binomial expansion, we have

$$(k+1)^p \equiv k^p + \binom{p}{1}k^{p-1} + \binom{p}{2}k^{p-2} + \cdots + \binom{p}{p-1}k + 1$$

$$\equiv k^p + 0 \times k^{p-1} + 0 \times k^{p-2} + \cdots + 0 \times k + 1 \equiv k^p + 1 \mod p.$$

(iii) Since $0^p \equiv 0$, repeated use of (ii) shows that $k^p \equiv k \mod p$ for all k with $0 \le k \le p - 1$ and so for all k.

(iv) If $k \not\equiv 0$, then (iii) gives

$$k^{p-1} \equiv k^{p-1} \times 1 \equiv k^{p-1} \times (k \times k^{-1}) \equiv (k^{p-1} \times k) \times k^{-1}$$

$$\equiv k^p \times k^{-1} \equiv k \times k^{-1} \equiv 1 \mod p.$$

∎

Exercise 4.4.6 *Let p be a prime number different from 2 and 5. Show that p divides $10^r - 1$ for infinitely many positive integers r. Hence show that p divides infinitely many of the integers*

$$11, \ 111, \ 1111, \ 11111, \ \ldots.$$

For the rest of this section and the next, we shall be occupied with 'square roots', first modulo a prime and then modulo more general integers.

Lemma 4.4.7 *Let p be an odd prime.*

(i) If $r^2 \equiv a^2$, then $r \equiv a$ or $r \equiv -a \mod p$.

(ii) If $u \not\equiv 0$, the equation $r^2 \equiv u \mod p$ either has no solution or has exactly two solutions (that is to say, exactly two solutions which are not equal modulo p). If $u \equiv 0$, then the equation has exactly one solution.

(iii) The sets

$$\{1 \le u \le p - 1 : u \equiv a^2 \mod p \text{ for some } a\}$$

and

$$\{1 \le u \le p - 1 : u \not\equiv a^2 \mod p \text{ for any } a\}$$

both contain $(p-1)/2$ elements.

Proof (i) Observe that, if $r^2 \equiv a^2$, then

$$0 \equiv r^2 - a^2 \equiv (r-a)(r+a)$$

and thus, by Lemma 4.4.3 (ii), either $r - a \equiv 0$ or $r + a \equiv 0$. Hence $r \equiv a$ or $r \equiv -a \mod p$.

(ii) Just use (i).

(iii) By (ii) each non-zero square has exactly two roots so

$$\{1 \leq u \leq p - 1 : u \equiv a^2 \mod p \text{ for some } a\}$$

contains $(p - 1)/2$ elements. ∎

Exercise 4.4.8 *What are the results corresponding to those of Lemma 4.4.7 when $p = 2$?*

Exercise 4.4.9 *Suppose that p is a prime and $p \neq 2$. Suppose further that $a \not\equiv 0 \mod p$. Show that the equation*

$$am^2 + bm + c \equiv 0 \mod p \qquad \bigstar$$

has a solution if and only if there exists a u with

$$u^2 \equiv b^2 - 4ac \mod p.$$

Show that, if such a u exists, the solutions of $\bigstar$ are given by

$$m \equiv 2^{-1}a^{-1}(-b \pm u) \mod p.$$

What can you say about solutions of $\bigstar$ in the case when $p = 2$?

The following result will form the basis for the next section.

Lemma 4.4.10 *(i) If p is a prime and $p \neq 2$ then, if m is any integer,*

$$m^{(p+1)/2} \equiv m \text{ or } m^{(p+1)/2} \equiv -m \mod p.$$

(ii) If p is a prime and $p \equiv 3 \mod 4$, then, if m is any integer,

$$(m^2)^{(p+1)/4} \equiv m \text{ or } (m^2)^{(p+1)/4} \equiv -m \mod p.$$

Proof (i) By Fermat's little theorem,

$$\left(m^{(p+1)/2} - m\right)\left(m^{(p+1)/2} + m\right) \equiv m^{p+1} - m^2 \equiv m^2 - m^2 \equiv 0$$

and thus, by Lemma 4.4.3 (ii),

$$m^{(p+1)/2} - m \equiv 0 \text{ or } m^{(p+1)/2} + m \equiv 0 \mod p.$$

(ii) Just apply (i). We need $p \equiv 3 \mod 4$ in order that $(p + 1)/4$ should be an integer. ∎

Exercise 4.4.11 *Suppose that p is a prime and $p \equiv 3$ mod 4.*

(i) By using Lemmas 4.4.10 and 4.4.7 (iii), or otherwise, show that the equation $u^2 \equiv -1$ mod p has no solution.

(ii) Suppose that $a \not\equiv 0$. By considering $(-a) \times a^{-1}$, or otherwise, show that, if the equation $u^2 \equiv a$ mod p has a solution, the equation $v^2 \equiv -a$ mod p has no solution.

(iii) By using Lemma 4.4.7 (iii), or otherwise, show that, if $a \not\equiv 0$, exactly one of the two equations $u^2 \equiv a$ mod p and $v^2 \equiv -a$ mod p has a solution.

Exercise 4.4.12 *Use Lemma 4.4.10 and Exercise 4.4.11 to discover if the equation*

$$u^2 \equiv a \quad \text{mod } 19$$

has a solution in the cases $a = 2, -2, 3, -3$ and to find the solutions if they exist.

Exercise 4.4.13 *By looking at the sequence a, a^2, a^4, ..., show that we can calculate a^{2^n} using n multiplications. By observing that any m with $0 \le m < 2^n$ can be written as*

$$m = w_0 + w_1 2 + w_2 2^2 + \cdots + w_{n-1} 2^{n-1}$$

with w_j taking the value 0 or 1, show that we can calculate a^m using at most $2n$ multiplications. (You may be able to do better.)

Use this idea to check whether the equation $u^2 \equiv 2$ mod 43 has any solutions.

In the next section, we discuss square roots in a slightly more general context. As a preliminary, the reader may wish to carry out her own investigation of some special cases.

Exercise 4.4.14 *(i) Find the solutions of*

$$u^2 \equiv a \quad \text{mod } 21$$

for all a with $0 \le a \le 20$.

(ii) Find the solutions of

$$u^2 \equiv a \quad \text{mod } 15$$

for all a with $0 \le a \le 14$.

4.5 Arithmetic modulo pq

Suppose that p and q are distinct primes. We already know that algebra modulo pq is not as simple as algebra modulo a prime since

$$p \not\equiv 0, \ q \not\equiv 0 \text{ yet } pq \equiv 0 \mod pq.$$

Exercise 4.5.1 *Use Euclid's algorithm to show that, if* $a \not\equiv 0 \mod p$ *and* $a \not\equiv 0 \mod q$, *we can find an integer* u *such that*

$$au \equiv 1 \mod pq.$$

(We shall write $u = a^{-1}$.)

However, we can still ask about solutions of

$$u^2 \equiv a \mod pq.$$

Our investigation is made easy by the Chinese Remainder theorem (Theorem 4.3.9) which tells us that $u^2 \equiv a \mod pq$ if and only if

$$u^2 \equiv a \mod p,$$
$$u^2 \equiv a \mod q.$$

Exercise 4.5.2 *Suppose that p and q are distinct odd primes.*
(i) Show that $u^2 \equiv a \mod pq$ has exactly one solution if $a \equiv 0 \mod pq$.
(ii) Show that $u^2 \equiv a \mod pq$ has has no solutions or has exactly two solutions if $a \equiv 0 \mod p$ but $a \not\equiv 0 \mod pq$ or if $a \equiv 0 \mod q$ but $a \not\equiv 0 \mod pq$.
(iii) If $a \not\equiv 0 \mod p$ and $a \not\equiv 0 \mod q$, show that $u^2 \equiv a \mod pq$ either has no solutions or has exactly four solutions.
(iv) Let $A = \{a : 1 \le a \le pq\}$ and write $|B|$ for the number of elements in a set B. Show that

$$|\{a \in A : u^2 \equiv a \mod pq \text{ has exactly two solutions}\}| = \frac{p+q-2}{2},$$

$$|\{a \in A : u^2 \equiv a \mod pq \text{ has exactly four solutions}\}| = \frac{(p-1)(q-1)}{4},$$

$$|\{a \in A : u^2 \equiv a \mod pq \text{ has no solutions}\}| = pq - \frac{(p+1)(q+1)}{4}.$$

Exercise 4.5.3 *Check that your results in Exercise 4.4.14 agree with the statements in Exercise 4.5.2.*

Exercise 4.5.4 *What are the results corresponding to Exercise 4.5.2 if $p = 2$ and q is an odd prime?*

Lemma 4.5.5 *Suppose that p and q are odd primes. Let g be an integer solution of*

$$g \equiv 1 \mod p,$$
$$g \equiv -1 \mod q.$$

If v is a solution of

$$u^2 \equiv a \mod pq,$$

then so are $-v$, gv *and* $-gv$. *Further, if* $a \not\equiv 0 \mod p$ *and* $a \not\equiv 0 \mod q$, *then* v, $-v$, gv *and* $-gv$ *are the four distinct solutions (modulo pq).*

Proof Observe that

$$g^2 \equiv 1 \mod p,$$
$$g^2 \equiv 1 \mod q$$

and so $g^2 \equiv 1 \mod pq$, whence

$$(gv)^2 \equiv g^2 v^2 \equiv v^2 \equiv a \mod pq$$

while, similarly, $(-gv)^2 \equiv (-v)^2 \equiv a$.

If $a \not\equiv 0 \mod p$ and $a \not\equiv 0 \mod q$, then

$$v^2 \not\equiv 0 \mod p \quad \text{and} \quad v^2 \not\equiv 0 \mod q$$

and so

$$v \not\equiv 0 \mod p \quad \text{and} \quad v \not\equiv 0 \mod q.$$

It follows that, if $va \equiv vb \mod pq$ then $a \equiv b \mod pq$. Since $1, -1, g$ and $-g$ are all unequal modulo pq, it follows that so are $v, -v, gv$ and $-gv$. ∎

Exercise 4.5.6 *State and prove the simpler result corresponding to that of Lemma 4.5.5 in the case when* $a \not\equiv 0 \mod p$ *and* $a \equiv 0 \mod q$.

When mathematicians began to operate with new number systems (and in particular with complex numbers) they invoked a 'principle of permanence of form' to the effect that, whenever an algebraic result held in one system, it would hold in all systems. Experience has shown that the principle must be restated as follows: 'as we extend and generalise a mathematical system the same results will continue to hold until we reach a point where they fail'. Here we have a spectacular failure of the useful principle that a quadratic equation has at most two roots.

In spite of this, the study of square roots modulo pq turns out to have a very interesting practical application. For the rest of this section we shall be considering large primes (by which we mean having a few hundred digits[12]) and calculations using computer programs able to handle integers of that size. We shall say that something can be computed rapidly if if it can be computed using a few million operations (additions, multiplications and so on).

Exercise 4.5.7 *(i) Show that, if we have two integers m and n of size about* 10^{300}, *we can compute their lowest common multiple using Euclid's algorithm in a few thousand operations.*

(ii) Show that, if we wish to apply the Chinese Remainder theorem in the form discussed in Lemma 4.3.10, with integers a few hundred digits long, then we can find solutions in a few thousand operations.

(iii) Show, using the idea of Exercise 4.4.13, that, if m, n and r are integers of size about 10^{300} *with* $r > 0$, *then we can compute* n^r *modulo m in a few thousand operations.*

Lemma 4.5.8 *Suppose that p and q are large primes such that*

$$p \equiv 3 \mod 4 \quad and \quad q \equiv 3 \mod 4.$$

If a is an integer we can compute the solutions of

$$u^2 \equiv a \mod pq$$

rapidly.

Proof Indeed, if we take p and q of the size suggested in Exercise 4.5.7, the observations in that exercise, together with the remarks which began this section, show that we can find the solutions with a few thousand computations. ∎

Lemma 4.5.8 says that it is easy to compute square roots modulo n if n factors into two known primes p and q (and p, $q \equiv 3 \mod 4$). How easy is it to find the primes p and q if we know how to take square roots? The following lemma, peculiar though it may seem at first sight, furnishes a partial answer.

Lemma 4.5.9 *Let* $n = pq$ *where p and q are unknown large primes. Suppose that we have access to an oracle who will give us one solution of the equation*

$$u^2 \equiv a \mod n$$

---

[12] The notion of a large prime varies with time but rather slowly. Just before the start of the computer age, a prime with ten digits, that is to say of size between 10^{10} and 10^{11}, would be considered large (see for example the beautiful essay by Borel in [37]). After sixty years of breakneck technological progress a prime with three hundred digits is considered large.

*(or tell us that no solution exists) for any integer a of our choosing. If we are
allowed to ask the oracle such a question once, then, with probability at least
1/2, we can determine p and q in a few thousand operations.*

Proof Choose an integer r with $1 \le r \le n - 1$ in such a way that each integer
in the range is equally likely to be chosen. Use Euclid's algorithm to find the
highest common divisor d of r and n. If $d \ne 1$, then $d = p$ or $d = q$ so we are
done.[13] If not, we know that r is not divisible by p or q.

We now compute $a \equiv r^2$ and ask the oracle to give us a v satisfying

$$v^2 \equiv a \mod n.$$

Using the notation of Lemma 4.5.5, we know that

$$v \equiv r, \ v \equiv -r, \ v \equiv gr, \ \text{or} \ v \equiv -gr \mod n$$

and, since each integer in our range was equally likely to be chosen, each of
these cases is equally likely. With probability $1/2$, $v \equiv r$ or $v \equiv -r$ and
the oracle has told us nothing that we did not already know. However, with
probability $1/2$, $v \not\equiv \pm r$, so $v \equiv \pm gr \mod n$. By interchanging p and q, if
necessary, we may suppose $v \equiv gr$. Since r is not divisible by p or q, we
can use Euclid's algorithm to find an integer r^{-1} such that $r^{-1}r \equiv 1$ and so
$g \equiv r^{-1}v \mod n$. We now know g.

Recall that

$$g \equiv 1 \mod p,$$
$$g \equiv -1 \mod q.$$

Thus $g - 1$ is divisible by p and $p = \gcd(g - 1, n)$. We can calculate p by
Euclid's algorithm. ∎

Lemma 4.5.10 *Let $n = pq$, where p and q are unknown large primes. Suppose that we have access to an oracle who will rapidly give us one solution of
the equation*

$$u^2 \equiv a \mod pq$$

*(or tell us that no solution exists) for any integer a of our choosing. If we are
allowed to ask the oracle such a question as many times as we wish, then, with
a probability of failure too small to matter in practice, we can determine p and
q rapidly.*

---

[13] In practice this outcome is so unlikely that we would not bother with this preliminary
computation.

Proof If we repeat the process of Lemma 4.5.9 64 times, say, each time picking r independently of our previous choices, then the probability that we will fail to factorise n is less than 2^{-64}, a probability which is certainly too small to matter in practice. We know that the oracle works rapidly and that we can complete each test rapidly, so the 64 trials can be completed rapidly.[14]   ■

Exercise 4.5.11 *Let* $n = pq$ *where* p *and* q *are unknown large primes. Suppose that we have access to an oracle who, with probability at least* 10^{-3}*, will rapidly give us one solution of the equation*

$$u^2 \equiv a \mod pq$$

(and otherwise will simply refuse to answer) for any integer a of our choosing. If we are allowed to ask the oracle such a question as many times as we wish then, with a probability of failure too small to matter in practice, we can determine p and q rapidly.

Although we shall not discuss how it is done here, there are rapid methods for determining whether a large integer is prime. It is also known that there is a reasonable chance that a large integer picked at random will, indeed, be prime.[15] It is thus very easy to find large primes p and q (with $p \equiv q \equiv 3$ mod 4 if desired) which are effectively chosen at random.

On the other hand, although people have been looking for the past 300 years, nobody has published a method for factoring large numbers n rapidly (even in the special case that n is the product of two primes p and q with $p \equiv q \equiv 3$ mod 4).

Rabin put these ideas together to come up with a remarkable secret code.

Suppose I wish people to be able to write to me and to be confident that no one else can read their messages. I secretly select two very large primes p and q with $p \equiv q \equiv 3$ mod 4. I keep the pair (p, q) secret, but I broadcast the public key $N = pq$. If someone wants to send me a message, they write it in binary code and split it into blocks of length m with $2^m < N < 2^{m+1}$. Each of these blocks is a number r_j with $0 \le r_j < N$. My correspondent computes s_j with $0 \le s_j \le N - 1$ such that $r_j^2 \equiv s_j$ modulo N and sends s_j. Since I know p and q, I can easily find the four square roots of s_j modulo N. One of

[14] If the reader looks at the matter in detail, she will see that our estimates are rather pessimistic and things will probably go rather faster than we implied.
[15] The prime number theorem tells us, in effect, that the probability that an integer m chosen at random between N and $2N$ will be prime is about $1/\log N$ when N is large. More sophisticated forms of the theorem tell us that a similar result holds even if we demand $m \equiv 3$ mod 4.

these roots will correspond to a sensible message and the other three will give garbage so I can easily decode the message.[16]

Suppose that someone else can decode messages sent by these means. Then they can extract square roots modulo N and so, by Lemma 4.5.10, they can factorise N, a task which so far as we know, is beyond the present capabilities of mathematicians armed with the fastest computers.

Exercise 4.5.12 *Consider the Rabin system with $N = 1333$ (needless to say, not a very large number). Find the four decodes corresponding to 183.*

The reader should not jump to the conclusion that we have achieved perfect secrecy. Even the best codes are like the lock on a safe. However good the lock is, the safe may be broken open by brute force, or stolen together with its contents, or a key holder may be persuaded by fraud or force to open the lock, or the presumed contents of the safe may have been tampered with before they go into the safe, or... A coding scheme is merely the cryptographic *element* of larger possible cryptographic *systems*. The planning of cryptographic systems requires not only mathematics but engineering, economics, psychology, humility and an ability to learn from past mistakes. Those who do not learn the lessons of history are condemned to repeat them.

The next exercise gives an example of a typical cryptographic mistake.

Exercise 4.5.13 *I announce to my extensive spy network that I shall be using Rabin's scheme with modulus N. My agent in X'Dofdro sends me a message m (with $1 \leq m \leq N - 1$) encoded in the requisite form. Unfortunately, my white cat eats the piece of paper on which the prime factors of N are recorded, so I am unable to decipher it. I therefore find a new pair of primes and announce that I shall be using the Rabin scheme with modulus $N' > N$. My agent now recodes the message and sends it to me again.*

The dreaded SNDO of X'Dofdro intercept both code messages. Show that they can find m. Can they decipher any other messages sent to me using only one of the coding schemes?

We shall not go into the practicalities of using Rabin and similar codes. However, the reader is entitled to ask what evidence there is that factorising the product of two large primes is a genuinely hard problem. A good answer will be found on the RSA Laboratories web site,[17] which contains a list of

---

[16] There will be problems if $r_j \equiv 0 \mod p$ or $r_j \equiv 0 \mod q$ or if two square roots correspond to reasonable messages. However, the probability of any of these things happening is so small that we can ignore the possibility.

[17] The ever shifting nature of the web makes it hard to give an address that is guaranteed to work. Try the *Wikipedia* article entitled *RSA Factoring Challenge*.

numbers of the type used in this type of code together with prizes available for factorising them.[18] In November 2005, the team of Bahr, Boehm, Franke and Kleinjung won 2000 dollars for factoring the 193 digit number

> 31074182404900437213507500358885679300373460228427
> 27545720161948823206440518081504556346829671723286
> 78243791627283380334154710731085019195485290073772
> 48227835257423864540146917366024776523466 09.

One of the factors was

> 16347336458092538484431338838650908598417836700330
> 9231218111085238933310010450815121211816751 1579.

The next number on the list, which carried a prize of 3000 dollars (unclaimed as of the middle of 2007 when the contest closed), is the 212 digit number

> 74037563479561712828046796097429573142593188889231
> 28908493623263897276503402826627689199641962511784
> 39958943305021275853701189680982867331732731089309
> 00552505116877063299072396380786710086096962537934
> 650563796359.

Organisations which use the Rabin and related systems rely on 'security through publicity'. Because the problem of cracking these codes is so notorious, any breakthrough is likely to be publicly announced.[19] Moreover, even if a breakthrough occurs, it is unlikely to be one which can be easily exploited by the average criminal. So long as the secrets covered by such codes need only be kept for a few days, rather than forever, the codes can be considered to be one of the strongest links in the security chain.

Exercise 4.5.14 *Consider the following variation on the Rabin coding scheme given on page 137. I secretly select two very large primes p and q with $p \equiv q \equiv 3 \mod 4$ and an integer b. I keep the pair (p, q) secret, but I broadcast the public key $N = pq$ and b. If someone wants to send me a message they write it in binary code and split it into blocks of length m with $2^m < N < 2^{m+1}$. Each of these blocks is a number r_j with $0 \leq r_j < N$. My correspondent computes s_j with $0 \leq s_j \leq N - 1$ such that $r_j^2 + b r_j \equiv s_j$ modulo N and sends s_j.*

---

[18] The prizes are no longer available but the challenges remain.
[19] And, if not, is more likely to be a government rather than a Mafia secret.

How do I decode the message? Is the new system as secure as the old? Give reasons for your answer.

4.6 Mr Jonas entertains

In *Martin Chuzzlewit*, Dickens relates how

> when the tea-tray was taken away, as it was at last, Mr Jonas produced a dirty pack of cards, and entertained the sisters with divers small feats of dexterity: whereof the main purpose of every one was, that you were to decoy somebody into laying a wager with you that you couldn't do it; and were then immediately to win and pocket his money. Mr Jonas informed them that these accomplishments were in high vogue in the most intellectual circles, and that large amounts were constantly changing hands on such hazards. And it may be remarked that he fully believed this; for there is a simplicity of cunning no less than a simplicity of innocence.

Let us suppose that, by dint of constant practice, Mr Jonas has reached the point where, given a pack of cards and a particular method of shuffling, he can execute that shuffle flawlessly as many times as we wish. Will the pack return to its original state and, if so, how many shuffles will it take?

We say that the top card is in the first place, the next card in the second place and so on. Let us write $Tr = s$ and $T^{-1}s = r$ if the shuffle takes the card in the rth place to the sth place. Then T and T^{-1} are functions from $X = \{1, 2, \ldots, n\}$ to itself, with T^{-1} representing the shuffle which reverses the effect of T. If we define $I : X \to X$ by $Ir = r$ (so that I is the shuffle which leaves everything the way it was, a particular favourite of Mr Jonas) then

$$TT^{-1} = T^{-1}T = I.$$

It is natural to write T^n for the shuffle corresponding to doing the shuffle T exactly n times and T^{-n} for the shuffle corresponding to doing the shuffle T^{-1} exactly n times. With these conventions,

$$T^0 = I,$$
$$T^m T^{-n} = T^{m-n} \text{ when } m, n \geq 0,$$

and, indeed,

$$T^m T^{-n} = T^{m-n} \text{ for all } m \text{ and } n.$$

We can now use a very general argument to show that, under the guidance of Mr Jonas, the pack will return to its original state.

Lemma 4.6.1 *There exists an $N \geq 1$ such that $T^N = I$.*

Proof Observe that there are $n!$ different ways of shuffling the pack. Thus if we write down the $n! + 1$ shuffles

$$T^0, T^1, T^2, T^3, \ldots, T^{n!}$$

two of them must be the same. In other words, we can find $0 \leq v < u \leq n!$ such that $T^u = T^v$ and so, writing $N = u - v$, we have

$$T^N = T^{u-v} = T^u T^{-v} = T^v T^{-v} = I$$

as required. ∎

We now turn to the more difficult question of how long it takes to return the pack to its original state. In other words, what is the least value of $N \geq 1$ for which $T^N = I$? We call this least value the *period*.[20] Lemma 4.6.1 tells us that the period $N \leq n!$ for any shuffle with a pack of n cards, but this is a very crude estimate.

Our investigations will be helped by a suitable notation. If we consider the card in position a_1, it will be moved to position a_2, the card in position a_2 will be moved to position a_3 and so on until we come to a card in position a_r which is moved to position a_1 (no earlier card having been moved to position a_1). We say that the shuffle gives rise to the cycle $(a_1 a_2 \ldots a_r)$.

Exercise 4.6.2 *With the notation just adopted, explain why we are sure that there will be an r such that the card in position a_r is moved to position a_1. If r is the smallest strictly positive[21] integer with this property, explain why we know that $a_j \neq a_k$ for all $1 \leq j < k \leq r$.*

Explain why the statement that the shuffle T gives rise to the cycle $(a_1 a_2 \ldots a_{r-1} a_r)$ is equivalent to the statement that the shuffle T gives rise to the shuffle $(a_2 a_3 \ldots a_r a_1)$.

Suppose that the shuffle T gives rise to the two cycles $(a_1 a_2 \ldots a_r)$ and $(b_1 b_2 \ldots b_k)$. Explain why either

$$\{a_1, a_2, \ldots, a_r\} \cap \{b_1, b_2, \ldots, b_k\} = \varnothing$$

---

[20] Many secret codes depend on some form of shuffling to produce a new 'key' each time. If the period of the shuffle is too short, then the same key will be repeated and the code becomes vulnerable to well-known methods of code breaking. The British convoy code during the early part of World War II was so heavily used that its period became too short and the Germans were able to break it. The reader is warned that, although a sufficiently long period is necessary for such a code to be secure, it is certainly not sufficient.

[21] That is, positive and non-zero.

or $r = k$,

$$\{a_1, a_2, \ldots, a_r\} = \{b_1, b_2, \ldots, b_r\},$$

and $b_s = T^u a_s$ *for all s and some u.*

Suppose that the shuffle T gives rise to the cycle $(a_1 a_2 \ldots a_r)$. If all the possible card positions appear within the cycle we stop. Otherwise, we pick a card position $b_1 \notin \{a_1, a_2, \ldots, a_r\}$ and write down the cycle starting with b_1. If the two cycles exhaust the positions we stop. If not we produce a third cycle and so on. We put the cycles together as

$$(a_1 a_2 \ldots a_r)(b_1 b_2 \ldots b_k)(c_1 c_2 \ldots c_l) \cdots$$

to express T as the product of cycles.

As an example, consider the shuffle T given as

$$
\begin{array}{cccccccccc}
u = & 1 & 2 & 3 & 4 & 5 & 6 & 7 & 8 & 9 \\
T(u) = & 2 & 4 & 5 & 1 & 3 & 6 & 9 & 7 & 8
\end{array}
$$

which is expressed as the product of cycles by

$$T = (124)(35)(6)(789).$$

As another example, observe that the shuffle

$$S = (1765)(432)(89)$$

may be given as

$$
\begin{array}{cccccccccc}
u = & 1 & 2 & 3 & 4 & 5 & 6 & 7 & 8 & 9 \\
S(u) = & 7 & 4 & 2 & 3 & 1 & 5 & 6 & 9 & 8.
\end{array}
$$

Exercise 4.6.3 *Consider a pack of* 17 *cards. Express each of these shuffles (where the card in position u goes to position $T_j(u)$) as a product of cycles.*

(a) The function defined by $T_1(u) \equiv 2u \mod 17$.

(b) The function defined by $T_2(u) \equiv u + 5 \mod 17$.

(c) The function defined by $T_3(u) \equiv 3u \mod 17$.

(d) Explain why the function defined by $T_4(u) \equiv u^2 \mod 17$ *does not give a shuffle.*

Consider a pack of 15 *cards. Find which of the following are shuffles and express them as a product of cycles.*

(e) The function defined by $S_1(u) \equiv 3u \mod 15$.

(f) The function defined by $S_2(u) \equiv u + 5 \mod 15$.

(g) The function defined by $S_3(u) \equiv u^3 \mod 15$.

We say that the cycle $(a_1 a_2 \ldots a_r)$ has length r. The next exercise shows that it is very easy to find the period of a shuffle from its expression as a product of cycles.

Exercise 4.6.4 *(i) Show that, if the shuffle T gives rise to the cycle $(a_1 a_2 \ldots a_r)$, then*

$$T^k a_j = a_j$$

for $1 \le j \le r$ if and only if k is divisible by r.

(ii) Show that, if T gives rise to cycles of length d_1, d_2, ..., d_{m-1} and d_m, then T has period the lowest common multiple of the d_s.

(iii) Use (ii) to find the periods of the shuffles described in Exercise 4.6.3.

Exercise 4.6.4 (ii) gives us some hold on the problem of determining the longest period for a shuffle with n cards.

Lemma 4.6.5 *The longest period of a shuffle with n cards is given by the maximum value of*

$$\mathrm{lcm}(d_1, d_2, \ldots, d_k)$$

subject to the conditions that k and d_1, d_2, ..., d_k are strictly positive integers with

$$d_1 + d_2 + \cdots + d_k = n.$$

We can make a further useful remark.

Exercise 4.6.6 *(i) Show that if u, $v \ge 2$, then*

$$uv \ge u + v.$$

(ii) Suppose that u and v are strictly positive integers with $\gcd(u, v) = 1$. Show that $\mathrm{lcm}(u, v) = uv$.

(iii) Suppose that d_1, d_2, ..., d_k are strictly positive integers. Show that we can find distinct primes p_1, p_2, ..., p_l and strictly positive integers $m(1)$, $m(2)$, ..., $m(l)$ such that

$$p_1^{m(1)} p_2^{m(2)} \cdots p_l^{m(l)} = \mathrm{lcm}(p_1^{m(1)}, p_2^{m(2)}, \ldots, p_l^{m(l)}) \ge \mathrm{lcm}(d_1, d_2, \ldots, d_k)$$

and

$$p_1^{m(1)} + p^{m(2)} + \cdots + p_l^{m(l)} \le d_1 + d_2 + \cdots + d_k.$$

(iv) Conclude that the longest period of a shuffle with n cards is given by the maximum value of

$$p_1^{m(1)} p_2^{m(2)} \cdots p_l^{m(l)}$$

where l is a strictly positive integer, p_1, p_2, …, p_l are distinct primes and $m(1)$, $m(2)$, …, $m(l)$ are strictly positive integers with

$$p_1^{m(1)} + p_2^{m(2)} + \cdots + p_l^{m(l)} \leq n.$$

I suspect that we cannot say much more without studying the distribution of the primes. However, when n is small, it is relatively easy to find the longest period by direct search.

Exercise 4.6.7 *Find the longest period of any shuffle of n cards for $1 \leq n \leq 12$. (So far as I know there is no particular pattern.) In each case write down a longest shuffle using cycle notation.*

We now look at two particularly important shuffles. In each case we look at packs of $2n$ cards.

Exercise 4.6.8 *(i) Mr Jonas is particularly adept at executing the 'perfect shuffle' in which the card in rth position in the pack moves to the $2r$th position for $1 \leq r \leq n$ and to the $2(r - n) - 1$th position for $n + 1 \leq r \leq 2n$. By using the cycle notation, find how many shuffles it takes him to return the pack to its initial state when $n = 1, 2, 3, 4, 5, 6, 7$? Are there any remarks about particular things for particular n that might be helpful to Mr Jonas? Remember that even a small amount of extra information can be useful.*

(ii) Why does Mr Jonas prefer a shuffle in which the card in rth position in the pack moves to the $2r - 1$th position for $1 \leq r \leq n$ and to the $2(r - n)$st position for $n + 1 \leq r \leq 2n$? (This is called an 'out-shuffle'. The shuffle described in (i) is called an 'in-shuffle'.) By using the cycle notation, find how many out-shuffles it takes to return the pack to its initial state when $n = 1, 2, 3, 4, 5, 6, 7$.

In order to find some underlying pattern in the periods of in-shuffles and out-shuffles we use an extension of Fermat's little theorem (Lemma 4.4.5) called the Euler–Fermat Theorem. First we need another consequence of Euclid's algorithm.

Exercise 4.6.9 *Use Euclid's algorithm to show that, if a and n are integers with $n \geq 2$ and $\gcd(a, n) = 1$, we can find an integer u such that*

$$au \equiv 1 \mod n.$$

(We shall write $u = a^{-1}$.)

We shall also need a consequence of the inclusion-exclusion formula.

Definition 4.6.10 *Let n be an integer with $n \geq 2$. Euler's totient function $\phi(n)$ is defined to be the number of integers r with $1 \leq r \leq n$ and $\gcd(r, n) = 1$.*

Exercise 4.6.11 *Suppose that* $n = p_1^{k_1} p_2^{k_2} \cdots p_u^{k_u}$ *with* $p_1, p_2, \ldots, p_u$ *distinct primes and* $k_j \geq 1$ *and let* $X = \{1 \leq r \leq n\}$.

(i) If $A_j = \{r \in X : p_j$ *divides* $r\}$ *and* $|B|$ *denotes the number of elements in the set* B, *show that, if* $1 \leq j(1) < j(2) < \cdots < j(s) \leq u$, *then*

$$|A_{j(1)} \cap A_{j(2)} \cap \cdots \cap A_{j(s)}| = \frac{n}{p_{j(1)} p_{j(2)} \cdots p_{j(s)}}.$$

(ii) Use the inclusion-exclusion formula of Lemma 2.3.9 to find $\left| \bigcup_{j=1}^{u} A_j \right|$ *and to show that*

$$\phi(n) = |X| - \left| \bigcup_{j=1}^{u} A_j \right| = p_1^{k_1} \left(1 - \frac{1}{p_1} \right) p_2^{k_2} \left(1 - \frac{1}{p_2} \right) \cdots p_u^{k_u} \left(1 - \frac{1}{p_u} \right).$$

(iii) Compute $\phi(n)$ *for* $2 \leq n \leq 12$.

***Theorem 4.6.12* [The Euler–Fermat Theorem]** *If* n *is a positive integer, then*

$$a^{\phi(n)} \equiv 1 \mod n$$

whenever $\gcd(a, n) = 1$.

Proof We write

$$A = \{1 \leq r \leq n : \gcd(r, n) = 1\}$$

and recall that the number of elements of A is $\phi(n)$. Observe that, if $r \in A$, then $\gcd(ar, n) = 1$, so $ar \equiv f(r) \mod n$ for some $f(r) \in A$.

If $r, r' \in A$ and $f(r) = f(r')$ then $ar \equiv ar' \mod n$ and so (since $\gcd(a, n) = 1$) $r \equiv r' \mod n$ and $r = r'$. Thus the collection of $f(r)$ with $r \in A$ is just a rearrangement of A and so

$$\prod_{r \in A} r \equiv \prod_{r \in A} f(r) \equiv \prod_{r \in A} ar \equiv a^{\phi(n)} \prod_{r \in A} r.$$

(Here, as usual, $\prod_{u \in U} u$, means the product of all $u \in U$.) Since $\gcd(r, n) = 1$ for all $r \in A$, we can apply Lemma 4.6.9 repeatedly to obtain

$$1 \equiv a^{\phi(n)} \mod n,$$

which is the desired result. ∎

Exercise 4.6.13 *Explain how the Euler–Fermat Theorem reduces to the little Fermat theorem when* n *is a prime.*

The next exercise will not be needed later but provides an interesting variation on the ideas we used to prove the Euler–Fermat Theorem.

Exercise 4.6.14 *(i) Let p be an odd prime and let* $A = \{r \; : \; 1 \le r \le p - 1\}$. *Show that we can find a subset B such that* $1, p - 1 \notin B$ *and, if* $u \ne 1, p - 1$, $u^{-1} \in B$ *if and only if* $u \notin B$. *Show that*

$$\prod_{r \in A} r \equiv 1 \times -1 \times \prod_{u \in B} u \prod_{u \in B} u^{-1} \equiv -1 \times \prod_{u \in B} uu^{-1} \equiv -1 \quad \mod p.$$

Deduce Wilson's[22] theorem that, if p is an odd prime, $(p - 1)! \equiv -1$ *mod p. (Although Wilson was the first to state the result, the first known proof is due to Lagrange.)*

(ii) Let n be an integer with $n \ge 2$ *Show that* $(n - 1)! \equiv -1$ *mod n if and only if n is prime. (You may need to consider the cases* $n = 2$ *and* $n = 4$ *separately.)*

(iii) Show that, if p is prime with $p \equiv 1$ *mod 4, then* $\big((p - 1)/2\big)!^2 \equiv -1$. *What can you say if p is prime with* $p \equiv -1$ *mod 4?*

(iv) Let p and q be distinct odd primes and let

$$A = \{1 \le r \le pq \; : \; \gcd(r, pq) = 1\}.$$

Show that

$$\prod_{r \in A} r \equiv 1 \quad \mod pq.$$

The Euler–Fermat Theorem enables us to say rather more about in-shuffles and out-shuffles.

Exercise 4.6.15 *(i) Show that the in-shuffle can be described using modular arithmetic by saying that the card in position r goes to position k where*

$$k \equiv 2r \quad \mod 2n + 1.$$

Explain why the pack returns to its original order after $\phi(2n + 1)$ *shuffles where* ϕ *is Euler's totient function (see Definition 4.6.10). Apply this result to a standard pack of 52 cards.*

(ii) Now consider the out-shuffle. Show that, if we ignore the first and last cards and renumber the remainder so that what was the $r + 1$*th card is now the rth card, then the effect of the out-shuffle can be described using modular arithmetic by saying that the card in position r goes to position k where*

---

[22] Rouse Ball's *History of the Study of Mathematics at Cambridge* [4] tells us that John Wilson (1741–93) '… was a good teacher and made his pupils work hard, but sometimes when they came for their lessons they found the door [closed] and "gone a fishing" written on the outside which Paley (who was one of them) deemed an addition of insult to injury, for he was himself very fond of the sport.' Wilson's subsequent career in law is described in the *Dictionary of National Biography*. (The first edition managed to omit his only lasting claim to fame!)

$$k \equiv 2r \quad \mathrm{mod}\ 2n - 1.$$

Explain why the pack returns to its original order after $\phi(2n - 1)$ shuffles. Apply this result to a standard pack of 52 cards.

(iii) Show that, in fact, out-shuffling returns a standard pack of 52 cards to it original state in 8 shuffles (making it a particularly useful shuffle for Mr Jonas and for stage magicians). Why is this consistent with the result of (ii)?

[Apparently, it is unknown for what values of n out-shuffling actually requires $\phi(2n - 1)$ shuffles to return the pack to its original state.]

(iv) Show that in-shuffling requires at least 52 shuffles to return the pack to its original order. (You should only need 26 easy calculations, or fewer, to show this. Cunning can replace computation but thinking of cunning tricks takes effort.)

In recent years, mathematicians like Aldous and Diaconis have made penetrating studies of card shuffling. They have shown that, provided your shuffling is not quite accurate, in-shuffling is a very good way of randomising the distribution of cards. In particular, Diaconis has shown that seven imperfect in-shuffles are sufficient to produce a 'well-shuffled' pack.

Even accomplished card sharps find it hard to 'keep control' of the entire pack with repeated shuffling. They concentrate on knowing the position of one or two cards and find that this extra knowledge is sufficient to give them a substantial advantage.[23]

---

[23] See Damon Runyon's *The Lacework Kid*.

5

A pack of cards

5.1 Find the largest

Suppose that we are helping some very cultured friends move house. They have already installed several walls of bookcases but left us to unpack the many crates of books which go on the bookcases. We decide to place the books in alphabetical order. This is not as easy as it seems, but we will be helped by various pieces of knowledge and guesswork. However badly jumbled the books have been in packing, there will probably be runs in near perfect alphabetical order. We expect that about a third of the authors will have initial letters A to G, about one third G to N and about one third N to Z. We know that most authors with initial letter W will have second letter A, E, H, I, O, R or Y and so on.

If we seek to mechanise the sorting process involved, then we can either attempt to identify and incorporate all these random and not very precise pieces of information or we must produce methods which make no use of them at all.

To make sure that we do not use extraneous information, let us consider the following model for sorting n cards bearing the numbers 1 to n. The cards are placed face down in front of you. You have an assistant who will look at any two cards that you indicate and tell you which of the two cards bears the largest number. You may then ask her either to replace the two cards in their original position or to swap them and replace them (or to place them at the top and bottom of the pile or something similar). This completes one operation. How can we sort the cards as quickly as possible (that is, with the fewest operations)?

Since we have no immediate ideas on how to attack this question, we attack a simpler one. How can you find the largest card as quickly as possible (that is, with the fewest operations)? It is not hard to come up with the following idea. Suppose the cards are laid out from right to left.

(1) Tell your assistant to examine the leftmost card (call it the first card from the left) and the card next to it (the second card from the left). If the first card from the left is bigger than the second card, the assistant should swap them. If not, she should leave them as they were. This ensures that the second card from the left is now bigger than the first card from the left.

(2) From (1), we know that the second card from the left is bigger than the first card from the left. Tell your assistant to examine the second and third cards from the left. If the second card from the left is bigger than the third card, the assistant should swap them. If not, she should leave them as they were. This ensures that the third card from the left is now bigger than both the first and second card from the left.

(3) From (2), we know that the third card from the left is bigger than the first and second cards from the left. Tell your assistant to examine the third and fourth cards from the left. If the third card from the left is bigger than the fourth card, the assistant should swap them. If not, she should leave them as they were. This ensures that the fourth card from the left is now bigger than the first, second and third cards from the left.

It is easy to write down the kth step.

(k) From ($k - 1$), we know that the kth card from the left is bigger than the jth from the left whenever $1 \le j \le k - 1$. Now tell your assistant to examine the kth and $k + 1$th cards from the left. If the kth card from the left is bigger than the $k + 1$th card, the assistant should swap them. If not, she should leave them as they were. This ensures that the $k + 1$th card from the left is bigger than the jth card from the left whenever $1 \le j \le k$.

After $n - 1$ steps the nth card from the left (that is to say the rightmost card) will be greater than the jth card from the left whenever $1 \le j \le n - 1$ (that is to say, greater than any other card in the pack) and we are done.

We have found a way to obtain the largest card in $n - 1$ operations. Could we do better?

Let each card be represented by a post and connect two posts by a wire whenever we ask our assistant to compare two cards. In order to know that a post represents the largest card it must be linked by a chain of wires (comparisons) to every other post. But, if we use fewer than $n - 1$ wires, then the posts cannot be linked together. Thus $n - 1$ comparisons are necessary and we cannot do better.

Exercise 5.1.1 *Suppose the n cards are numbered with distinct real numbers. What changes, if any, are required in our discussion above?*

Suppose the n cards are numbered with real numbers which need not be all distinct. What changes, if any, are required in our discussion above?

We return to the more difficult question of sorting all the cards later in the chapter.

Exercise 5.1.2 *Describe a method of sorting all the n cards into order. Roughly how many operations does it take? Do you think it is possible to do better? We saw that it is impossible to find the largest card in fewer than n − 1 operations. Can you produce a similar lower bound for sorting all the n cards into order?*

Exercise 5.1.3 *(A brain-teaser.) Show how to find the largest and smallest card in a pack of 2n cards using 3n − 2 operations.*

5.2 Records

We are fascinated by records. When a new sport is introduced, we get a plethora of records, but the longer a sport has been practised, the rarer records become. It is said that, at first, records reflect raw ability, then training and finally the limits of what can be achieved by combining natural ability and training.

This may well be true, but it is interesting to look at a pack of cards and see how it behaves in the matter of records. As before, we require an assistant. This time, we ask her to shuffle a pack of n cards numbered with distinct real numbers and then turn over the cards one at a time saying 'record', if the number revealed is bigger than any she has previously seen, and 'no record', if not.

We shall attempt to answer the following two questions.

(a) What is the probability of a record when the kth card is turned over and how does it depend (if at all) on what has happened before?

(b) On average, how many records do you expect with a pack of n cards?

Exercise 5.2.1 *Guess the answer to (b) when $n = 5$, $n = 10$, $n = 100$ and $n = 1000$.*

We start by asking what the probability is that the last card dealt gives a record. This is the same as asking what the probability is that the last card dealt is the largest in the pack. This probability is clearly $1/n$ and independent of the earlier announcements by our assistant.

What about the probability that the kth card is a record? Suppose that, after the kth card has been dealt, we tell our assistant to throw away the rest of the pack. This cannot change the announcement of a record or non-record for the kth card, but reduces the problem to asking whether the last card of a well-shuffled pack of k cards, is a record and we know, from the previous paragraph that this probability is $1/k$ independent of any earlier announcements by our assistant.

We now know that the probability of the kth card giving a record is $1/k$, independent of announcements for any other cards. What is the expected number of records if we deal n cards? There is a standard technique for attacking problems of this nature.

Exercise 5.2.2 *We continue with the discussion above. Set $X_k = 1$, if the kth card is a record, and $X_k = 0$, if not. Explain why*

$$Y = X_1 + X_2 + \cdots + X_n$$

is the total number of records. Compute $\mathbb{E}X_j$ and hence show that

$$\mathbb{E}Y = 1 + \frac{1}{2} + \frac{1}{3} + \cdots + \frac{1}{n}.$$

There is also a standard technique for finding the approximate value for sums like that given in Exercise 5.2.2. Figure 5.1 is practically self explanatory, but we give a more formal argument in the next exercise.

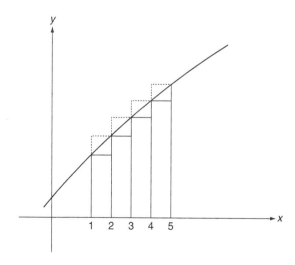

Figure 5.1. Comparing integrals and sums.

Exercise 5.2.3 *(i) Suppose that f is a well-behaved increasing function. Explain why*

$$f(n) \leq f(x) \leq f(n+1) \text{ for } n \leq x \leq n+1$$

and deduce that

$$f(n) \leq \int_n^{n+1} f(x)\,dx \leq f(n+1).$$

By summing the inequalities, deduce that

$$\sum_{n=1}^{N-1} f(n) \leq \int_1^N f(x)\,dx \leq \sum_{n=2}^N f(n).$$

Show also that, if $f(x) \geq 0$ for all x,

$$\int_1^N f(x)\,dx \leq \sum_{n=1}^N f(n) \leq \int_1^{N+1} f(x)\,dx$$

(ii) Restate the argument of (i) in words using Figure 5.1.

(iii) State and prove a result similar to (i) which applies to well-behaved decreasing functions.

(iv) Use (iii) to show that

$$\log N \leq \sum_{n=1}^N \frac{1}{n} \leq 1 + \log N$$

for $N \geq 1$.

We have thus shown that the expected number of records when n cards are dealt one after the other is approximately $\log n$.

Exercise 5.2.4 *Let $b > 1$. Suppose that we deal a pack with $[bn]$ cards.[1] Show, by using the ideas of Exercise 5.2.2 and Exercise 5.2.3, that the expected number of records between the nth and the last card is approximately $\log b$ when n is large.*

The arguments above show that, if we make a long series of measurements, then we may expect records to become rarer unless there is a long-term trend for the quantity measured to increase. This fits in with what we see when we consider records in sport, but, no doubt, many other factors must be taken into account. (For example, in pole vaulting, technological improvements in the pole have, essentially, produced a new sport.).

---

[1] We use $[x]$ to denote the integral part of x, that is to say, the largest integer m with $m \leq x$.

In many places, we have records of average temperatures for over 100 years. If someone claims that the average temperature is not increasing, they should be prepared to bet at odds of one hundred to one that the average temperature next year will not be a record.

It is surprising that such simple arguments should give the information that they do. However, we produced this simplicity by ignoring everything about records except the fact that they were records. If we build a sea wall, we are interested not so much in the probability of a record sea height during the lifetime of the sea wall, but in the actual height when it occurs. To the old saying 'prophecy is difficult, particularly when it concerns the future' we may add 'prophecy is difficult when it concerns events which have not occurred before'. The great physicist Lorentz headed the group which advised the Dutch government on how high to build its sea walls.

5.3 How to choose a restaurant

As you walk back to your car along the Manly beach front you will pass a kilometre of fish restaurants. You are determined never to backtrack (so, once you have passed a restaurant, you cannot go back to it), but you wish to eat in the best restaurant. What is your optimum strategy and what is the probability that you will eat in the best restaurant if you adopt it?

To answer this question, we return to the discussion of records in the previous section. Recall that we asked our assistant to shuffle a pack of n cards numbered with distinct real numbers and then turn over the cards one at a time saying 'record' if the number revealed is bigger than any she has previously seen and 'no record' if not. Your object is now to stop her when she has announced the largest card.

We make the following remarks.

(1) If you fail to stop before the nth card, you will have to chose the nth card.

(2) If you stop on an earlier card, that card should be a record card (since any card which is not a record cannot be the largest card).

(3) As we saw earlier, the statement that the kth card is a record is simply the statement that the kth card is the largest of the first k cards dealt. Thus your strategy cannot depend on the announcements of record or non-record made about the early cards.

It follows that your strategy can only depend on where you are in the deal (on the first card, second card, third card and so on) and must have form 'stop if a record is declared on the kth card if $k \in E$ "the permitted stopping set", otherwise do not stop unless you are at the nth card'. Our question reduces to finding the set E.

Table 5.1. *Choosing from four restaurants*

1	1	1	1	1	1	2	2	3	4	3	4	2	2	3	4	3	4	2	2	3	4	3	4
2	2	3	4	3	4	1	1	1	1	1	1	3	4	2	2	4	3	3	4	2	2	4	3
3	**4**	2	2	**4**	**3**	**3**	**4**	2	2	**4**	**3**	1	1	1	1	1	1	**4**	**3**	**4**	**3**	2	2
4	3	**4**	3	2	2	**4**	3	**4**	3	2	2	**4**	**3**	**4**	**3**	2	2	1	**1**	1	**1**	**1**	**1**

If $k \in E$, then we have decided that, if the kth restaurant we come to is better than all the previous ones, then we wish to eat there. Suppose that we reach the kth restaurant but, disappointingly, it turns out not to be better than all the previous ones. We now pass on to $k + 1$th restaurant and discover that it is better than all the previous ones. With fewer restaurants to come, we must surely choose to eat at that restaurant. Thus, if $k \in E$, it follows that $k + 1 \in E$ and so E consists of all k greater than or equal to some fixed m. Our rule is, thus, 'stop if a record is declared on the kth card, if $k \geq m$, otherwise do not stop unless you are at the nth card'.

Exercise 5.3.1 *(i) Table 5.1 shows the $4! = 4 \times 3 \times 2 \times 1 = 24$ different ways of dealing a pack of 4 cards labelled 1, 2, 3 and 4. Check that the numbers in boldface represent the cards we choose if we follow the rule 'stop if a record is declared on the kth card if $k \geq m$, otherwise do not stop unless you are at the nth card' with $m = 3$. Check, by counting, that the probability that we obtain the largest card is $5/12$.*

Obtain the same result by calculating the probability that either the third card is 4 or the fourth card is 4 and the third card is not 3.

(ii) Carry out the arguments of part (i) with $m = 1$, $m = 4$ and $m = 2$. What is the best choice for m?

(iii) Find the best choice for m when there are three cards in the pack.

We can make some simple estimates to guide our choice of m. Write $X_1 = x_1$ if the largest card is the x_1th card to be turned over, $X_2 = x_2$ if the second largest card is the x_2th card to be turned over and so on.

Lemma 5.3.2 *Consider the system discussed in this section.*
(i) If $n = 2u - 1$ and $m = u$, then

$$\Pr(\textit{stop at largest card}) > 1/4.$$

(ii) If $n = 2u$ and $m = u + 1$, then

$$\Pr(\textit{stop at largest card}) > 1/4.$$

(iii) If $m - 1 \geq 3n/4$, then

$$\Pr(\textit{stop at largest card}) \leq 1/4.$$

(iv) If $m \le n/80$, then

$$\text{Pr}(\textit{stop at largest card}) \le 1/4.$$

Proof (i) We have

$$\text{Pr(stop at largest card)} \ge \text{Pr}(X_1 = u) + \text{Pr}(X_1 > u, \, X_2 < u)$$
$$= \frac{1}{2u-1} + \frac{u-1}{2u-1} \times \frac{u-1}{2u-2}$$
$$= \frac{1}{2u-1}\left(1 + \frac{u-1}{2}\right) = \frac{u+1}{2(2u-1)} > \frac{1}{4}$$

as required.

(ii) Left as an exercise for the reader.

(iii) We have

$$\text{Pr(stop at largest card)} \le \text{Pr}(X_1 \ge m) = \frac{n-m+1}{n} \le \frac{1}{4}$$

when $m - 1 \ge 3n/4$.

(iv) (Better estimates are possible, but we just need one to be going on with.) Observe that, for us to stop at the largest card, at least one of the events

$$A_1 = \{X_1 < X_2, X_3, X_4, X_5\},$$
$$A_2 = \{X_2 < u\}, \; A_3 = \{X_3 < u\}, \; A_4 = \{X_4 < u\}, \; A_5 = \{X_5 < u\}$$

must happen. Thus

$$\text{Pr(stop at largest card)} \le \text{Pr}(A_1 \cup A_2 \cup A_3 \cup A_4 \cup A_5)$$
$$\le \text{Pr}(A_1) + \text{Pr}(A_2) + \text{Pr}(A_3) + \text{Pr}(A_4) + \text{Pr}(A_5)$$
$$< \frac{1}{5} + \frac{1}{80} + \frac{1}{80} + \frac{1}{80} + \frac{1}{80} = \frac{1}{4}.$$

∎

In the next lemma, we replace estimates by an exact (though not particularly transparent) formula.

Lemma 5.3.3 *Consider the system discussed in this section.*

(i) Let A_j be the event that $X_j \ge m$ and A_j^c be the event that $X_j < m$. Let B_j be the event that $X_1 < X_2, X_3, \ldots, X_j$. In order that we stop at the largest card, one of the following disjoint events[2] must occur

$$A_1 \cap A_2^c, \; A_1 \cap A_2 \cap A_3^c \cap B_2, \; A_1 \cap A_2 \cap A_3 \cap A_4^c \cap B_3,$$
$$\ldots, \; A_1 \cap A_2 \cap \cdots \cap A_j \cap A_{j+1}^c \cap B_j, \ldots$$

---

[2] Recall that two events C and D are disjoint if $C \cap D = \varnothing$, that is to say, at most one of the two events can happen.

(ii) The events $A_1 \cap A_2 \cap \cdots \cap A_j \cap A_{j+1}^c$ and B_j are independent. We have

$$\Pr(B_j) = \frac{1}{j}.$$

(iii) If $m \geq 2$, we have

$$\Pr(\textit{stop at largest card}) = \frac{n-m+1}{n} \times \frac{m-1}{n-1}$$
$$+ \frac{1}{2} \times \frac{n-m+1}{n} \times \frac{n-m}{n-1} \times \frac{m-1}{n-2}$$
$$+ \frac{1}{3} \times \frac{n-m+1}{n} \times \frac{n-m}{n-1} \times \frac{n-m-1}{n-2} \times \frac{m-1}{n-3} + \cdots$$

If $m = 1$

$$\Pr(\textit{stop at largest card}) = \frac{1}{n}.$$

Proof (i) If $X_1 < m$, then we will not stop at the largest card. If $X_k \geq m$ for $1 \leq k \leq j$ but $X_{j+1} < m$, then we will stop at the largest card if and only if $X_1 < X_k$ for $2 \leq k \leq j$.

(ii) If all we know is that $X_k \geq m$ for $1 \leq k \leq j$ but $X_{j+1} < m$, then, since the cards are well-shuffled, each order of X_1, X_2, ..., X_j remains equally likely, so the events $A_1 \cap A_2 \cap \cdots \cap A_j \cap A_{j+1}^c$ and B_j are independent and

$$\Pr(B_j) = \frac{1}{j}.$$

(iii) The case $m = 1$ is immediate so we concentrate on the case $m \geq 2$. By (ii)

$$\Pr(A_1 \cap A_2 \cap \cdots \cap A_j \cap A_{j+1}^c \cap B_j) = \frac{1}{j} \Pr(A_1 \cap A_2 \cap \cdots \cap A_j \cap A_{j+1}^c).$$

We observe that the event A_1 occurs if we place the largest card in one of $n - m + 1$ positions out of a total of n possibilities. Once the largest card has been placed so that A_1 occurs, the event A_2 occurs if we place the largest card in one of $n - m$ remaining positions out of a total of $n - 1$ possibilities. More generally, once the largest $k - 1$ cards have been placed so that $A_1 \cap A_2 \cap \ldots$ $\cap A_{k-1}$ occurs, then A_k occurs if we place the kth largest card in one of $n + 1 - m - k$ remaining positions out of a total of $n - k$ possibilities. Finally, once the largest j cards have been placed so that $A_1 \cap A_2 \cap \cdots \cap A_j$ occurs, then A_{j+1}^c occurs if we place the $j + 1$th largest card in one of $m - 1$ positions out of a total of $n - j - 1$ possibilities.

Thus

$$\Pr(A_1 \cap A_2 \cap \cdots \cap A_j \cap A_{j+1}^c)$$
$$= \frac{n-m+1}{n} \times \frac{n-m}{n-1} \times \cdots \times \frac{n+1-m-k}{n-k} \times \frac{m-1}{n-k-1}$$

and

$$\Pr(A_1 \cap A_2 \cap \cdots \cap A_j \cap A_{j+1}^c \cap B_j)$$
$$= \frac{1}{j} \times \frac{n-m+1}{n} \times \frac{n-m}{n-1} \times \cdots \times \frac{n+1-m-j}{n-k} \times \frac{m-1}{n-j-1}.$$

We know from part (i) that

Pr(stop at largest card)

$$= \Pr\left((A_1 \cap A_2^c) \cup (A_1 \cap A_2 \cap A_3^c \cap B_2) \cup (A_1 \cap A_2 \cap A_3 \cap A_4^c \cap B_3) \cup \cdots\right)$$
$$= \Pr(A_1 \cap A_2^c) + \Pr(A_1 \cap A_2 \cap A_3^c \cap B_2)$$
$$+ \Pr(A_1 \cap A_2 \cap A_3 \cap A_4^c \cap B_3) + \cdots$$

so the required formula follows. ∎

Exercise 5.3.4 *Carry out sufficient calculations to to find the best choice of m when $n = 5$.*

One of the standard tricks of a mathematician faced with a complicated formula involving n is to see what happens as $n \to \infty$. We know, from Lemma 5.3.2, that the best choice of m will satisfy $n/80 < m < 3n/4$, so it makes sense to look at $t = m/n$.

Lemma 5.3.5 *Consider the system discussed in this section with $m = nt$ and $0 < t < 1$. If n is large*

$$\Pr(\text{stop at largest card}) \approx -t \log t.$$

Proof Observe that

$$\frac{1}{j} \times \frac{n-m+1}{n} \times \frac{n-m}{n-1} \times \cdots \times \frac{n+1-m-j}{n-k} \times \frac{m}{n-j-1}$$
$$= \frac{1}{j} \times \frac{n-tn+1}{n} \times \frac{n-tn}{n-1} \times \cdots \times \frac{n+1-tn-j}{n-k} \times \frac{tn}{n-j-1}$$
$$= \frac{1}{j} \times ((1-t) + n^{-1}) \times \frac{1-t}{1-n^{-1}} \times \cdots \times \frac{1-t-(j-1)n^{-1}}{1-kn^{-1}}$$
$$\times \frac{t}{1-(j+1)n^{-1}} \approx \frac{1}{j}(1-t)^j t.$$

Using the Taylor series for $\log(1-x)$ (see Exercise A.9), we obtain

$$\text{Pr(stop at largest card)} \approx t(1-t) + \frac{t(1-t)^2}{2} + \frac{t(1-t)^3}{3} + \cdots$$

$$= t\left((1-t) + \frac{(1-t)^2}{2} + \frac{(1-t)^3}{3} + \cdots\right)$$

$$= t\left(-\log\left(1 - (1-t)\right)\right) = -t\log t$$

as stated.[3]   ■

Exercise 5.3.6 *In the proof of Lemma 5.3.5, we obtained the result*

$$\text{Pr(}stop\ at\ largest\ card\text{)} \approx t(1-t) + \frac{t(1-t)^2}{2} + \frac{t(1-t)^3}{3} + \cdots$$

by first obtaining the general result of Lemma 5.3.3 and then allowing n to become large. Here is an alternative direct argument.

Explain why, when n is large compared with u and v, the probability that u given cards are dealt before the tnth card and v given cards are dealt after the tnth card is approximately $t^u(1-t)^v$. Use this to obtain the result stated in the previous paragraph.

Exercise 5.3.7 *Use standard calculus techniques to show that $-t\log t$ takes its largest value for $0 < t < 1$ when $t = e^{-1}$.*

We can now answer the problem posed in this section, when n is large, by combining Lemma 5.3.5 and Exercise 5.3.7.

Lemma 5.3.8 *If n is large, the strategy which gives the greatest probability of stopping at the highest card is 'do nothing while roughly the first ne^{-1} cards are dealt and then stop at the first time a record is declared or at the last card'. The probability of stopping at the highest card is then approximately e^{-1}.*

You should walk a distance $1/e$ back to the car and then choose any restaurant which is better than any you have previously seen. Since a function does not change very quickly close to a maximum, $-t\log t$ does not change very much close to its maximum, so the rule 'walk about $1/3$ of the distance back to the car' will work pretty well. We obtained our rule for the case when the number n of cards or restaurants was large, but a closer look at the kind of

---

[3] If this was a text on analysis, we would need to proceed considerably more cautiously, but the idea of the proof would remain the same.

approximations we made shows that the rule will work well even when n is small. The strategy is a natural one though I find it surprising that it is so effective.

Exercise 5.3.9 *(i) Show that, if we use the strategy 'wait to the mth card dealt, then stop at a record' we will choose the last card and that card will not be the largest if and only if the largest card is dealt before the card m.*

If n is large and we choose $m \approx ne^{-1}$, show that the probability that we stop at the last card is approximately e^{-1} and the probability that we neither choose the last card nor the largest is approximately $1 - 2e^{-1}$.

(ii) Now suppose that $m = tn - 1$ and k is an integer with $k \geq 2$. Show that the probability that we stop at the last card is t and the probability that none of the k largest cards are turned over before the mth card is $(1 - t)^m$. Deduce that

$$\Pr(\text{stop at one of the } k \text{ largest cards}) \geq 1 - \left(t + (1 - t)^k\right).$$

Without indulging in orgies of calculation, show that, given any $\epsilon > 0$, we can find a k (depending on ϵ) and a strategy which guarantees that we stop at one of the k largest cards with probability at least $1 - \epsilon$.

Anyone proposing to use the ideas above for the selection of a spouse probably overestimates both their ability to estimate the number of suitable partners they are likely to meet and their attractiveness to the opposite sex.

5.4 Back to sorting

Let X be a set. Suppose that we play the following game. You choose an $x \in X$ and allow me three questions to discover x. I must ask questions to which the answer is 'yes' or 'no' and you must answer such questions truthfully. What is the largest size of X which will let me always win?

Observe that I can only ask questions of the form 'Does $x \in A$?'. If you reply 'yes', I know that $x \in A$. If you reply no, I know that $x \in A^c = X \setminus A$, the complement of A. If my three questions are 'Does $x \in A_j$?' for $j = 1, 2, 3$, then, after you have replied, I will know that x is in one of

$$A_1 \cap A_2 \cap A_3, \ A_1 \cap A_2 \cap A_3^c, \ A_1 \cap A_2^c \cap A_3, \ A_1 \cap A_2^c \cap A_3^c,$$
$$A_1^c \cap A_2 \cap A_3, \ A_1^c \cap A_2 \cap A_3^c, \ A_1^c \cap A_2^c \cap A_3, \ A_1^c \cap A_2^c \cap A_3^c$$

and that is all I know. In order that I may be certain of winning, each of these 8 sets can contain at most one point. Since the union of these 8 sets is X and since they are disjoint, I know that (writing $|B|$ for the number of elements of B)

$$
\begin{aligned}
|X| = {} & |A_1 \cap A_2 \cap A_3| + |A_1 \cap A_2 \cap A_3^c| |A_1 \cap A_2^c \cap A_3| + |A_1 \cap A_2^c \cap A_3^c| \\
& + |A_1^c \cap A_2 \cap A_3| + |A_1^c \cap A_2 \cap A_3^c| + |A_1^c \cap A_2^c \cap A_3| \\
& + |A_1^c \cap A_2^c \cap A_3^c| \\
\leq {} & 1 + 1 + 1 + 1 + 1 + 1 + 1 + 1 = 8.
\end{aligned}
$$

Thus I can only be sure of finding the element x using 3 yes/no questions if X contains 8 elements or fewer. Similarly, I can only be sure of finding an element $y \in Y$ using N yes/no questions if Y contains 2^N elements or fewer.

Exercise 5.4.1 *(i) Consider the 8 sequences*

$$000, \ 001, \ 010, \ 011, \ 100, \ 101, \ 110, \ 111.$$

Give three yes/no questions which will enable me to find any specified sequence. By adapting this idea, or otherwise, show how to give three yes/no questions which will enable me to find any member of set consisting of 8 or fewer members. Show how to give n yes/no questions which will enable me to find any member of set consisting of 2^n or fewer members.

(ii) In the middle of 2005, the Oxford English Dictionary *contained about half a million 'head words'.*[4] *Show that it is possible to identify a particular word with 20 yes/no questions and describe how you would go about this task.*

What does this have to do with sorting? Recall that, in our statement of the problem, a pack of n cards bearing the numbers 1 to n is shuffled and the cards then placed face down in front of you. You have an assistant who will look at any two cards that you indicate and tell you which of the two cards bears the largest number. You may then ask her either to replace the two cards in their original position or to swap them and replace them (or take some other specified action of this kind). This completes one operation.

Let X be the set of $n!$ different possible orders for the shuffled pack. Once you have the pack in order, you can reverse the swaps that your assistant made in order to produce the original order $x \in X$. Each time you ask your assistant 'Which of the two cards is higher?' you are actually asking 'If I take the original shuffled pack and perform the various swaps I asked you to do, will the card in the first position I specify be larger than the card in the second position?' and this is a yes/no question. If it requires at least N yes/no questions of any type to be sure of discovering x, it will certainly require N yes/no questions of

---

[4] But no entry for that important word 'counterexample'.

this particular type to be sure of discovering x. Thus it will require at least N operations to be sure of sorting your pack of n cards.

We know that X contains $n!$ elements and that we can only be sure of identifying an $x \in X$ with N yes/no questions if $2^N \geq |X| = n!$. We have proved the following lemma.

Lemma 5.4.2 *If we can guarantee to sort a pack of n cards in N operations, then $2^N \geq n!$.*

The inequality $2^N \geq n!$ may be rewritten $N \log 2 \geq \log n!$, but in order to estimate N we need a good estimate for $\log n!$. Fortunately the equality

$$\log n! = \log 1 + \log 2 + \log 3 + \cdots + \log N$$

suggests using the ideas of Exercise 5.2.3. These give us a version of Stirling's formula.

Lemma 5.4.3 [Stirling's formula] *(i) If $n \geq 2$ then*

$$n \log n - (n - 1) \leq \log n! \leq (n + 1) \log(n + 1) - n.$$

(ii) We have

$$\frac{\log n!}{n \log n} \to 1$$

as $n \to \infty$.

Proof (i) Since log is increasing,

$$\int_1^n \log x \, dx \leq \sum_{r=2}^n \log r = \sum_{r=1}^n \log r \leq \int_2^{n+1} \log x \, dx \leq \int_1^{n+1} \log x \, dx.$$

By integrating by parts,

$$\int_1^b \log x \, dx = \int_1^b 1 \times \log x \, dx$$
$$= \left[x \log x \right]_1^b - \int_1^b \frac{x}{x} \, dx$$
$$= b \log b - \int_1^b 1 \, dx = b \log b - (b - 1).$$

Putting the two formulae together, we obtain the result.

(ii) This is an immediate consequence of (i). ∎

Exercise 5.4.4 *Although it is helpful to have the full force of Lemma 5.4.3, all we really need for this section is that*

$$An \log n \geq n! \geq Bn \log n$$

for some A, B > 0 and all sufficiently large n. This can be proved more simply.
  (i) If n is even, explain why

$$n^n \geq n! \geq (n/2)^{n/2}$$

and deduce that

$$n \log n \geq \log n! \geq \frac{1}{4} n \log n$$

for n ≥ 4.
  (ii) Obtain a similar result for all n.

Using either Lemma 5.4.3 or Exercise 5.4.4, we have shown that we cannot sort n cards in much fewer than $n \log n$ operations.

Lemma 5.4.5 *There exists a constant K such that, if we can guarantee to sort a pack of n cards in N operations then $N \geq Kn \log n$.*

Can we actually sort this fast? When human beings sort things, they often do a 'rough sort' into a number of more or less equal piles and then sort the piles. To show that this is a good strategy, we return to our assistant but give her the additional power of being able to indicate the middle card (by value) of any pile. (More precisely, given a pile of m cards, she can indicate the card such that $[(m-1)/2]$ cards have lower value and $(m-1) - [(m-1)/2]$ have higher value.)

If we have a pack of 2^m cards, we ask our assistant to find the middle card and then examine all the other cards, placing them in two piles, according to whether their value is higher or lower than the middle card. We then ask her to place the middle card on the smaller pile giving us two packs of 2^{m-1} cards. We say that this takes 2^m operations (one operation to find the middle card and $2^m - 1$ comparisons).[5] If we write a_m for the number of operations required to sort a pack of 2^m cards then we have shown that

$$a_m = 2^m + 2a_{m-1}$$

(since we still have to sort two packs of 2^{m-1} cards).

Exercise 5.4.6 *Take $a_1 = 2$ and compute a_j for $j = 2, 3, 4, 5$. Show, by induction, that*

$$a_m = m2^m$$

for all m ≥ 2.

---

[5] The reader may prefer a different operation count, but this will only change the arithmetic and not the spirit of our argument.

We have shown that, if our assistant has the power to find the middle card, we can sort a pack of 2^m cards in $m2^m$ operations.

Exercise 5.4.7 *Explain why this means that, if our assistant has the power to find the middle card and $2^{m-1} < n \leq 2^m$, we can sort a pack of n cards in $m2^m$ operations.*

Deduce that we can sort a pack of n cards in

$$\frac{2}{\log 2} n(1 + \log n)$$

operations.

This argument falls into the 'If we had some ham, we could have some ham and eggs, if we had some eggs' class since, usually, the only way to find the middle card is to have a sorted pack.

However, Hoare[6] had the following brilliant idea. Suppose that our assistant simply picks a card at random and deals the cards into two piles according to whether they are bigger or smaller than the randomly chosen card. Sometimes the randomly chosen card will be a very high or very low value and we will get a very small pile and a very large pile (which is almost as hard to sort as our original pack) but, with probability about $1/2$, the larger of the two new piles will contain fewer than $3/4$ of the number of cards in our original pack and this is almost as good as an exact division into two equal piles. The law of large numbers tells us that we can expect to be lucky quite often and so, with high probability, we will get something very much like the result produced by the 'middle card' procedure.

We shall show that, on average, Hoare's idea of 'quicksort' runs as fast as we can reasonably hope. The mathematics that follows may seem complicated, but it is important that the reader understands that it is the initial idea which matters and that, once we are convinced that the idea will work, it merely requires perseverance to provide a proof.

Let us write e_n for the expected number of operations required to sort n cards using quicksort. If we start with such a pack and choose a card at random, then it requires $n - 1$ comparisons to produce two piles, one containing k cards, say, consisting of those cards which are less than our randomly chosen card together with the card itself, and one consisting of $n - k$ cards consisting of

---

[6] Classicist, philosopher, linguist, industrial and academic computer scientist and all-round clever man. His reminiscences [28] include the observation that '... there are two ways of constructing a software design. One way is to make it so simple that there are *obviously* no deficiencies and the other way is to make it so complicated that there are no *obvious* deficiencies.'

those cards which are greater than our randomly chosen card. If we count the choice of the random card as one operation, then

total number of operation required to sort our n cards

$= n +$ total number of operation required to sort the pile of k cards

$+$ total number of operation required to sort the pile of $n - k$ cards

and so

total expected number of operation required to sort our n cards

$= n + e_k + e_{n-k}.$

Now k can take each of the values $1, 2, \ldots, n$ with probability $1/n$, so

$e_n =$ expected number of operations required to sort our n cards

$$= \sum_{k=1}^{n} \frac{1}{n} (\text{expected number of ops req to sort } n \text{ cards if } k\text{th card chosen})$$

$$= \sum_{k=1}^{n} \frac{1}{n} (n + e_k + e_{n-k}).$$

Thus

$$e_n = n + \frac{1}{n}\big((e_1 + e_{n-1}) + (e_2 + e_{n-2}) + \cdots + (e_{n-1} + e_1) + (e_n + e_0)\big)$$

and so, since $e_0 = e_1 = 0$,

$$e_n = \frac{n^2}{n-1} + \frac{2}{n-1}(e_2 + \cdots + e_{n-1}) \qquad\qquad \bigstar$$

for all $n \geq 3$.

Equation $\bigstar$ looks very complicated, but we do not want to solve it. We only want to show that $e_n \leq Kn \log n$ for some constant K, and the obvious way forward is induction using the kind of estimate obtained in the next lemma.

Lemma 5.4.8 *If $n \geq 2$,*

$$2\log 2 + 3\log 3 + \cdots + (n-1)\log(n-1) \leq \frac{n^2 \log n}{2} - \frac{n^2}{4}.$$

Proof Once again, we use the ideas of Exercise 5.2.3. If we set $f(x) = x \log x$, then f is a well-behaved increasing function and

$$\sum_{r=2}^{n-1} f(r) \leq \int_3^n f(x)\,dx,$$

that is to say,

$$2 \log 2 + 3 \log 3 + \cdots + (n-1) \log(n-1) \le \int_3^n x \log x \, dx.$$

Integration by parts gives

$$\int_3^n x \log x \, dx = \left[\frac{x^2}{2} \log x \right]_3^n - \int_3^n \frac{x}{2} \, dx$$

$$= \left[\frac{n^2}{2} \log n - \frac{9}{2} \log 3 \right] - \left[\frac{n^2}{4} - \frac{9}{2} \right]$$

$$\le \frac{n^2}{2} \log n - \frac{n^2}{4}.$$

■

Since one operation is required to sort two cards, we have $e_2 = 1$ and we can start the induction.

Exercise 5.4.9 *We know that $e_2 = 1$ and that ★ holds. Use induction to show that*

$$e_n \le 4n \log n$$

for all $n \ge 2$.

Different ways of counting the number of operations and of organising the calculations will give different estimates, but it is clear that the expected number of operations for quicksort is within a constant multiple of the best possible.

The average number of operations is not the same as the worst possible number. If we are very unlucky, then we will divide our pack of n cards into a pile of $n-1$ cards and a pile of 1 card,[7] taking n operations to do this. If our bad luck is repeated each time, we will end up having to take

$$n + (n-1) + (n-2) + \cdots + 1 \approx n^2/2$$

operations, which is very slow indeed.

We are very unlikely to be this unlucky.

---

[7] Indeed, the way we have described the treatment of the randomly chosen card, we could end up with a pile of n cards and be no better off than we started. An extra instruction to our assistant will avoid this.

Exercise 5.4.10 *Let* (Ω, Pr) *be a probability space and* X *a random variable with* $X(\omega) \geq 0$ *for all* $\omega \in \Omega$. *Show, by using the ideas of Lemma 2.5.3, or otherwise, that*

$$\mathrm{Pr}(X \geq a) \leq a^{-1}\mathbb{E}X$$

whenever $a > 0$. *How does this relate to the sentence preceding this exercise?*

Exercise 5.4.11 *Suppose that, to make things easier, we always use the bottom card of any pile as the 'sorting card'. Explain why, if we are presented with a properly shuffled deck, this will make no difference. Observe that, if the pack is already sorted, then this version of quicksort will take about* $n^2/2$ *operations! Thus, if we wish to sort a partially ordered pack in this way, our first act will be to shuffle the pack thoroughly!*

We can get much better estimates than those given in Exercise 5.4.10 which show that, for a well-shuffled pack, quicksort is very unlikely to be slow. However, the proof of the pudding is in the eating, and quicksort is probably the most used sorting algorithm for computers.[8] To see why it is preferred to other methods (even some which are guaranteed to be fast in all cases), we would need to look in detail at how it would be implemented on a computer rather than talking airily about packs of cards and assistants.

Exercise 5.4.12 *In the middle of 2005, Google claimed to index about* 8×10^9 *pages. How long would be required to sort that number of items using a program which took* $n^2/2$ *operations if each operation took* 10^{-9} *seconds? How long would be required for a program which took* $4n \log n$ *operations?*

5.5 Shortest paths

One of the reasons why sorting algorithms and methods for finding the largest and smallest of a collection of objects are so important is that they are often used as components of other algorithms.

In this final section, we look at some rather more intricate problems involving repeated comparisons. We need the result of the next exercise.

Exercise 5.5.1 *Why is finding the least of* n *numbers essentially the same problem as finding the largest of* n *numbers?*

---

[8] Since sorting is so important, this may make quicksort the most used algorithm of all.

There are many sites on the Internet which tell us things like the shortest route by road,[9] or the cheapest route by air between two specified towns.[10] How do they work?

The next exercise shows that they cannot examine every possible route.

Exercise 5.5.2 *Suppose that we have n airports and there are flights between every two airports. Explain why there are $(n-2)!$ routes between two specified airports which visit every other airport exactly once.*

A little thought gives us a better method.

Exercise 5.5.3 *Get hold of a map which shows the shortest route by road between some given town (London, say) and many other towns. If one is not available try and construct one yourself. Is there any pattern that you notice? Can you give a reason for the pattern?*

I hope that you notice that a map of shortest routes to London looks like a tree or, even more suggestively, a river with tributaries.

Why does it look like this? Two answers come to mind. The first, which is not directly useful, is that 'water seeks the easiest way to the sea' and we are seeking the shortest way to London. The second is that, once two rivers join, they never separate,[11] and the same is true of shortest routes to London. Suppose that the shortest route from A to C passes through B. Then the shortest route from A to C must coincide with the shortest route from B to C for that part which goes from B to C.

This gives us a way of finding shortest routes. Suppose that we have n towns $A_1, A_2, A_3, \ldots, A_n$ and we wish to find the shortest route from A_1 to A_n. Instead of concentrating on this single problem, we try to find the shortest route from any A_j to A_1. (Of course, the moment we find the result in the particular case $j = n$ we can stop.) First find a town which is the shortest direct distance from A_1. Suppose (by renaming the towns) that this is A_2. Clearly, any route from A_2 to A_1 via any further town will be at least as long, so we have found a shortest route from A_2 to A_1. Now find a town (other than A_1 or A_2) such that either the direct route to A_1 or the route to A_1 via A_2 is the shortest. Suppose (by renaming the towns) that this is A_3. Clearly, any route from A_3 to A_1 via any other route will be at least as long, so we have found a shortest route from

---

[9] There is a minor problem here since there may be several routes of the same length and so no *unique* shortest route. Sometimes, when I talk about 'the shortest route', I should more correctly say 'a shortest route'. The reader should convince herself, in such cases, that it is easy to deal with the case when the shortest route is not unique.

[10] And, of course, the satellite navigation equipment in a car is forever calculating shortest routes.

[11] Not quite true (think of islands in a river and of the Nile delta) but near enough.

A_3 to A_1. Now find a town (other than A_3, A_2 or A_1) such that either the direct route to A_1 or the route to A_1 via A_2 or the route to A_1 formed by taking the direct route to A_3 and then taking the shortest route to A_1 is the shortest.

If we continue in this way, we see that at the rth step we will have r towns $A_1, A_2, \ldots, A_r$, say, for which we know the shortest way to A_1. We now look at all the

$$b(k, i) = \text{distance from } A_k \text{ to } A_i + \text{length shortest route from } A_i \text{ to } A_1$$

with $r + 1 \leq k \leq n$ and $1 \leq i \leq r$. The smallest $b(k, i)$ (strictly speaking, any smallest $b(k, i)$) will correspond to a town A_k for which a shortest route to A_1 consists in travelling to A_i and then taking the shortest route from A_i to A_1.

Observe that at the rth step we have to examine $r \times (n - r)$ possibilities and that we may have to perform $n - 1$ steps.

Exercise 5.5.4 *Show that* $r(n - r) \leq n^2/4$.

Exercise 5.5.5 *(i) The following problem goes back at least a thousand years. A peasant must row a wolf, a goat and a cabbage across a river in a boat that will only carry one passenger at a time. If he leaves the wolf with the goat, then the wolf will eat the goat. If he leaves the goat with the cabbage, then the goat will eat the cabbage. The cabbage represents no threat to the wolf nor the wolf to the cabbage. By considering possible paths (some involving two boat trips and some one) between $\{P, W, G, C\}$ (all on starting bank), $\{P, W, G\}$ (peasant, wolf and goat on starting bank, cabbage on final bank), $\{P, W, C\}$, $\{P, G, C\}$, $\{P, C\}$ and $\varnothing$ (all crossed) find the smallest number of trips that the peasant must make.*

(ii) This problem appears in a book by Tartaglia (famous, as we mention in Appendix B, for finding the solution of the cubic). Here is how it appears in Bachet's Problèmes Plaisants et Délectables *published in 1612.*

Two boon companions have 8 pints of wine which they wish to share equally. The wine is in an 8 pint jug and all they have to help them are two other jugs of capacity 5 and 3 pints. How can they do this only using these three jugs?[12]

Formulate the problem as a shortest path problem and proceed as far as you wish with the solution.

[From the point of view of this section, problem (ii) has (and problem (i) can be made to have) the special feature that all distances are 0, 1 or ∞. This can be exploited to give faster algorithms. However, the point of the question is to

---

[12] The onward march of civilisation saw the problem reappear in the 1995 film *Die Hard 3* in connection with disarming a bomb.

show that there is a uniform way of solving these problems,[13] *not to find the best uniform way.*]

Exercise 5.5.6 *If the reader tries our shortest path algorithm on a few examples, she will find that most of the steps change nothing.*

It is not the business of this book to give the best algorithms for doing things.[14] *However, if the reader is prepared to do the work, she may be interested to see how Dijkstra speeded things up by using better book-keeping*[15] *to eliminate many of the unnecessary steps of our crude method.*

We describe the rth stage of Dijkstra's algorithm.

At the beginning of the rth stage there are $r - 1$ 'old settled towns', one 'new settled town' and $n - r$ 'unsettled towns'. In each settled town there is a signpost showing the next town on a shortest route to A_1 and the total distance to A_1 along that shortest route. In each unsettled town there is a signpost showing the direction to an old settled town such that the shortest route to A_1 through that particular old settled town is no greater than the shortest route to A_1 through any other old settled town. (If there is no direct road to a settled town, the signpost points up in the air.) The signpost also records the distance to A_1 by the route suggested. (If the signpost points upwards, we take the distance to be infinite.)

During the rth step, each unsettled town examines the shortest route to A_1 via the new settled town. If this is shorter than the distance presently recorded on the signpost, the town changes the direction of the signpost and the distance recorded appropriately. If not, it leaves things as they are. We now look at all the unsettled towns and choose one with the shortest distance on its signpost. This becomes the new settled town for the $r + 1$th stage and the new settled town for the rth stage joins the old settled town for the $r + 1$th stage.

(i) What modifications are required at the first and last stage?

(ii) Show that the algorithm works.

---

[13] The charming book of O'Beirne [47] contains a discussion of both of the types of problem given in this exercise. He gives good ways for human beings (as opposed to machines) to attack them. In his historical remarks he explains why the problem of part (i) may have been among 'some examples of subtlety in Arithmetic for your enjoyment' sent by Alcuin to his imperial pupil the Emperor Charlemagne and how solving a problem along the lines of (ii) convinced Poisson that he should be a mathematician!

[14] Textbooks along the lines of *Twenty Algorithms to Teach Your Pet Monkey* perform a useful function, but this is not such a book.

[15] It is unlikely that he would approve of my presentation. In his essay *On the cruelty of really teaching computing science* he wrote: 'It is the prevailing educational practice, ... to present every thing that could be an exciting novelty as something as familiar as possible. ... The educational dogma seems to be that everything is fine as long as the student does not notice that he is learning something really new; more often than not, the student's impression is indeed correct'.

(iii) Explain why this algorithm only requires at most 2n comparisons at each stage and at most $2n^2$ comparisons in all. (As usual, you can obtain slightly better bounds with a little thought.)

(iv) Guess an appropriate n for the route finding equipment in a car and consider the difference made by using Dijkstra's algorithm rather than the one we thought up.

Although we started out by thinking about the d_{ij} as ordinary distances, it turns out to be useful to stretch the notion of distance to its utmost. The first generalisation is to allow $d_{ij} \neq d_{ji}$ (though, for the moment, we continue to take $d_{ij} \geq 0$). If d_{ij} is the time by bicycle from A_i to A_j with A_i at the top of a hill and A_j at the bottom, then $d_{ij} \neq d_{ji}$.

Exercise 5.5.7 *Check that our algorithm continues to work in this case.*

If we use our algorithm, it is almost as quick to work out all the shortest paths to a particular town A_1 as to work out the shortest path between A_1 and a single specified town. What happens if we try to work out the shortest paths between every pair of paths?

We adopt the convention that, if there is no path of a certain type between two points, we say that the distance between those two points via paths of that type is ∞. If a is a real number, we set

$$\min\{a, \infty\} = a \text{ and } \min\{\infty, \infty\} = \infty.$$

Lemma 5.5.8 *Consider points A_1, A_2, ..., A_n. Suppose that the 'distance' from A_i to A_j via a route of type $\mathcal{P}$ is p_{ij} and the 'distance' from A_j to A_k via a route of type $\mathcal{Q}$ is q_{jk}. Then if r_{ik} is the minimum 'distance' from i to k obtained by first following a route of type $\mathcal{P}$ from A_i to A_j and then a route of type $\mathcal{Q}$ from A_j to A_k we have*

$$r_{ik} = \min_{1 \leq j \leq n} (p_{ij} + q_{jk}).$$

Proof We just need to check that this is consistent with our conventions about ∞. ∎

If we write P for the $n \times n$ array (or *matrix*) of the p_{ij}, Q for the array of the q_{ij} and R for the array of r_{ij} we can use the suggestive notation

$$R = P \bullet Q$$

to mean $r_{ik} = \min_{1 \leq j \leq n}(p_{ij} + q_{jk})$.

Exercise 5.5.9 *(i) Show that, if P and Q are $n \times n$ matrices, then we can compute $P \bullet Q$ using at most n^3 comparisons.*

(ii) Show that, if U, V and W are $n \times n$ matrices, then

$$U \bullet (V \bullet W) = (U \bullet V) \bullet W.$$

Exercise 5.5.10 Let U, V and W be $n \times n$ matrices. Are the following always true or sometimes false? Give a proof or a counterexample as appropriate. (If A and B are matrices, we write $B = A^T$ if $b_{ij} = a_{ji}$.)
(i) $U \bullet V = V \bullet U$.
(ii) If U and V are symmetric (that is to say $U^T = U$ and $V^T = V$), then $U \bullet V = V \bullet U$.
(iii) $(U \bullet V)^T = V^T \bullet U^T$.
(iv) $(U + V) \bullet W = U \bullet W + V \bullet W$.

Since $\bullet$ is a form of multiplication, it is natural to write $P^{[2]} = P \bullet P$, $P^{[3]} = P^{[2]} \bullet P$ and, more generally, $P^{[m+1]} = P^{[m]} \bullet P$ for $m \geq 1$, where we set $P^{[1]} = P$.

Exercise 5.5.11 Suppose we have n points A_i and the direct distance from A_i to A_j is d_{ij}. (We take $d_{ij} = \infty$ if there is no direct route from A_i to A_j. We take $d_{ii} = 0$.) Let D be the $n \times n$ matrix with entries d_{ij} and let $d_{ij}^{[m]}$ be the (i, j)th entry in the matrix $D^{[m]}$.

(i) Show, by induction on m, or otherwise, that $d_{ij}^{[m]}$ is the shortest distance from A_i to A_j by a route passing through at most m towns (including the final town but not including the first).

(ii) We have assumed implicitly that $d_{ij} \geq 0$ for all i, j. Explain why, under this assumption, $D^{[m]} = D^{[n-1]}$ for all $m \geq n-1$. Explain why, if $d_{ij}^{[n-1]} = \infty$, there is no route of any kind from A_i to A_j. Explain why, if $d_{ij}^{[n]} \neq \infty$, $d_{ij}^{[n]}$ is the length of the shortest path from A_i to A_j.

At first sight, since each '$\bullet$ multiplication' requires about n^3 comparisons, it looks as though the computation of $D^{[n-1]}$ will require about n^4 comparisons which is rather a lot. However, we can use a simple trick to cut down on our work. Since

$$A^{[2m]} = A^{[m]} \bullet A^{[m]} \text{ and so } A^{[2^{r+1}]} = A^{[2^r]} \bullet A^{[2^r]},$$

we only need N '$\bullet$ multiplications' to compute $A^{[2^N]}$.

Exercise 5.5.12 (i) Explain why, if $2^{N-1} \leq n \leq 2^N$ we have $A^{[2^N]} = A^{[n]}$.
(i) Show that if $2^{N-1} \leq n$, then

$$N \leq 1 + \frac{\log n}{\log 2}.$$

(ii) Show that there is a constant K such that we only need $Kn^3 \log n$ comparisons to find the length of the shortest path between all pairs of n points.

This algorithm is due to Floyd.[16]

Floyd's algorithm has a particular advantage. Suppose that we allow negative distances. More formally, we allow d_{ij} to be a positive or negative real number or ∞. (As before, we take $d_{ij} = \infty$ if there is no direct path from A_i to A_j. We still demand $d_{ii} = 0$.) We wish to find the minimum value of

$$d_{i(1)i(2)} + d_{i(2)i(3)} + \cdots + d_{i(r-1)i(r)}$$

with $i(1) = i$ and $i(r) = j$. The first problem is that there need be no minimum.

Exercise 5.5.13 *Let $d_{11} = d_{22} = 0$ and $d_{12} = 1$ and $d_{21} = -2$. Compute*

$$d_{i(1)i(2)} + d_{i(2)i(3)} + \cdots + d_{i(r-1)i(r)}$$

where $i(2u) = 2$ and $i(2u - 1) = 1$.

Definition 5.5.14 *Consider points A_1, A_2, ..., A_n. Suppose that the direct 'distance' from A_i to A_j is d_{ij}. We say that $A_{j(1)}$, $A_{j(2)}$, ..., $A_{j(s)}$ is a distance decreasing cycle if $j(s) = j(1)$ and*

$$d_{j(1)j(2)} + d_{j(2)j(3)} + \cdots + d_{j(s-1)j(s)} < 0.$$

Lemma 5.5.15 *Consider points A_1, A_2, ..., A_n. Suppose that the direct 'distance' from A_i is d_{ij}. If there is no distance decreasing cycle, then (if there is any path joining A_i to A_j) there is a path of minimum length.*

Proof The condition that there are no distance decreasing cycles tells us that, given any path from A_i to A_j in which some town is visited twice, we can find a shorter path (or one of the same length) in which each town is visited once (since leaving out loops will not increase the length of the path). But there are only a finite number of paths in which each town is visited once, so there must be one path joining A_i to A_j of minimum length.   ■

Exercise 5.5.16 *(i) Show that, if there is a length decreasing cycle, then there are two towns which can be joined by paths of arbitrarily large negative length.*

(ii) Show that, if there is a path connecting any two towns and there is a length decreasing cycle, then any two towns can be joined by paths of arbitrarily large negative length.

---

[16] As with the other results in this section the version given is the simplest but not the best.

Exercise 5.5.17 *Suppose we have four towns* $A_1, \ldots, A_4$ *that* $d_{ii} = 0$ *for all* i, $d_{12} = d_{23} = 1$, $d_{14} = 4$, $d_{43} = -4$ *and* $d_{ij} = \infty$ *otherwise. Show that there is a shortest path from* A_1 *to* A_3, *but that the method described on page 167 will fail to find it.*

Floyd's algorithm copes easily with this problem.

Exercise 5.5.18 *Suppose we have* n *points* A_i *and the direct 'distance' from* A_i *to* A_j *is* d_{ij} *(where* d_{ij} *may be positive, negative or infinite). Let* D *be the* $n \times n$ *matrix with entries* d_{ij} *and let* $d_{ij}^{[n]}$ *be the* (i, j)th *entry in the matrix* $D^{[n]}$.

(i) Explain why the system has length a decreasing cycle if and only if $d_{kk}^{[n]} < 0$ *for some* k.

(ii) Suppose that $d_{kk}^{[n]} \geq 0$ *for all* k. *Explain why* $d_{kk}^{[n]} = 0$ *for all* k. *Explain why, if* $d_{ij}^{[n]} = \infty$, *there is no route of any kind from* A_i *to* A_j. *Explain why, if* $d_{ij}^{[n]} \neq \infty$, $d_{ij}^{[n]}$ *is the length of the shortest path from* A_i *to* A_j.

As an application, consider the problem of changing money. Suppose we have n currencies and the exchange rate between currencies is p_{ij} units of currency j in return for one unit of currency i. If we start with one unit of currency $i(1)$, exchange it for currency $i(2)$, exchange the result for currency $i(3)$ and so on ending up in currency $i(r)$ then we will have

$$p_{i(1)i(2)}\, p_{i(2)i(3)} \cdots p_{i(r-1)i(r)}$$

units in the final currency. If we have money in currency 1 and want money in currency n, then we want a chain of currencies with $i(1) = 1$ and $i(r) = n$ which maximises that product.

As we have done before, we change a problem on multiplication into one on addition by taking logarithms. Our object is to maximise

$$\log\left(p_{i(1)i(2)}\, p_{i(2)i(3)} \cdots p_{i(r-1)i(r)}\right) = \log p_{i(1)i(2)} + \log p_{i(2)i(3)} + \cdots$$
$$+ \log p_{i(r-1)i(r)}$$
$$= -\left((-\log p_{i(1)i(2)}) + (-\log p_{i(2)i(3)}) + \cdots + (-\log p_{i(r-1)i(r)})\right)$$

and this is the same problem as seeking to find the shortest path from 'town' A_1 to A_n if the 'distance' between A_i and A_j is $d_{ij} = -\log p_{ij}$ for $i \neq j$.

Since $d_{ij} = -d_{ji}$, some of the d_{ij} will be negative so we have to use some method like Floyd's and bear in mind that there may not be a shortest path. However, if our algorithm tells us that there is no shortest path, it also tells us that there exists a distance decreasing cycle, that is to say that we can find $i(1)$, $i(2), \ldots, i(r)$ such that $i(1) = i(r)$ and

$$0 > d_{i(1)i(2)} + d_{i(2)i(3)} + \cdots + d_{i(r-1)i(r)}$$
$$= \log \left(p_{i(1)i(2)} p_{i(2)i(3)} \cdots p_{i(r-1)i(r)} \right)$$

and so with

$$p_{i(1)i(2)} p_{i(2)i(3)} \cdots p_{i(r-1)i(r)} > 1.$$

By going through the indicated sequence of exchanges we can end up with more money than we started with — a perfect illustration of arbitrage.[17]

Exercise 5.5.19 *In the nineteenth century, many countries used gold coins for large units of currency and silver coins for small change. France and several other countries used a system in which a gold coin could be exchanged for $15\frac{1}{2}$ times its weight in silver coins. In the United States you could exchange precious metal for its equivalent weight in coins, elsewhere you could the same but you would lose slightly on the transaction. Until 1834 the United States used an exchange ratio 1 : 15 (so that a gold coin could be exchanged for 15 times its weight in silver coins) and people complained that there were very few gold coins in circulation. In 1834 the United States changed the ratio to 1 : 16 and this was followed by a scarcity of small change. Explain this.[18]*

Exercise 5.5.20 *(i) Suppose we have a set of towns A_1, A_2, ..., A_n. Suppose that all distances are positive and that, if $i \neq j$, at least one of d_{ij} or d_{ji} must take the value ∞ (so all routes are one way). Suppose we want the longest route from A_1 to A_n. What conditions are required to make this problem sensible? How would you solve this problem in the case when it is sensible? (You are asked to provide a reasonable method and not to worry if it is best possible.[19])*

(ii) It is not usual to build a house from the roof down. More generally, there are certain tasks which have to be completed before others can be started. How can you use the ideas of (i) to find out how long it will take to build a house? Can you identify tasks which, if they take longer than expected, will delay the completion of the building?

Exercise 5.5.21 *In this exercise we look at at a slightly different type of problem. Suppose we have n towns A_i and the price of building a road between the*

---

[17] And one of the oldest. The word arbitrage comes from the French term for this kind of exchange.

[18] Like many simple stories, this one becomes more complicated when looked at more closely.

[19] When the problem is sensible, it has rather special properties which can exploited to produce much faster algorithms than the kind suggested by the previous discussion.

town A_i and the town A_j is d_{ij}, where $d_{ij} = d_{ji} \geq 0$ for all $i \neq j$. We want a cheapest set of roads such that it is possible to travel from every town to every other town by some route.

(i) Find a cheapest solution when $n = 6$ and $d_{12} = d_{23} = d_{34} = d_{45} = d_{46} = 1$ while $d_{ij} = 10$ for all other $i \neq j$.

(ii) By using induction on n, or otherwise, show that the following algorithm produces a cheapest path. As a first step, find the smallest value of d_{ij} and choose two towns A_r and A_s such that d_{rs} takes this value. Build a road between A_r and A_s and call the linked towns the city. At the mth step we have a city of m linked towns. Find a town A_k closest to the city (that is find j, k such that A_k is not in the city, A_j is in the city and $d_{kj} \leq d_{KJ}$ whenever A_K is not in the city and A_J is). Build a road joining k to j and add A_k to the city. Stop when all towns are in the city.

(iii) Find a rough estimate of the number of comparisons needed for a unsophisticated application of the method of (ii).

(iv) By using induction on n, or otherwise, show that the following algorithm also produces a cheapest path. At each stage look for the shortest unbuilt road which will not produce a circuit and build it. (Note that during the process we will usually have several disconnected groups of linked towns.)

(v) Suppose that you are given a list of the possible $n(n-1)/2$ pairs of towns in order of the costs of joining them (something like (A_3, A_7) has the cheapest link, (A_2, A_5) the next cheapest, and so on). Adapt the algorithm of (iv) to produce a cheapest path.

(vi) Find a rough estimate of the expected number of comparisons needed if we first perform quicksort and then use the algorithm of (iv).

(vii) In the 'travelling salesman problem' we consider a salesperson who wishes to travel to each of n towns, visiting each one exactly once and returning to where she started. The cost of travelling between the town i and town j is d_{ij} where $d_{ij} = d_{ji} \geq 0$ for all $i \neq j$. She wishes to find a cheapest route. Find a cheapest route if the d_{ij} are as in (i), taking care to show that you have a cheapest route.

Nobody knows of any algorithm for solving this problem in An^m operations for any A or any m for any reasonable meaning of operations. Most people think there is no such algorithm, but no one can prove this. A resolution of this impasse would be a major mathematical advance.[20]

---

[20] And would entitle the solver to one of the million dollar prizes offered by the Clay Mathematics Institute.

6

Other people

6.1 Marrying

So far in this book, we have considered the question of finding the best outcome for a single person under a fixed set of rules. What is the best way to bet, given the appropriate odds and probabilities? Should I take out an annuity? What is the shortest route from A to B?

Life becomes much more complicated when there are many people with different goals and the action of one person changes the rules for the others. In this chapter we shall see that, even in these circumstances, mathematics can sometimes provide insight. We shall also see that problems arise which lie outside the province of the mathematician.

We start by looking at problems of the following type. Suppose we wish to form $2n$ children into pairs. If we match Amber with Bertha and Caroline with Delia but Amber prefers Caroline to Bertha while Caroline prefers Amber to Delia, then the pairing is unstable since Amber and Caroline would both prefer to break up with their present partners and form a pair together. If, however, Amber prefers Caroline to Bertha but Caroline prefers Delia to Amber this particular event will not happen (though there may be other ways in which the pairing is unstable).

Our problem is the following.

The Kindergarten Problem *Is it always possible to arrange $2n$ children in stable pairs (i.e. so there are not two children in different pairs who would prefer each other to their present partner)?*

The following problem is closely related and requires even less explanation.

The Marriage Problem *Is it always possible to arrange n ladies and n gentlemen in stable pairs (i.e. match each lady with a gentleman in such a way that there do not exist one lady and one gentleman in different pairs who would prefer each other to their present partner)?*

Exercise 6.1.1 *Without reading further, attempt to guess whether the two problems are soluble. If you guess that one is soluble but the other not, try and explain why one should be soluble but the other not.*

The discussion of the kindergarten problem turns out to be quite easy. Let us write $B > C > D$ to mean that B is preferred to C and C is preferred to D. Consider the table of preferences.

A's preferences	$B > C > D$
B's preferences	$C > A > D$
C's preferences	$A > B > D$

(It turns out that poor old D's preferences do not matter.)

If we start by matching A to B and C to D, then the pairing will break down because B and C prefer each other to their present partners. We try the pairing (B, C), (A, D) but this will break down because C and A prefer each other to their present partners, . . .

Exercise 6.1.2 *Check that each of the suggested pairings break down in the way indicated by the arrow.*

$$\begin{matrix} (A, B) \\ (C, D) \end{matrix} \longrightarrow \begin{matrix} (B, C) \\ (A, D) \end{matrix} \longrightarrow \begin{matrix} (C, A) \\ (B, D) \end{matrix} \longrightarrow \begin{matrix} (A, B) \\ (C, D) \end{matrix}$$

Since every pairing breaks down, there is no solution to the kindergarten problem in this case.

Exercise 6.1.3 *Write down a set of preferences for A, B, C and D for which the kindergarten problem is soluble. Write down a solution and explain why it is a solution.*

However, Gale and Shapley showed that the marriage problem is always soluble.[1] They proved their result directly by giving an algorithm to solve the problem. Here it is.

We throw a party and, at a certain point, invite ladies to stay in the ball room and the gentlemen go outside.

Each gentleman enters in turn. He proposes to his preferred lady. If accepted, he stays with her as her fiancé. If rejected, he proposes to his next preferred. If accepted, he stays with her as her fiancé. If rejected he proposes to his next preferred and so on until he is accepted.

---

[1] They announced this remarkable result in a very readable paper [22].

There are two further rules.

(1) If a lady has no fiancé she must accept any proposal. The next gentleman now enters the room and begins a round of proposals.

(2) If a lady with a fiancé receives a proposal she chooses whichever of her current fiancé or the proposer she prefers. If she accepts the new proposer she terminates her previous engagement and the newly rejected swain proceeds to offer his heart to his next preferences until accepted.

If all the gentlemen have fiancées the process stops.

Exercise 6.1.4 *Suppose the gentlemen are Albert, Bertram, Charles and David and the ladies are Josephine, Katherine, Louise and Mary. Suppose their preferences are as follows.*

A's preferences	$j > k > l > m;$	*j's preferences*	$D > B > C > A$
B's preferences	$j > m > k > l;$	*k's preferences*	$B > A > C > D$
C's preferences	$j > k > m > l;$	*l's preferences*	$B > C > A > D$
D's preferences	$j > k > l > m;$	*m's preferences*	$A > B > C > D$

Apply the algorithm, continuing until it stops.

When we applied the algorithm in Exercise 6.1.4, it terminated. We need to show that this will always happen. To see this, observe that each man will make at most n proposals, since he never proposes to a lady who has rejected him, and so there can be at most n^2 proposals in total.

Now that we know that the algorithm terminates, we need to show that it terminates at a solution for the stable marriage problem. Observe that, when the algorithm terminates, each gentleman has been turned down by any lady whom he prefers to his present fiancée. Since turning him down, the ladies may have changed fiancés but, since each new fiancé is preferred to the old, the ladies who have turned him down will still prefer their present fiancé to him. Thus, if we marry off the pairs, every lady will prefer their husband to any errant husband who prefers them to their own wife.

Thus we have a complete solution to the marriage problem ... or do we?

Exercise 6.1.5 *Is there any aspect of the mathematical problem that we have failed to cover?*

Exercise 6.1.6 *Suppose that we stop our algorithm at some earlier point at which there are r pairs but everybody else has no partner. Show that we have a solution of the stable marriage problem for the r pairs.*

Exercise 6.1.7 *Consider the following suggested algorithm for the kindergarten problem.*

We choose n children and tell the remaining n children to leave the classroom.

Each of the children from outside enters in turn. They go to their preferred partner. If accepted, they form a pair. If rejected, they go to their next preferred. If accepted, they form a pair. If rejected, they go to their next preferred and so on until they are accepted.

There are two further rules.

(1) If a child has no partner they must accept any proposal. The next child from outside now enters the room and begins a round of proposals

(2) If a member of a pair receives a proposal, they choose whichever of their current partner or the proposer they prefer. If they accept the new proposer, the rejected child then proposes to their next choice.

If all the children are in a pair, the process stops.

Show that the process need not terminate. Explain why the argument we used to show that the marriage algorithm must terminate does not apply. Show that, if the process does terminate, the pairing it produces is stable.

Although it looks as though our discussion is complete, we should recall that, as we said before and will say again, mathematical theorems are like legal contracts. They say exactly what they say and not what we believe they say. The reader should do at least part (i) of the next exercise.

Exercise 6.1.8 *(i) Suppose the gentlemen are Albert and Bertram and the ladies are Josephine and Katherine. Suppose their preferences are as follows.*

$$A\text{'s preferences} \quad j > k; \quad j\text{'s preferences} \quad B > A$$
$$B\text{'s preferences} \quad k > j; \quad k\text{'s preferences} \quad A > B.$$

Find a stable solution by applying the algorithm. Now find a stable solution by applying the algorithm with the roles of the ladies and gentlemen interchanged. (So the ladies propose to the gentlemen.)

Which procedure is preferred by the gentlemen? Which procedure is preferred by the ladies?

(ii) Show that, in the general case, if each gentleman has a different first choice of lady, then the ladies' preferences do not affect the result of our algorithm.

(iii) Suppose that the gentlemen are Albert, Bertram, Charles and David and the ladies are Josephine, Katherine, Louise and Mary. Suppose their preferences are as follows.

$$A\text{'s preferences} \quad j > k > l > m; \quad j\text{'s preferences} \quad B > A > C > D$$
$$B\text{'s preferences} \quad k > j > l > m; \quad k\text{'s preferences} \quad A > B > C > D$$
$$C\text{'s preferences} \quad l > m > j > k; \quad l\text{'s preferences} \quad D > C > A > B$$
$$D\text{'s preferences} \quad m > l > j > k; \quad m\text{'s preferences} \quad C > D > A > B$$

Show that there are at least 4 *stable arrangements.*

(iv) Show that, if n is even, we can specify preferences for n ladies and n gentlemen in such a way that there are at least $2^{n/2}$ *stable solutions to the marriage problem. Show that, if n is odd, we can specify preferences for n ladies and n gentlemen in such a way that there are at least* $2^{(n-1)/2}$ *stable solutions to the marriage problem.*

(v) Show that we can specify preferences for n ladies and n gentlemen in such a way that there is only one stable solution to the marriage problem.

It turns out that our algorithm produces the most favourable result for the gentlemen.

Lemma 6.1.9 *In any stable solution to the marriage problem, each gentleman will be married to a lady whom he likes at most as much as the one he receives under our algorithm.*

At first reading, it is more important to understand the meaning of Lemma 6.1.9 than to worry about the details of the proof.

To prove Lemma 6.1.9, we recall that when our algorithm finishes, each lady prefers the husband she has to all the gentlemen she has previously rejected. The lemma is thus equivalent to the following statement.

Lemma 6.1.10 *If a gentleman A is rejected by a lady a during the execution of our algorithm, then there is no stable solution in which A and a are married.*

We prove Lemma 6.1.10 inductively.

Lemma 6.1.11 *Suppose that we apply our algorithm. If the kth rejection involves a lady a rejecting a gentleman A then there is no stable solution in which A and a are married.*

Proof Suppose that the result is true for all $k \leq m$. Suppose that at the $m + 1$th rejection Agatha rejects Albert. Then Agatha must have rejected Albert in favour of someone else, say Bertram. (Either Albert proposed and she preferred to stick with Bertram or Bertram proposed and she let go of Albert to take up Bertram.) Thus Agatha prefers Bertram to Albert and Bertram prefers Agatha to all those ladies who have not yet rejected him.

Suppose there is a stable solution in which Agatha and Albert are married. Then Bertram must be married to a lady, Belinda say, whom he prefers to Agatha (otherwise Agatha and Bertram would prefer each other to their actual partners). The last sentence of the previous paragraph tells us that, in our algorithm, Belinda must have rejected Bertram before the $m + 1$th rejection. By the inductive hypothesis, this means that there is no stable solution in which Bertram and Belinda are married. We have arrived at a contradiction with the second sentence of this paragraph and so shown that the result is true for $k = m + 1$.

A similar, but simpler, argument proves the result for $k = 1$ and completes the inductive argument. ∎

Exercise 6.1.12 *Write down the argument for $k = 1$.*

Exercise 6.1.13 *At first glance, the result of our algorithm might appear to depend on the order in which the gentlemen enter the ball room. Explain why Lemma 6.1.9 shows that this is not the case.*

It will come as no surprise that our algorithm gives the worst possible result for the ladies.

Lemma 6.1.14 *In any stable solution to the marriage problem, each lady will be married to a gentleman whom she likes at least as much as the one she receives under our algorithm.*

Proof Suppose that, under our algorithm, Agatha is married to Albert and, in some other stable solution, Agatha is married to Bertram and Albert to Belinda. If Agatha prefers Albert to Bertram, then, in order to ensure stability, Albert must prefer Belinda to Agatha contradicting Lemma 6.1.9, which says that our algorithm always produces the best result for the gentlemen. ∎

When there are many solutions to the stable marriage problem, it is natural to ask for the 'fairest solution'. A look at Exercise 6.1.8 (iii) should convince the reader that there is unlikely to be any satisfactory answer to this question.[2] Some stable solutions are better for some people than others, but there is no mathematical way of determining whether a solution in which Josephine gets her first choice and Louise her second is preferable to one in which Louise gets her first choice and Josephine her second.

The present author does not believe that our discussion bears any relation to how people choose marriage partners or that we would be any happier if it did.[3] However, it does bear some relation to situations like application to university. Here, each candidate applies to several universities, but can attend only one

---

[2] Knuth discusses these matters and other interesting developments in his book [35], a particularly happy marriage of subject and author.

[3] The song

> If you were the twenty-fourth girl on my list,
> And I was your sixteenth boy

does not have the right ring to it.

of them. The candidates can rank the universities in order of preference and the universities can rank the candidates in order of preference. We say that an assignment of students to universities is stable if there does not exist a student a at university A and a student b at university B such that a would prefer to be at B and B would prefer student a to student b.

Suppose, for the moment, that the total number of places offered equals the number of candidates applying and that candidates prefer going to any of the universities to not going to any university. If a university A offers k places, we split that university into k shadow universities $A_1, A_2, \ldots, A_k$ each with the same order of preference among the students. The students rank these shadow universities in such a way that, if they prefer university A to university B, they prefer the shadow university A_r to the shadow university B_s. Any stable solution for the marriage problem, with the shadow universities playing the part of gentlemen and the students playing the part of ladies, will give a stable assignment of students to universities.

In general, there are more students who want to go to university than there are available places. We deal with this by introducing a fictitious university which all the students place at the bottom of their list of preferences and which offers exactly enough places so that the total number of places equals the number of students.

Exercise 6.1.15 *Show that, whichever order of preference among students we assign to the fictitious university, we get the same stable solutions.*

Exercise 6.1.16 *(i) Explain how to modify our methods if some candidates do not wish to go to some universities under any circumstances and some universities will not take some candidates under any circumstances (even if this leaves places unfilled).*

(ii) Sometimes, universities offer a certain fixed number of places with scholarships and a certain number without. Explain how to modify our methods to cope with this complication.

The Gale–Shapley algorithm is used by American medical schools to assign internships and has even, I believe, survived legal challenge. Why is it not used more frequently?

Consider 100 universities each offering 100 places. For the algorithm to operate, each university must rank all 10 000 students in order although, ultimately, it will only take 100. We should not be surprised if they are unwilling to do this.[4]

---

[4] However, this is not the final word on the matter. A modified version of the Gale–Shapely algorithm has been introduced for the very large New York school system [1].

If universities need to sort through very large numbers of students compared with the number of places on offer, they frequently use some automatic ranking procedure such as the rank of each candidate in a national exam. In this case, the objection of the previous paragraph ceases to hold and all the universities will rank students in the same order. However, the problem of finding a stable assignment becomes much less interesting.

Exercise 6.1.17 *Suppose that all the ladies in the marriage problem have the same order of preference among the gentlemen. Show that there is only one stable solution and that it may be attained by the most favoured gentleman choosing his favourite among all the ladies, then the second most favoured choosing his favourite among the remainder and so on.*

Explain the corresponding method for universities choosing candidates if all universities rank the candidates in the same order.

Several countries use more or less close approximations to the method discussed in Exercise 6.1.17.

Exercise 6.1.18 *In our discussion of the stable marriage problem, we have assumed that the ladies and gentlemen have strict preferences (so that Agatha either prefers Arnold to Bertie or Bertie to Arnold but cannot like them both equally). Suppose that we allow non-strict preferences (so that Agatha can like both Arnold and Bertie equally). Show that we can still match each lady with a gentleman in such a way that there do not exist one lady and one gentleman in different pairs who would strictly prefer each other to their present partner. Does this remain true if we replace 'strictly prefer each other to' by 'like each other at least as much as'?*

We discuss the number of proposals required by the marriage algorithm in Theorem 9.3.19 and Exercise 9.3.20.

6.2 Voting

A society of mathematicians wishes to have dinner at a restaurant. The proprietor tells them that he can give them a special cheap rate if they all order the same dishes. All the mathematicians want the same first and second courses, but they may disagree about the dessert.

The proprietor hands them the dessert menu and explains that not all the items may be available but, if the mathematicians place the dishes in order of preference, he will give them their favourite among the items available. The mathematicians decide to establish their order of preference by voting. They decide that each participant will write down their order of preference. They agree to follow the following rules.

(1) Each participant, when presented with a choice between A and B, will prefer A to B or will prefer B to A.

(2) If someone prefers A to B and B to C, then they will prefer A to C. (This is called transitivity.)

They want a set of rules which takes everybody's complete list of preferences and delivers 'the society's list of preferences'. They agree that the society's list of preferences should obey the following rules.

(1') If everybody prefers A to B, then the society prefers A to B. (We call this the unanimity rule.)

(2') If the society prefers A to B and B to C, then the society prefers A to C. (Transitivity.)

(3') The question of whether the society prefers A to B should not depend on its views about a third matter C. (This is called the principle of indifference to irrelevant alternatives.[5])

Thus, if the society prefers apple pie to plum duff, then the subsequent announcement, by the proprietor, that profiteroles are also available should not change the society's preference for apple pie over plum duff.

The mathematicians now sit down to try and work out an appropriate voting method.

Exercise 6.2.1 *Try to think of some appropriate method. Remember that there are several options and several voters. Think about whether your methods satisfy the conditions (1') to (3'). (If you cannot come to a conclusion, do not worry, this is what we are about to discuss.)*

Exercise 6.2.2 *Suppose that there are only two options A and B. Explain why conditions (2') and (3') are automatically satisfied. Show that the rule 'the society prefers A to B if at least half the voters prefer A to B, otherwise the society prefers B to A' satisfies (1').*

To see why we should expect problems, observe that, if there are n options, then there are $n!$ ways of placing the desserts in order of preference (we can choose the favourite in n ways, the second favourite in $n - 1$ ways and so on). If there are m mathematicians, there are $(n!)^m$ different voting patterns which could give rise to $n!$ different 'society's orders'. Thus a voting method can be considered as a function $f : X \to Y$ from a set X containing $(n!)^m$ points to a set Y consisting of $n!$ points. When n and m are large, Y contains many points but X contains very many more.

---

[5] Or independence of irrelevant alternatives.

Table 6.1. *A difficult vote*

	A	B	C	D	E	F
first preferences:	a	a	b	b	c	c
second preferences:	b	c	c	a	a	b
third preferences:	c	b	a	c	b	a

It is also clear that, for some voting patterns, the choice of the society's order of preference will appear arbitrary. As an example, suppose that there are three desserts a, b and c and six voters A to F with the preferences shown in Table 6.1. For this distribution of preferences any voting method must break the symmetry of the voter's choices.

For these and other reasons, voting methods tend to be rather complicated. The *Cambridge University Statutes and Ordinances* take three pages to describe its single transferable vote system.[6]

In order to understand their chosen voting method, the mathematicians program the rules into a computer (a 'voting machine') and see what the machine produces in response to various choices of lists. *Note that these lists do not correspond to the actual lists of the mathematicians but are simply intended to help understand how the rules operates.* In what follows, we assume that the society's voting method obeys all the rules we have set out, and try to see what the 'voting machine' will do.

First we look for the smallest set of people who can get their way on some issue. The answer is rather surprising.

Lemma 6.2.3 *If there are at least three options, then there exist two options A and B and a voter x such that, if x prefers A to B and everybody else prefers B to A, the society prefers A to B.*

Proof For every pair of options C and D, we know that if everybody prefers C to D, then the society prefers C to D (unanimity rule). Thus there must exist a smallest set of people,[7] call it $\mathcal{E}(C, D)$, such that if the voters in $\mathcal{E}(C, D)$ prefer C to D and everybody else prefers D to C the society prefers C to D.

Choose A and B so that $\mathcal{E}(A, B)$ is as small as possible. We claim that $\mathcal{E}(A, B)$ contains exactly one person.

---

[6] The Mathematics Faculty asked for exemption on the grounds that it did not understand the rules, but the University central authorities refused to budge.

[7] That is to say, a set containing the fewest people. If there are several such sets, we choose one of them.

Suppose not. Let $y \in \mathcal{E}(A, B)$ and consider an option $C \neq A, B$. What will be the result if the voters have the following preferences?

y's preferences	$A > B > C$
$\mathcal{E}(A, B) \setminus \{y\}$'s preferences	$C > A > B$
everybody else's preferences	$C > B > A$

If the society prefers A to C, then, since y prefers A to C and everybody else prefers C to A, we see that $\mathcal{E}(A, C)$ contains only one person. This contradicts our two assumptions that $\mathcal{E}(A, B)$ is as small as possible and $\mathcal{E}(A, B)$ contains more than one person.

Thus, with the given preferences, the society prefers C to A. But, with the given preferences, the voters in $\mathcal{E}(A, B)$ (including y) prefer A to B and everybody else prefers B to A. Thus the society prefers A to B. By transitivity, since the society prefers C to A and A to B, it follows that the society prefers C to B. But only members of $\mathcal{E}(A, B) \setminus \{y\}$ prefer C to B and everybody else prefers B to C. Thus $\mathcal{E}(C, B)$ contains fewer people than $\mathcal{E}(A, B)$ and we have a contradiction.

We have shown that $\mathcal{E}(A, B)$ has exactly one member. We call that member x to complete the proof. ∎

Exercise 6.2.4 *Consider the voting systems you thought of for Exercise 6.2.1. In each case, either show that Lemma 6.2.3 holds or show that one of the conditions (1′) to (3′) does not hold.*

The next lemma shows that Lemma 6.2.3 is a very powerful result.

Lemma 6.2.5 *Suppose there exist two options U and V and a voter x such that, if x prefers U to V and everybody else prefers V to U, the society prefers U to V.*

(i) If W is an option and $W \neq U, V$, then, if x prefers U to W, the society prefers U to W regardless of the preferences of everybody else.

(ii) If W is an option and $W \neq U, V$, then, if x prefers W to V, the society prefers W to V regardless of the preferences of everybody else.

Proof (i) Let the set of people (apart from x) who prefer U to W be $\mathcal{F}$ and the set of people who prefer W to U be $\mathcal{G}$. We set out the preferences in the following table.

x's preferences	$U > W$
$\mathcal{F}$'s preferences	$U > W$
$\mathcal{G}$'s preferences	$W > U$

By the principle of indifference to irrelevant alternatives, the society's preferences between U and W will be the same, whatever its members' views on V. In particular the society's preferences will be the same if the preferences of its members are those shown in the following table.

x's preferences	$U > V > W$
$\mathcal{F}$'s preferences	$V > U > W$
$\mathcal{G}$'s preferences	$V > W > U$

We observe that, with these preferences, x prefers U to V and everybody else prefers V to U so, by hypothesis, the society prefers U to V. We also note that everybody prefers V to W, so by the unanimity rule, the society prefers V to W. Since the society prefers U to V and V to W, we see, by transitivity, that society prefers U to W.

(ii) Left to the reader as a strongly recommended exercise. ∎

The next lemma is a trivial consequence.

Lemma 6.2.6 *Suppose there exist two options U and V and a voter x such that, if x prefers U to V, then the society prefers U to V regardless of the preferences of everybody else.*

(i) If W is an option and $W \neq U$, V, then if x prefers U to W, the society prefers U to W regardless of the preferences of everybody else.

(ii) If W is an option and $W \neq U$, V, then if x prefers W to V, the society prefers W to V regardless of the preferences of everybody else.

Proof The hypothesis of Lemma 6.2.6 is stronger than the hypothesis of Lemma 6.2.5. ∎

By using Lemmas 6.2.5 and 6.2.6 repeatedly we can parlay Lemma 6.2.3 into even more remarkable results.

Lemma 6.2.7 *Suppose there are at least three options and A, B and x are as in Lemma 6.2.3.*

(i) If x prefers A to B, then the society prefers A to B, regardless of the preferences of everybody else.

(ii) If x prefers B to A, then the society prefers B to A, regardless of the preferences of everybody else.

(iii) Suppose C is any option. If x prefers C to A, then the society prefers C to A, regardless of the preferences of everybody else. If x prefers A to C, then the society prefers C to A, regardless of the preferences of everybody else. The same results hold with A replaced by B.

(iv) Suppose C and D are any distinct options. If x prefers C to D, then the society prefers C to D, regardless of the preferences of everybody else.

Proof (i) Suppose that x prefers A to B. Let C be an option with $C \neq A, B$. By the principle of indifference to irrelevant alternatives, the society's preferences between A and B will be unchanged whatever its members views on C. Thus we may assume that x prefers A to C and C to B. By Lemma 6.2.5 (i) with $U = A$, $V = B$ and $C = W$, it follows that the society prefers A to C and, by Lemma 6.2.5 (ii), the society prefers C to B. Since the society prefers A to C and C to B it follows, by transitivity, that the society prefers A to B.

(ii) Let C be an option with $C \neq A, B$. By Lemma 6.2.5 (i) with $U = A$, $V = B$ and $W = C$, we know that if x prefers A to C, then the society prefers A to C regardless of the preferences of everybody else. Using Lemma 6.2.5 (i) again, but with $U = A$, $V = C$ and $W = B$, it follows that if x prefers B to A, then the society prefers B to A regardless of the preferences of everybody else and this is the required result.

(iii) Observe that $C \neq A$. If $C = B$, we know from (i) that if x prefers C to A, then the society prefers C to A, regardless of the preferences of everybody else. If $C \neq B$, then we know from part (ii) that if x prefers B to A, the society prefers B to A, regardless of the preferences of everybody else. Using Lemma 6.2.5 (ii) with $U = B$, $V = A$ and $W = C$, we see that if x prefers C to A, the society prefers C to A, regardless of the preferences of everybody else. Thus, if C is a possible option, and x prefers C to A, the society prefers C to A, regardless of the preferences of everybody else.

The remaining statements follow similarly.

(iv) If either of C or D is A or B, the result has already been proved directly in (iii). If not then, by the principle of indifference to irrelevant alternatives, we may assume that x prefers C to A, A to B and B to D. By earlier parts of this lemma, it follows that the society prefers C to A, A to B and B to D and so, by transitivity, the society prefers C to D. ■

Thus we can replace the book of rules by a single rule 'x gets to choose'. We have proved Arrow's impossibility theorem[8] (sometimes called Arrow's paradox).

---

[8] I cannot resist retelling an anecdote from Arrow's wartime service in the Weather Division of the US Army Airforce. His group subjected the prevailing weather prediction techniques to statistical test against a simple technique based on historical averages for the date in question. Finding that the prevailing techniques were not significantly more reliable, several junior officers sent a memo to the general in charge suggesting that the unit be disbanded. After a succession of such memos, the general's secretary is reported to have replied brusquely on his behalf 'The general is well aware that your division's forecasts are worthless. However, they are required for planning purposes'.

Theorem 6.2.8 *If there are three or more options, the only voting system which obeys conditions (1′), (2′) and (3′) (the unanimity rule, transitivity and the principle of indifference to irrelevant alternatives) is one in which the society's choice is the same as a particular individual.*

More briefly, if there are three or more options, the only voting system which obeys conditions (1′), (2′) and (3′) is a dictatorship!

In order to produce a coup de theatre, I did not announce the theorem before proving it. This is not the right way do things, but I hope that the reader will be sufficiently intrigued by the result to go back over the various steps in the proof.[9]

In my opinion, the key steps are Lemma 6.2.3 (there is one preference between some pair of options and one person x such that, if x has that preference and everybody else the other then x gets his way) and Lemma 6.2.5 (x can use the principle of indifference to show that there is a class of preferences where he gets his way regardless of other people's choices).

6.3 Preferring

When Arrow proved his theorem it created a great stir, but time has reduced its impact. It certainly shows that any voting system must produce an unsatisfactory result for some possible distribution of votes. (Unsatisfactory because of use of random methods to decide a result, failure to consider all options, failure of transitivity, failure to obey the principle of indifference to irrelevant alternatives and so on.) However, this fact was already known for all the standard systems. As we pointed out earlier, it is difficult to produce any satisfactory outcome for the distribution of preferences given in Table 6.1.

Sometimes people say that Arrow's theorem shows the impossibility of democracy, but this is only true if we interpret democracy in a very narrow sense. When small groups are run democratically, they usually only have to choose one course of action rather than to rank all possible courses of action. It is often true that such groups seek to make decisions by consensus rather than voting, even if this requires long discussion. We can list some of the advantages[10] of democratic decision-making for small groups.

---

[9] 'Allow me,' said Mr Gall. 'I distinguish the picturesque and the beautiful, and I add to them, in the laying out of grounds, a third and distinct character, which I call *unexpectedness*.'
   'Pray sir,' said Mr Milestone 'by what name do you distinguish this character, when a person walks round the grounds for a second time?' [Peacock, *Headlong Hall*.]

[10] The disadvantages of small-group democracy are not our concern. However, when I have an operation I prefer decisions to be made by the surgeon, rather than by the vote of all those present.

(1) The group has to take joint responsibility for its decisions.

(2) Since they had a part in making the decision, the members of the group may be more willing to implement it.

(3) Discussion may improve both the quality of the decisions and the understanding of that decision.

None of these advantages has much to do with the conditions of Arrow's theorem.

Large-scale democracy[11] is even further removed from the rarefied heights of the previous section. Again, we list some of the advantages.[12]

(1) It allows for peaceful changes of government.

(2) It allows everybody to express their views. The losers may be satisfied by the fact that they were allowed to state their case even if they lost it.

More simply, the purpose of democratic elections is to 'get the scoundrels out' and to 'let everybody have a good shout' but not to 'draw up a list of society's preferences'.

It is also true that even a very partial and imperfect democratic system changes the relation between governors and governed. A French visitor to England in the mid eighteenth century wrote that

> The proudest of Englishmen will converse familiarly with the meanest of his countrymen; he will take part in their rejoicings ... It is true that persons of higher rank find the common people necessary to realise their ambitious designs, and it is not uncommon, at elections, and those for members of parliament especially, to see the lowest of citizens receiving letters from the most illustrious candidates, in which, in the most polite way possible, they solicit the favour of their votes; and when these agree to their request, they are not long in receiving a letter, in which the candidate expresses his gratitude in the warmest terms. Have we not lately seen the Duchess of Devonshire lavishing, on such an occasion, not only gold but kisses?
>
> *(Quoted in [52], Chapter 7)*

If we look round the various modern democracies, we see a great variety of voting methods (that is to say, electoral systems) which appear to function to the reasonable satisfaction of the various electorates. It is worth remarking that, although it may be possible to choose a voting system which favours one particular option over others for a particular set of voters at a particular time, it is very hard to control future options. Provided that a voting scheme is fixed, the belief that it may favour your schemes today can be balanced by the hope that it will favour my schemes tomorrow.

---

[11] We shall be thinking about *representative democracy* in which the population elects a small group of representatives to make decisions on their behalf.

[12] The fact that large-scale democracies are historically very rare suggests that there may be disadvantages, but, again, these do not concern us here.

Arrow's theorem hinges on the fact that every pattern of voting has to be allowed for and conditions $(1')$ to $(3')$ must apply whatever patterns are considered. If a group of people share similar aims and are not too concerned with the effect of every decision (provided that, in general, things are going reasonably) they are unlikely to be worried by Arrow's theorem. If large groups of voters have irreconcilable aims, then the democratic process may collapse. The most notable examples are, perhaps, the American Civil War and the destruction of the Weimar Republic, but there are other examples both great and small.[13] I know of no case that can be plausibly associated with Arrow's theorem.

Nonetheless, Arrow's theorem is a very severe blow to philosophers, economists, political theorists and people like the present author who wish to talk about society as a whole. We are used to making statements like 'the British prefer change to come locally' or 'Europe looks to the past and the US looks to the future'. We treat societies as having desires and preferences in the same way as individuals. But Arrow's theorem shows that there is no reasonable way of aggregating the preferences of the individuals which make up a society into 'super-preferences' which we can assign to that society. Perhaps we should not be surprised by this: a society of cyclists does not itself ride a bicycle, so we should not expect a society made up of people with preferences to have preferences. Whenever we feel impelled to make grandiloquent assertions about the wishes of large groups of people, we should remember Arrow's theorem and moderate our voices.[14]

Up to now, in our discussion of marrying and voting, we have taken it as given that if we prefer A to B and B to C we must prefer A to C. Should we have accepted this without discussion?

Suppose that we have three fair six-sided dice.[15] These are: A with faces marked $(2, 2, 4, 4, 9, 9)$, B with faces marked $(3, 3, 5, 5, 7, 7)$ and C with faces marked $(1, 1, 6, 6, 8, 8)$. I allow you to choose which ever die you prefer and

---

[13] The great university library of Louvain was destroyed during the deliberate burning of Louvain by the German army in 1914. It was reconstituted as a monument to civilised values after the war. In 1940, during the German invasion, it was again destroyed by fire. It was again reconstituted as a monument to civilised values. After 1968, it was decided that Flemish speaking and French speaking students could not share the same university. A new university was built at Louvain-la-Neuve and the library was split. Documents and books with an odd registration number stayed in Louvain, those with even registration numbers moved to Louvain-la-Neuve.

[14] The reader who wishes to go further into these matters could start with the very nice book by Brams [8].

[15] I use the mnemonic that the markings are the rows of the magic square

$$
\begin{array}{ccc}
2 & 9 & 4 \\
7 & 5 & 3. \\
6 & 1 & 8
\end{array}
$$

you allow me to choose whichever of the remaining two I prefer. We then throw each die once and the one whose die shows the higher number wins. Which die should you choose?

Suppose you choose A and I choose B. The following table shows the winner for the various combinations of throws.

	2	4	9
3	B	A	A
5	B	B	A
7	B	B	A

We see that the probability that B beats A is $5/9$.

Exercise 6.3.1 *Check that C beats B with probability $5/9$ and C beats A with probability $5/9$.*

Once we have got over our surprise, we recall the many occasions in football when team A has beaten team B, team B has beaten team C and team C has beaten team A. Rather more to the point we recall the children's game of Scissors, Paper, Stone. The two players simultaneously put out their right hands, two extended fingers represent scissors, an open hand paper, and a clenched fist stone. Scissors cut paper, paper wraps stone and stone blunts scissors.

Exercise 6.3.2 *The dice paradox above is a descendant of a paradox of Steinhaus and Trybula published in 1959. Here is their example. Consider independent random variables X, Y, Z with*

$$\Pr(X = 1) = p, \ \Pr(X = 4) = 1 - p, \ \Pr(Y = 2) = 1,$$
$$\Pr(Z = 0) = 1 - p, \ \Pr(Z = 3) = p.$$

Calculate $\Pr(X > Y)$, $\Pr(Y > Z)$, $\Pr(Z > X)$ and find the value of p which maximises $\min \{ \Pr(X > Y), \Pr(Y > Z), \Pr(Z > X) \}$. Deduce that it is possible to find independent random variables U, V, W such that

$$\Pr(U > V) = \Pr(V > W) = \Pr(W > U) = \tau \text{ where } \tau = \frac{\sqrt{5} - 1}{2}.$$

The surprise arises because $\tau > 1/2$.

They also showed (see [64]) that, if U, V and W are independent random variables, then

$$\min \{ \Pr(U > V), \Pr(V > W), \Pr(W > U) \} \leq \tau,$$

but I do not know of any simple way of proving this.

Suppose that you are driving down a three-lane motorway when you see a traffic jam in the distance. You have to choose one lane while a very visible red car chooses another. You will be happy if you come to the end of the jam before the red car and unhappy if the red car reaches the end before you. Suppose the time in the jam is given by U, V and W for the three lanes. What will happen if you choose after the red car? What will happen if the driver of the red car has the same feelings about you as you do about him and there is ample opportunity to change lanes several times before arriving at the jam?

Exercise 6.3.3 *Here are a couple more examples of non-transitive dice.*

(i) Efron produced a very nice set of four fair six-sided dice A with faces marked $(0, 0, 4, 4, 4, 4)$, *B with faces marked* $(3, 3, 3, 3, 3, 3)$, *C with faces marked* $(2, 2, 2, 2, 6, 6)$ *and D marked marked* $(1, 1, 1, 5, 5, 5)$. *Show that A beats B, B beats C, C beats D and D beats A, each with probability* $2/3$.

(ii) Consider the following three fair six-sided dice: A with faces marked $(5, 6, 7, 8, 9, 18)$, *B with faces marked* $(2, 3, 4, 15, 16, 17)$, *C with faces marked* $(1, 10, 11, 12, 13, 14)$. *Check that each of the integers* 1 *to* 18 *occurs exactly once on a face. Show that A beats B with probability* $21/36$, *B beats C with probability* $21/36$ *and C beats A with probability* $25/36$.

Exercise 6.3.4 *The first person to study voting systems seriously was the French mathematician Condorcet.*[16] *He discovered the first voting paradox. Consider three voters A, B and C with the following preferences.*

	A	B	C
first preferences:	a	b	c
second preferences:	b	c	a
third preferences:	c	a	b

Show that the majority of the voters prefer a to b, b to c and c to a.

Show that, if these voters are first asked to vote between two alternatives and then vote between the winner of the first vote and the remaining option, then the remaining option will aways win. This kind of phenomenon accounts for the fact that assemblies may spend as much time debating the order in which issues are addressed as in debating the issues themselves.

It is well established that when people are asked their preferences between pairs of objects they will often produce cycles preferring A_1 to A_2, A_2 to $A_3, \ldots$ and A_n to A_1. (I prefer to eat a chocolate rather than an apple, I prefer to eat two chocolates rather than one, three chocolates rather than two, $\ldots$ and

[16] Condorcet went on to take an important part in the French Revolution. When still more radical politicians took over, he was outlawed. He went into hiding for eight months during which he wrote his *Esquisse d'un Tableau Historique des Progrès de l'Esprit Humain* proclaiming the equality of the sexes and the infinite perfectibility of the human race. He was found dead in his cell two days after his capture.

fifty chocolates rather than forty-nine but, frankly, I prefer an apple to having to eat fifty chocolates.) Mathematicians say that these are inconsistent preferences, but this is no answer to someone who, after careful reflection, maintains a set of cyclic preferences.[17]

Sometimes it is reasonable to expect consistent preferences, and the theorems of the previous two sections apply. Sometimes it is not reasonable to expect consistent preferences, and they do not.

The idea of non-cyclic preferences is closely linked to that of utility functions (see Section 3.5). If someone has a utility function, that is to say they can assign a real number (the 'utility') to every possible outcome of an action and always prefer the outcome with the largest utility, then it is clear that they cannot have cyclic preferences. Conversely, if someone can always decide which of two possible outcomes they prefer,[18] and does not have cyclic preferences, then it is plausible that they will have a utility function.[19] A selfish person might have a utility function depending on their personal pleasure, an unselfish person a utility function depending of the happiness of others and so on. Presented with a choice of actions, such individuals will seek to maximise the expected value of their utility function.

From this point of view, Arrow's theorem tells us that, even if the individuals in a group have utility functions, there may be no way of aggregating the individual utility functions into a utility function for the group in a way consistent with the conditions of Arrow's theorem.

---

[17] My father used to quote the scholar who annotated Goethe's 'She of all women I loved the most' with the words 'Here Goethe was wrong'.

[18] This is not a trivial condition. Those who remember Chesterton's parody of a newspaper editorial beginning 'Whatever we may think of the rights and wrongs of the vivisection of pauper children, we shall all agree that it should only be done, in any event, by fully qualified practitioners' (*The Flying Inn*) may feel that sometimes we have a duty not to choose.

[19] At its simplest, we can imagine asking such a person 'Would you rather have 1 dollar or an ice cream?' 'Would you rather have 2 dollars or an ice cream?' and so on. More sophisticated versions of this argument may be found in [67].

7

Simple games

7.1 Scissors, Paper, Stone

In the previous chapter we recalled the game of Scissors, Paper, Stone. In it, the two players simultaneously put out their right hands, two extended fingers represent scissors, an open hand paper, and a clenched fist stone. Scissors cut paper, paper wraps stone and stone blunts scissors.

How should we play this game? The answer depends on our opponent. If our opponent is a small child[1] we may observe that it never uses stone or that it never repeats the weapon it used in the previous round. You can use this information to ensure that you win more times than you lose.

Exercise 7.1.1 *Explain why, if you know that your opponent will choose from two specified weapons, you can arrange so that you never lose and sometimes win.*

It is more interesting to consider what we should do if faced by an opponent cleverer than ourselves.[2] Whatever plans we make, we must expect them to be anticipated. Under these circumstances, it makes sense to play at random, choosing each weapon with probability 1/3 independent of what has gone before. (We could, for example, throw a die and play scissors if the die shows 1 or 2, paper if the die shows 3 or 4 and stone if the die shows 5 or 6.) If our opponent plays stone, then with probability 1/3 we play scissors and lose, with probability 1/3 we play paper and win and with probability 1/3 we play stone and draw. Similar arguments apply if our opponent plays scissors or paper. Thus, if we follow the random strategy outlined, our chance of winning equals

---

[1] Some of my kindlier readers may be playing to lose rather than to win in this situation, but the principle remains the same.

[2] 'Across the gulf of space, minds that are to our minds as ours are to those of the beasts that perish, intellects vast and cool and unsympathetic regarded this earth with envious eyes.' [Wells *The War of the Worlds*]

our chance of losing. Since the game is symmetric and our opponent is cleverer than we are, we cannot do better than this.

Armed with this insight, we tackle a slightly more complicated game.

Example 7.1.2 *'Rhinestone' Roland is lounging on the deck of a Mississippi paddle steamer reading an old-fashioned but amusing[3] and informative book entitled* The Compleat Strategyst *[70] when he is approached by 'Cards' Collins who suggests a game of Matching Pennies. Roland says it is too hot for violent exercise. 'Well then', Collins replies, 'let us just lie here and speak the words "heads" or "tails" – and, to make it interesting, I'll give you \$30 when I call tails and you call heads and \$10 when it's the other way around. And – just to make it fair, you give me \$20 when they match.'*

We first look at this game from the point of view of Roland. Roland strongly suspects that Collins is cleverer than he is, so he decides to play a random strategy, calling heads with probability p and tails with probability $1 - p$. If Collins calls heads, then Roland loses 20 dollars if he calls heads and gains 10 if he calls tails. His expected winnings, if Collins calls heads, are thus

$$e_H(p) = -20p + 10(1 - p) = 10 - 30p.$$

If Collins calls tails, then the same kind of reasoning shows that Roland's expected winnings are

$$e_T(p) = 30p - 20(1 - p) = 50p - 20.$$

Roland believes that Collins is clever enough to guess whatever p he has chosen and will make whichever call minimises Roland's expected winnings. Thus Collins will call heads if $e_H(p) < e_T(p)$ and tails if $e_H(p) > e_T(p)$. If $e_H(p) = e_T(p)$, it does not matter which call Collins makes. Since

$$e_T(p) - e_H(p) = 80p - 30,$$

this means the following.

(A) If $p < 3/8$, Collins calls tails and Roland's expected winnings are $e(p) = e_T(p) = 50p - 20$.

(B) If $p > 3/8$, Collins calls heads and Roland's expected winnings are $e(p) = e_H(p) = 10 - 30p$.

(C) If $p = 3/8$, then Roland's expected winnings are $e(3/8) = e_H(3/8) = e_T(3/8) = -5/4$.

Roland chooses p to maximise his expected winnings $e(p)$ under the assumption that Collins can guess whatever p he chooses. By drawing a diagram,

---

[3] It points out that the Chinese version of 'Scissors cut paper, paper wraps stone, stone blunts scissors' is 'Man eats rooster, rooster eats worm, worm eats man'.

or observing that $e(p)$ increases as p increases from 0 to 3/8 and decreases as p increases from 3/8 to 1, we see that he will choose $p = 3/8$ and his expected winnings are then $-5/4$ (that is to say, he must expect to lose on average one dollar and 25 cents each time he plays) whatever Collins does.

Now let us change sides and look at this game from the point of view of Collins. Just as Roland thinks Collins is cleverer, so Collins thinks Roland is cleverer. He, too, decides to play a random strategy calling heads with probability q and tails with probability $1 - q$. His expected winnings are thus

$$f_H(q) = 20q - 30(1 - q) = 50q - 30,$$

if Roland calls heads and

$$f_T(q) = -10q + 20(1 - q) = 20 - 30q$$

if Roland calls tails. Since Collins believes that Roland is clever enough to guess whatever q he has chosen, he believes that Roland will make whichever call minimises Collins's expected winnings. Since

$$f_H(q) - f_T(q) = 80q - 50$$

this means the following.

(A) If $q > 5/8$, Roland calls tails and Collins's expected winnings are $f(q) = f_T(q) = 20 - 30q$.

(B) If $q < 5/8$, Roland calls heads and Collins's expected winnings are $f(q) = f_H(q) = 50q - 30$.

(C) If $q = 5/8$, then it does not matter what Roland does and Collin's expected winnings are $f(5/8) = f_H(5/8) = f_T(5/8) = 5/4$.

Collins chooses q to maximise his expected winnings $f(q)$ under the assumption that Roland can guess whatever q he chooses. By drawing a diagram, or observing that $f(q)$ increases as q increases from 0 to 5/8 and decreases as q increases from 5/8 to 1, we see that he will choose $q = 5/8$ and his expected winnings are then 5/4 (that is to say, he must expect to gain on average one dollar and 25 cents each time he plays) whatever Roland does.

A remarkable fact about the chosen strategies is that, if you were to go to Collins and tell him 'Roland is totally unable to guess what you will do but will choose heads 3/8th of the time', then Collins could not improve his strategy of choosing heads 5/8th of the time. Similarly, if you were to tell Roland 'Collins is totally unable to guess what you will do but will choose heads 5/8th of the time', then Roland could not improve his strategy of choosing heads 3/8th of the time.

Exercise 7.1.3 *Check the statements just made.*

Here is another example along the same lines.

Example 7.1.4 *I own a stand at the stadium where a cup final will be played. One of the semi-finals involves Foxton Athletic whose supporters wave a teddy bear and Shelford Dynamos whose supporters wear green rosettes. If Foxton get through to the final, I will be able to sell any number of teddy bears at €20 a bear but will have to throw away any green rosettes. If Shelford get through, I will be able to sell any number of rosettes at €5 but but will have to sell the teddy bears to a toy shop for €5 per bear. Bears cost €10 and rosettes €1. I must place any order for rosettes and bears before the result of the semi-final is known.*

Being a pessimist, I believe that whatever decision I make, fate will ensure the worst possible outcome. I therefore decide to imitate the Southern gentlemen of the previous example. I send €10 to the manufacturers with instructions to send me a teddy bear with probability p and otherwise to send me 10 green rosettes. If Foxton win, my expected profit is

$$e_F(p) = -10 + 20p,$$

but if Shelford win, my expected profit is

$$e_S(p) = -10 + 5p + 50(1 - p) = 40 - 45p.$$

Assuming that fate will arrange for Shelford to win if $e_F(p) > e_S(p)$ and for Foxton to win if $e_S(p) > e_F(p)$, my expected profit will be

$$e(p) = \min\{e_F(p), e_S(p)\} = \begin{cases} e_F(p) = -10 + 20p & \text{if } p \le 10/13 \\ e_S(p) = 40 - 45p & \text{if } p \ge 10/13. \end{cases}$$

Since $e(p)$ is maximised at $10/13$, I choose $p = 10/13$ and my expected profit is $e(10/13) = 70/13$ euros.

Exercise 7.1.5 *Check the statements just made.*

A little reflection shows that I do not have to indulge in the peculiar procedure I have just outlined to obtain an expected profit of €70/13 per €10 sent to the manufacturers. If I send money to the manufacturers and instruct them that $10/13$ is to be spent on teddy bears and $3/13$ on rosettes, I shall make a guaranteed profit of €70/13 per €10 sent to the manufacturers. I will then have followed exactly in the footsteps of the bettor described in Section 1.1 by placing a bet to maximise my winnings, whichever team wins the semi-final.

We close this section with a warning example.

Example 7.1.6 *In a training exercise, Colonel Schröder must take his troops through one of two passes. If he uses the first pass and it is undefended, it will take his troops 4 hours to reach their destination, but if it is defended, it will take them 6 hours to reach their destination. If he uses the second pass and it is undefended, it will take his troops 2 hours to reach their destination, but if it is defended it will take them 3 hours to reach their destination. Lieutenant Lukáš is told to delay Colonel Schröder as long as possible but is only given enough troops to defend one pass. Both men know all these facts.*

Inspection of the problem shows that there is a common sense solution. It is interesting to see what happens if we apply the methods of this section.

For the usual reasons, Colonel Schröder decides to use a random strategy and take the first pass with probability p and the second with probability $1 - p$. If Lukáš decides to defend the first pass the expected time it will take Colonel Schröder's troops to reach their destination will be

$$e_1(p) = 6p + 2(1 - p) = 2 + 4p$$

and if he decides to defend the second, the expected time will be

$$e_2(p) = 4p + 3(1 - p) = 3 + p.$$

Assuming that Lukáš can guess the chosen p and makes the correct decision, the expected time for Colonel Schröder's troops to reach their destination will be

$$e(p) = \max\{e_1(p), e_2(p)\} = \begin{cases} e_1(p) = 2 + 4p & \text{if } p \leq 1/3 \\ e_2(p) = 3 + p & \text{if } p \geq 1/3. \end{cases}$$

However, unlike our previous examples, $e(p)$ is an increasing function of p over the entire range $0 \leq p \leq 1$. Since Colonel Schröder wishes to minimise $e(p)$, he will take $p = 0$. Thus he will always take the second pass.

Exercise 7.1.7 *(i) Explain why it is clear without going through the argument just given that Colonel Schröder will always take the second pass. What will Lieutenant Lukáš do?*

(ii) Confirm that, if Lieutenant Lukáš uses the arguments of this section, he will indeed do what you say he will.

7.2 Scissors, Paper

In the previous section, we looked at special cases of the following game played between two players Rowena and Calum. Rowena has two options called row 1 and row 2 and Calum has two options called column 1 and column 2. Each player makes their choice and places it in an envelope. The envelopes are then opened. If Rowena has chosen row i and Calum has chosen column j, then Calum pays the amount a_{ij} (which may be negative) to Rowena. The object of Rowena is to maximise the expected sum paid to her and the object of Calum is to minimise that sum. Each player is sufficiently afraid of the other to adopt a random strategy and to assume that the other player will guess exactly what that strategy is.

This game is called a two-person, zero-sum, 2×2 game. It is called two-person because there are two players, zero-sum because one player's winnings are exactly equal to the other player's losses and 2×2 because it is associated with the matrix

$$\begin{pmatrix} a_{11} & a_{12} \\ a_{21} & a_{22} \end{pmatrix}.$$

If we start to analyse this general game, it soon becomes clear that the outcome depends on the relative sizes of the a_{ij}. We can reduce the number of cases to be considered by the following simple argument. First, observe that the game is essentially unaltered if we interchange the names of Rowena's options (that is to say, interchange the two rows). Similarly we can interchange the two columns. In this way we can ensure that a_{11} is the largest of the a_{ij}. The nature of the game is also unaltered if we exchange the roles of Rowena and Calum (that is, replace a_{ij} by a_{ji}) so we can ensure that $a_{12} \geq a_{21}$. Thus, we may assume that

$$a_{11} \geq a_{12} \geq a_{21} \text{ and } a_{11} \geq a_{22}.$$

Let us look at the game from Rowena's point of view. She is playing a random strategy, choosing row 1 heads with probability p and row 2 with probability $1 - p$. Her expected winnings are

$$e_1(p) = a_{11}p + a_{21}(1 - p) = a_{21} + (a_{11} - a_{21})p$$

if Calum chooses column 1 and

$$e_2(p) = a_{12}p + a_{22}(1 - p) = a_{22} + (a_{12} - a_{22})p$$

if Calum chooses column 2. She seeks to maximise

$$e(p) = \min\{e_1(p), e_2(p)\}.$$

There are two cases.

First case If $a_{12} \geq a_{22}$, then both $e_1(p)$ and $e_2(p)$ are increasing functions of p as p runs from 0 to 1 and so $e(p)$ must also be increasing. If Rowena chooses

$p = 1$, she will maximise her expected winnings. Her expected winnings will be a_{12}.

Second case If $a_{12} < a_{22}$, then, automatically, $a_{21} < a_{22}$ and

$$0 < a_{22} - a_{21} \le (a_{11} - a_{21}) + (a_{22} - a_{12}) = (a_{11} + a_{22}) - (a_{12} + a_{21})$$

so, writing[4]

$$\hat{p} = \frac{a_{22} - a_{21}}{(a_{11} + a_{22}) - (a_{12} + a_{21})} = 1 - \frac{a_{11} - a_{12}}{(a_{11} + a_{22}) - (a_{12} + a_{21})},$$

we have $0 < \hat{p} < 1$. Now

$$e_1(p) - e_2(p) = -(a_{22} - a_{21}) + \big((a_{11} - a_{21}) + (a_{22} - a_{12})\big)p$$

$$\begin{cases} < 0 & \text{if } 0 \le p < \hat{p}, \\ = 0 & \text{if } p = \hat{p}, \\ > 0 & \text{if } \hat{p} < p \le 1 \end{cases}$$

and so

$$e(p) = \begin{cases} e_2(p) = a_{22} + (a_{12} - a_{22})p & \text{if } 0 \le p < \hat{p}, \\ e_1(\hat{p}) = e_2(\hat{p}) & \text{if } p = \hat{p}, \\ e_1(p) = a_{21} + (a_{11} - a_{21})p & \text{if } \hat{p} < p \le 1. \end{cases}$$

It follows that $e(p)$ increases as p increases from 0 to $\hat{p}$ and decreases as p increases from $\hat{p}$ to 1, so Rowena will choose $p = \hat{p}$ with expected winnings

$$e_1(\hat{p}) = a_{11}\hat{p} + a_{21}(1 - \hat{p})$$

$$= \frac{a_{11}(a_{22} - a_{21})}{(a_{11} + a_{22}) - (a_{12} + a_{21})} + \frac{a_{21}(a_{11} - a_{12})}{(a_{11} + a_{22}) - (a_{12} + a_{21})}$$

$$= \frac{a_{11}a_{22} - a_{12}a_{21}}{(a_{11} + a_{22}) - (a_{12} + a_{21})}.$$

Exercise 7.2.1 *Before doing the algebra above (and on the various occasions during the algebra when he lost track of things), the author drew several diagrams.*

Sketch the graphs of $e_1(p)$, $e_2(p)$ and $e(p)$ as p runs from 0 to 1 on the same diagram, first in the case $a_{12} < a_{22}$ and then, in a new diagram, in the case $a_{12} > a_{22}$. Explain why it is obvious that in the first case, Rowena should choose $p = 1$ and, in the second case, she should choose $p = \hat{p}$ where $\hat{p}$ can be characterised graphically.

Now examine Calum's choices graphically. Explain how you expect the algebra to run.

---

[4] Not surprisingly, mathematicians pronounce $\hat{p}$ as 'p hat'.

Now we consider Calum's point of view. He plays a random strategy choosing column 1 with probability q and column 2 with probability $1 - q$. His expected losses are

$$f_1(q) = a_{11}q + a_{12}(1 - q) = a_{12} + (a_{11} - a_{12})q$$

if Rowena chooses row 1 and

$$f_2(q) = a_{21}q + a_{22}(1 - q) = a_{22} + (a_{21} - a_{22})q$$

if Rowena chooses row 2. He seeks to minimise

$$f(q) = \max\{f_1(q), f_2(q)\}.$$

As before, there are two cases, though the first looks slightly different from Calum's side.

First case   If $a_{12} \geq a_{22}$, then

$$f_1(q) - f_2(q) = (a_{11} - a_{21})q + (a_{12} - a_{22})(1 - q) \geq 0 + 0 = 0$$

and so $f_1(q) \geq f_2(q)$ for all q with $0 \leq q \leq 1$. We have

$$f(q) = f_1(q).$$

Since f_1 is increasing, Calum will minimise his losses by choosing $q = 0$. His expected losses will be a_{12}.

Second case   If $a_{12} < a_{22}$ then, automatically, $a_{21} < a_{22}$ and

$$0 < a_{22} - a_{12} \leq (a_{11} - a_{21}) + (a_{22} - a_{12}) = (a_{11} + a_{22}) - (a_{12} + a_{21})$$

so, writing

$$\hat{q} = \frac{a_{22} - a_{12}}{(a_{11} + a_{22}) - (a_{12} + a_{21})} = 1 - \frac{a_{11} - a_{21}}{(a_{11} + a_{22}) - (a_{12} + a_{21})},$$

we have $0 < \hat{q} < 1$. Thus

$$f_1(q) - f_2(q) = -(a_{22} - a_{12}) + \big((a_{11} - a_{12}) + (a_{22} - a_{21})\big)q$$

$$\begin{cases} < 0 & \text{if } 0 \leq q < \hat{q}, \\ = 0 & \text{if } q = \hat{q}, \\ > 0 & \text{if } \hat{q} < q \leq 1 \end{cases}$$

and so

$$f(q) = \begin{cases} f_2(q) = a_{22} + (a_{21} - a_{22})q & \text{if } 0 \leq q < \hat{q}, \\ f_1(\hat{q}) = f_2(\hat{q}) & \text{if } q = \hat{q}, \\ f_1(q) = a_{12} + (a_{11} - a_{12})q & \text{if } \hat{q} < q \leq 1. \end{cases}$$

It follows that $f(q)$ decreases as q increases from 0 to $\hat{q}$ and increases as q increases from $\hat{q}$ to 1, so Calum will choose $q = \hat{q}$ with expected losses

$$f_1(\hat{q}) = a_{11}\hat{q} + a_{12}(1 - \hat{q})$$

$$= \frac{a_{11}(a_{22} - a_{12})}{(a_{11} + a_{22}) - (a_{12} + a_{21})} + \frac{a_{21}(a_{11} - a_{21})}{(a_{11} + a_{22}) - (a_{12} + a_{21})}$$

$$= \frac{a_{11}a_{22} - a_{12}a_{21}}{(a_{11} + a_{22}) - (a_{12} + a_{21})}.$$

In order to make our notation run smoothly we take $\hat{p} = 1$ and $\hat{q} = 0$ when $a_{12} \geq a_{22}$.

The next result shows why we must have $e(\hat{p}) = f(\hat{q})$.

Lemma 7.2.2 *(i) If Rowena uses a random strategy with $p = \hat{p}$, then her expected winnings against any random strategy adopted by Calum will be $e(\hat{p})$.*

(ii) If Calum uses a random strategy with $q = \hat{q}$, then his expected losses against any random strategy adopted by Rowena will be $f(\hat{q})$.

Proof (i) Suppose that Calum chooses column 1 with probability q and column 2 with probability $(1 - q)$. Then the probability that Rowen chooses row 2 and Calum chooses column 1 is $(1 - \hat{p})q$ and similar results hold for the other combinations of rows and columns. The expected winnings of Rowena, in these circumstances, will be

$$a_{11}\hat{p}q + a_{12}\hat{p}(1 - q) + a_{21}(1 - \hat{p})q + a_{22}(1 - \hat{p})(1 - q)$$
$$= \big(a_{11}\hat{p} + a_{21}(1 - \hat{p})\big)q + \big(a_{11}\hat{p} + a_{21}(1 - \hat{p})\big)(1 - q)$$
$$= e_1(\hat{p})q + e_2(\hat{p})(1 - q)$$
$$= e(\hat{p})q + e(\hat{p})(1 - q) = e(\hat{p})$$

as stated.

(ii) As for (i). ∎

Thus, even if Calum knows for certain that Rowena is going to employ a random strategy with $p = \hat{p}$, he cannot improve on his original random strategy with $q = \hat{q}$. The result holds with the roles of Calum and Rowena reversed.

Exercise 7.2.3 *Explain why parts (i) and (ii) of Lemma 7.2.2 immediately imply that $e(\hat{p}) = f(\hat{q})$.*

Exercise 7.2.4 *(i) If $a_{11} = a_{12} = a_{21} = a_{22}$, explain why any choice of q is as good for Calum as any other.*

(ii) Investigate under what circumstances there is no unique best value of q for Calum. (Use graphs rather than algebra.)

(iii) If Calum does not have a unique best possible q, does it follow that Rowena will not have a unique best possible choice of p?

7.3 Can we generalise?

It is not surprising that, if Albert and Bertha play a game, then Albert will have a strategy P^* which maximises his minimum return against all possible strategies of Bertha, nor that Bertha will have a strategy Q^* which minimises her maximum expected loss against all possible strategies of Albert. However, the fact that P^* maximises Albert's minimum return against all possible strategies of Bertha does not mean that P^* maximises Albert's return against Q^* so, if Albert knows that Bertha is using Q^*, his best policy may be to use another strategy $P^{**} \neq P^*$. If Bertha knows that Albert is using P^{**}, her best policy may now be to switch from Q^* to Q^{**} and so on.

The remarkable fact about the two-person, zero-sum, 2×2 game discussed in the previous section is that this does not happen. We found a strategy $\hat{P}$ for Rowena which maximised her minimum expected return against all possible strategies of Calum and a strategy $\hat{Q}$ for Calum which minimised his maximum expected losses against all possible strategies of Rowena. We then discovered that if Rowena plays $\hat{P}$, Calum could not do better[5] than play $\hat{Q}$ and if Calum played $\hat{Q}$, Rowena could not do better than play $\hat{P}$.

Can we generalise our results about two-person, zero-sum, 2×2 games to more complicated games? The natural way forward is to relax each of the three conditions in turn leading to the study of

(a) k-person, zero-sum, $2 \times 2 \times \cdots \times 2$ games,

(b) two-person, non-zero-sum, 2×2 games, and

(c) two-person, zero-sum, $n \times m$ games.

We shall see that both k-person and non-zero-sum games raise new mathematical and non-mathematical issues so, for the time being we concentrate on two-person, zero-sum, $n \times m$ games.

Let us start by asking what a two-person, zero-sum, 3×3 game should look like. Our discussion of the 2×2 case suggests the following. There are two players Rowena and Calum. Rowena has three options called row 1, row 2 and row 3 and Calum has three options called column 1, column 2 and column 3. Each player makes their choice and places it in an envelope. The envelopes are then opened. If Rowena has chosen row i and Calum has chosen column j, then Calum pays the amount a_{ij} (which may be negative) to Rowena. The

---

[5] In fact, Lemma 7.2.2 shows that, if Rowena plays $\hat{P}$, then it does not matter what Calum plays and, if Calum plays $\hat{Q}$, then it does not matter what Rowena plays. However, as we shall see, this stronger result does not carry over when we consider more general games.

object of Rowena is to maximise the expected sum paid to her and the object of Calum is to minimise that sum. The game is thus associated with the 3×3 matrix

$$A = \begin{pmatrix} a_{11} & a_{12} & a_{13} \\ a_{21} & a_{22} & a_{23} \\ a_{31} & a_{32} & a_{33} \end{pmatrix}.$$

We shall say that Rowena adopts strategy $\mathbf{p}$ if she chooses row i with probability p_i and that Calum adopts strategy $\mathbf{q}$ if he chooses row j with probability q_j. If Rowena adopts strategy $\mathbf{p}$ and Calum adopts strategy $\mathbf{q}$, then the expected gain for Rowena is

$$\begin{aligned} e(\mathbf{p}, \mathbf{q}) = {} & a_{11} p_1 q_1 + a_{12} p_1 q_2 + a_{13} p_1 q_3 \\ & + a_{21} p_2 q_1 + a_{22} p_2 q_2 + a_{23} p_2 q_3 \\ & + a_{31} p_3 q_1 + a_{32} p_3 q_2 + a_{33} p_3 q_3 \end{aligned}$$

which may be written more briefly as

$$e(\mathbf{p}, \mathbf{q}) = \sum_{i=1}^{3} \sum_{j=1}^{3} a_{ij} p_i q_j = \sum_{i=1}^{3} \sum_{j=1}^{3} p_i a_{ij} q_j$$

or, still more briefly, in matrix notation,[6] as

$$e(\mathbf{p}, \mathbf{q}) = \mathbf{p}^T A \mathbf{q}.$$

The question we wish to ask is the following. Do there exist $\hat{\mathbf{p}}$ and $\hat{\mathbf{q}}$ obeying the following conditions?

(1) $\min_{\mathbf{q}} e(\hat{\mathbf{p}}, \mathbf{q}) \geq \min_{\mathbf{q}} e(\mathbf{p}, \mathbf{q})$ for all $\mathbf{p}$. (Thus $\mathbf{p} = \hat{\mathbf{p}}$ maximises Rowena's minimum expected return against all possible strategies of Calum.)

(2) $\max_{\mathbf{p}} e(\mathbf{p}, \hat{\mathbf{q}}) \geq \max_{\mathbf{p}} e(\mathbf{p}, \mathbf{q})$ for all $\mathbf{q}$.

(3) $e(\hat{\mathbf{p}}, \hat{\mathbf{q}}) = \max_{\mathbf{p}} e(\mathbf{p}, \hat{\mathbf{q}})$. (Thus $\mathbf{p} = \hat{\mathbf{p}}$ maximises Rowena's expected return against Calum when he takes $\mathbf{q} = \hat{\mathbf{q}}$.)

(4) $e(\hat{\mathbf{p}}, \hat{\mathbf{q}}) = \min_{\mathbf{q}} (\hat{\mathbf{p}}, \mathbf{q})$.

If the answer to this question is yes, we say that the game has a solution and that the solution is for Rowena to play $\mathbf{p} = \hat{\mathbf{p}}$ and for Calum to take $\mathbf{q} = \hat{\mathbf{q}}$. (More briefly, we say that the solution is $(\hat{\mathbf{p}}, \hat{\mathbf{q}})$.) I hope that the previous sections have convinced the reader that, unless she believes

(a) that her opponent is playing badly, and

(b) that she *understands* the way in which her opponent is playing badly,

the solution, if it exists, represents the correct way for her to play the game.

---

[6] If you are familiar with this notation you will recall that we need to treat our vectors as column vectors. If you are not familiar with this notation, do not worry, we shall not use it.

Exercise 7.3.1 *Extend the discussion above to the $n \times m$ case where Rowena has n options and Calum has m options.*

If we look at conditions (1) to (4) long enough, we may be led to make the following remark.

Lemma 7.3.2 *If the two-player, zero-sum, $n \times m$ game with associated matrix A has the solution $(\hat{\mathbf{p}}, \hat{\mathbf{q}})$, then*

$$\max_{\mathbf{p}} \min_{\mathbf{q}} e(\mathbf{p}, \mathbf{q}) = e(\hat{\mathbf{p}}, \hat{\mathbf{q}}) = \min_{\mathbf{q}} \max_{\mathbf{p}} e(\mathbf{p}, \mathbf{q}).$$

Proof Condition (1) gives

(1)′ $\min_{\mathbf{q}} e(\hat{\mathbf{p}}, \mathbf{q}) = \max_{\mathbf{p}} \min_{\mathbf{q}} e(\mathbf{p}, \mathbf{q})$

and condition (2) gives

(2)′ $\max_{\mathbf{p}} e(\mathbf{p}, \hat{\mathbf{q}}) = \min_{\mathbf{q}} \max_{\mathbf{p}} e(\mathbf{p}, \mathbf{q})$,

so, using (3) and (4) above, we deduce that,

$$\max_{\mathbf{p}} \min_{\mathbf{q}} e(\mathbf{p}, \mathbf{q}) = \min_{\mathbf{q}} e(\hat{\mathbf{p}}, \mathbf{q}) = e(\hat{\mathbf{p}}, \hat{\mathbf{q}}) = \max_{\mathbf{p}} e(\mathbf{p}, \hat{\mathbf{q}}) = \min_{\mathbf{q}} \max_{\mathbf{p}} e(\mathbf{p}, \mathbf{q}),$$

and we have the stated result.   ∎

This result should be contrasted with the following very general observation.

Lemma 7.3.3 *If $E(\mathbf{p}, \mathbf{q})$ is any real-valued function of $\mathbf{p}$ and $\mathbf{q}$, then we have*

$$\max_{\mathbf{p}} \min_{\mathbf{q}} E(\mathbf{p}, \mathbf{q}) \leq \min_{\mathbf{q}} \max_{\mathbf{p}} E(\mathbf{p}, \mathbf{q})$$

Proof Observe that, by definition,

$$\min_{\mathbf{q}} E(\mathbf{p}, \mathbf{q}) \leq E(\mathbf{p}, \mathbf{u})$$

for all $\mathbf{u}$. It follows that

$$\max_{\mathbf{p}} \min_{\mathbf{q}} E(\mathbf{p}, \mathbf{q}) \leq \max_{\mathbf{p}} E(\mathbf{p}, \mathbf{u})$$

for all $\mathbf{u}$ and so, by definition,

$$\max_{\mathbf{p}} \min_{\mathbf{q}} E(\mathbf{p}, \mathbf{q}) \leq \min_{\mathbf{u}} \max_{\mathbf{p}} E(\mathbf{p}, \mathbf{u}).$$

Rewriting the last inequality we obtain

$$\max_{\mathbf{p}} \min_{\mathbf{q}} E(\mathbf{p}, \mathbf{q}) \leq \min_{\mathbf{q}} \max_{\mathbf{p}} E(\mathbf{p}, \mathbf{q}),$$

as stated.   ∎

Lemma 7.3.3 gives us another way of characterising solutions.

Lemma 7.3.4 *A two-player, zero-sum, $n \times m$ game with associated matrix A has the solution $(\hat{\mathbf{p}}, \hat{\mathbf{q}})$ if and only if there exists a v with*
  (i) $e(\hat{\mathbf{p}}, \mathbf{q}) \geq v$ for all $\mathbf{q}$, and
  (ii) $e(\mathbf{p}, \hat{\mathbf{q}}) \leq v$ for all $\mathbf{p}$.
  Further, if (i) and (ii), hold then $v = e(\hat{\mathbf{p}}, \hat{\mathbf{q}})$.

Proof Suppose first that $(\hat{\mathbf{p}}, \hat{\mathbf{q}})$ is a solution. If we set $v = e(\hat{\mathbf{p}}, \hat{\mathbf{q}})$ then Lemma 7.3.2 tells us that

$$\max_{\mathbf{p}} \min_{\mathbf{q}} e(\mathbf{p}, \mathbf{q}) = v = \min_{\mathbf{q}} \max_{\mathbf{p}} e(\mathbf{p}, \mathbf{q}).$$

Condition (1) tells us that

$$\min_{\mathbf{q}} e(\hat{\mathbf{p}}, \mathbf{q}) = \max_{\mathbf{p}} \min_{\mathbf{q}} e(\mathbf{p}, \mathbf{q}) = v$$

and so

$$e(\hat{\mathbf{p}}, \mathbf{q}) \geq v$$

for all $\mathbf{q}$. Condition (ii) follows similarly.

We now prove the converse. Suppose that conditions (i) and (ii) hold. We note first that by conditions (i) and (ii)

$$e(\hat{\mathbf{p}}, \hat{\mathbf{q}}) \geq v \geq e(\hat{\mathbf{p}}, \hat{\mathbf{q}})$$

and so $v = e(\hat{\mathbf{p}}, \hat{\mathbf{q}})$. We now use Lemma 7.3.3 to obtain

$$\max_{\mathbf{p}} \min_{\mathbf{q}} e(\mathbf{p}, \mathbf{q}) \geq \min_{\mathbf{q}} e(\hat{\mathbf{p}}, \mathbf{q}) \geq v$$

$$\geq \max_{\mathbf{p}} e(\mathbf{p}, \hat{\mathbf{q}}) \geq \min_{\mathbf{q}} \max_{\mathbf{p}} e(\mathbf{p}, \mathbf{q}) \geq \max_{\mathbf{p}} \min_{\mathbf{q}} e(\mathbf{p}, \mathbf{q}).$$

It follows that

$$\max_{\mathbf{p}} \min_{\mathbf{q}} e(\mathbf{p}, \mathbf{q}) = \min_{\mathbf{q}} e(\hat{\mathbf{p}}, \mathbf{q}) = e(\hat{\mathbf{p}}, \hat{\mathbf{q}}) = v$$

$$= \max_{\mathbf{q}} e(\mathbf{p}, \hat{\mathbf{q}}) = \min_{\mathbf{q}} \max_{\mathbf{p}} e(\mathbf{p}, \mathbf{q})$$

and conditions (1), (2), (3) and (4) can be read off together with the value of v. ∎

Exercise 7.3.5 *Give an example of a matrix*

$$H = \begin{pmatrix} h_{11} & h_{12} \\ h_{21} & h_{22} \end{pmatrix}$$

such that

$$\max_{i} \min_{j} h_{ij} \neq \min_{i} \max_{j} h_{ij}.$$

Suppose that someone tells us that an $n \times m$ game has a solution and that the solution is $(\hat{\mathbf{p}}, \hat{\mathbf{q}})$. A modification of Lemma 7.3.4 gives an easy way of checking whether the statement is true.

Lemma 7.3.6 *A two-player, zero-sum, $n \times m$ game with associated matrix A has the solution $(\hat{\mathbf{p}}, \hat{\mathbf{q}})$ if and only if there exists a v with*

(i) $\sum_{i=1}^{n} \hat{p}_i a_{ij} \geq v$ *for all j, and*

(ii) $\sum_{j=1}^{m} a_{ij} \hat{q}_j \leq v$ *for all i.*

Further, if (i) and (ii) hold, then $v = \sum_{i=1}^{n} \sum_{j=1}^{m} \hat{p}_i a_{ij} \hat{q}_j$.

Proof If A has the solution $(\hat{\mathbf{p}}, \hat{\mathbf{q}})$ then Lemma 7.3.4 tells us that

(i') $e(\hat{\mathbf{p}}, \mathbf{q}) \geq v$ for all $\mathbf{q}$.

Now take $q_k = 0$ for $k \neq j$ and $q_j = 1$ to obtain

(i) $\sum_{i=1}^{n} \hat{p}_i a_{ij} \geq v$.

We obtain condition (ii) similarly.

Now suppose, conversely, that (i) and (ii) hold. Then, using (i),

$$e(\hat{\mathbf{p}}, \mathbf{q}) = \sum_{i=1}^{n} \sum_{j=1}^{m} \hat{p}_i a_{ij} q_j = \sum_{j=1}^{m} q_j \sum_{i=1}^{n} \hat{p}_i a_{ij} \geq \sum_{j=1}^{m} q_j v = v.$$

for all $\mathbf{q}$. Similarly, (ii) gives

$$e(\mathbf{p}, \hat{\mathbf{q}}) \leq v$$

for all $\mathbf{p}$. Applying Lemma 7.3.4, we see that $(\hat{\mathbf{p}}, \hat{\mathbf{q}})$ is a solution to the game and

$$v = e(\hat{\mathbf{p}}, \hat{\mathbf{q}}) = \sum_{i=1}^{n} \sum_{j=1}^{m} \hat{p}_i a_{ij} \hat{q}_j$$

as stated.   ∎

Remark After reading the past few pages the reader may well have a dazed impression of lots of maxima of minima and minima of maxima engaged in a baffling formal minuet. The ideas involved cannot be understood by reading someone else's proof. Rather, you have work out the proofs yourself (referring back to the proofs in the book when you get stuck). After you have done this several times over a period, the strangeness will vanish and you will see the proofs as 'merely routine verification'.

It is also the case that results like Lemma 7.3.6 become much easier to understand once you have seen them used. We shall give a good example of the use of Lemma 7.3.6 in the next section.

7.4 Morra

The game of Morra is played between two people. Each player extends a number of fingers and simultaneously guesses the number of fingers their opponent will extend.[7] If one player guesses right and the other wrong the correct guess wins. If both players guess right or both players guess wrong the result is a draw.

The game goes back as far as Roman times[8] and gave rise to the Latin saying 'He is so trustworthy you could play Morra with him in the dark'. The game is often loud and enjoyable but was, at one time, banned in Italy because of the resulting fights. We shall assume that our players are trustworthy and know each other to be trustworthy.

Exercise 7.4.1 *Suppose that the players can extend one or two fingers and the loser gives the winner 1 gold piece.*

Write out the 4×4 pay-off matrix $(a_{[rs][pq]})$ (with entries representing Row's winnings) for this game. I give the part of the matrix to help check that you are on the right track. The pair $[rs]$ means 'extend r fingers and guess s fingers'.

	[11]	[12]
[11]	0	1
[12]	−1	0

Guess the best strategy for this game and use Lemma 7.3.6 to show that your guess is correct.

Generalise to the case where the players can extend $1, 2, \ldots$, or n fingers.

As Exercise 7.4.1 shows, the simple form of Morra is about as mathematically interesting as Scissors, Paper, Stone. We shall look at a more interesting version where the loser pays the winner a sum in gold pieces corresponding to the total number of fingers extended.

Exercise 7.4.2 *Suppose that, in this new version, the players can extend one, two or three fingers. Check that the associated matrix is that given in Figure 7.1. (Recall $[rs]$ means extend r and guess s. The entries show Row's winnings.)*

We observe that the game is symmetric, so, if $\hat{\mathbf{p}}$ is a best strategy for one player, it must be a best strategy for the other. Similarly, the v of Lemma 7.3.6 must take the value 0. Lemma 7.3.6 now takes the following form.

---

[7] Or the total number of fingers shown. Why does this amount to the same thing?

[8] And, if you are prepared to exercise your imagination on vase and wall paintings which show two people holding up fingers, much further.

	[11]	[12]	[13]	[21]	[22]	[23]	[31]	[32]	[33]
[11]	0	2	2	-3	0	0	-4	0	0
[12]	-2	0	0	0	3	3	-4	0	0
[13]	-2	0	0	-3	0	0	0	4	4
[21]	3	0	3	0	-4	0	0	-5	0
[22]	0	-3	0	4	0	4	0	-5	0
[23]	0	-3	0	0	-4	0	5	0	5
[31]	4	4	0	0	0	-5	0	0	-6
[32]	0	0	-4	5	5	0	0	0	-6
[33]	0	0	-4	0	0	-5	6	6	0

Figure 7.1. The Morra matrix

Lemma 7.4.3 *The game of Morra with the matrix given in Figure 7.1 has the solution* $(\hat{\mathbf{p}}, \hat{\mathbf{p}})$ *if and only if*

$$a_{[rs][11]}\hat{p}_{[11]} + a_{[rs][12]}\hat{p}_{[12]} + a_{[rs][13]}\hat{p}_{[13]}$$
$$+a_{[rs][21]}\hat{p}_{[21]} + a_{[rs][22]}\hat{p}_{[22]} + a_{[rs][23]}\hat{p}_{[23]}$$
$$+a_{[rs][31]}\hat{p}_{[31]} + a_{[rs][32]}\hat{p}_{[32]} + a_{[rs][33]}\hat{p}_{[33]} = 0$$

for each strategy $[rs]$.

Exercise 7.4.4 *(i) Use Lemma 7.4.3 to show that playing each of the nine possible moves with probability* $1/9$ *is not an optimal strategy.*

(ii) If you were playing against an opponent playing the strategy outlined in (i), what strategy would you adopt and why?

(iii) Try and guess some good strategies for Morra. Use Lemma 7.4.3 to see if your choices are optimal.

(iv) If your choices are not optimal, find appropriate counter strategies and explain why they are the right ones to use.

The next exercise shows that our version of Morra is solvable and gives the solution.

Exercise 7.4.5 *(i) Use Lemma 7.4.3 to show that the game of Morra (with the matrix given in Figure 7.1) is solvable with solution* $(\hat{\mathbf{p}}, \hat{\mathbf{p}})$ *where*

$$(\hat{p}_{[11]}, \hat{p}_{[12]}, \hat{p}_{[13]}, \hat{p}_{[21]}, \hat{p}_{[22]}, \hat{p}_{[23]}, \hat{p}_{[31]}, \hat{p}_{[32]}, \hat{p}_{[33]})$$
$$= (0, 0, 5/12, 0, 1/3, 0, 1/4, 0, 0).$$

(ii) Suppose that you play the random strategy of (i) (i.e. the random strategy with associated probabilities $\hat{\mathbf{p}}$*).[9] Show that, if your opponent only uses* [12],

---

[9] The traditional way of doing this is to look at your digital watch and note the integer n shown by the seconds display. If $0 \le n \le 24$, extend one finger and guess three fingers, if

[13], [22], [31] *and* [32] *your expected winnings are zero but that if, at any time, they play another choice your expected winnings are at least* $1/12$.
(iii) *Is it always true that, if your opponent only plays* [13], [22] *and* [31] *with known probabilities, then there is no way you can play to make your expected winnings positive?*

The reader will, I hope, be suitably impressed by the fact that there is a best strategy for such a complicated game. However, this will not stop her asking where the solution came from. The answer that it came from Williams's book [70] may strike her as honest but unsatisfactory, so I shall now discuss the matter further.

In this chapter we have shown that every two-person, zero-sum 2×2 game is soluble (i.e. has best strategies) and shown how to find a best strategy. In Lemma 7.3.6 we gave a necessary and sufficient condition for a pair of strategies to be a solution for a two-person, zero-sum $n \times m$ games.

In the next chapter we will outline a proof that every two-person, zero-sum $n \times m$ game has a pair of strategies satisfying the conditions of Lemma 7.3.6. Thus every two-person, zero-sum $n \times m$ game is soluble. Unfortunately, our proof only tells us that the required pair exists, and gives no way of finding them.

This mirrors the history of the subject. The first proof (by von Neumann in 1928) that suitable strategies exist also gave no way of finding them. The algebraic and geometric difficulties of a direct attack on any particular $n \times m$ game when $n, m \leq 4$ are not quite as formidable as they may look to the reader, particularly if one can leave the numerical work to other people. However, as n and m increase, the number of computations in a direct attack increases so rapidly as to make it impractical even with a fast computer.

Around 1950, Dantzig invented a new method called the simplex method for maximisation subject to a large numbers of linear inequalities and showed that it could be applied to finding solution pairs for large games. The interested reader can consult books like [21].

Exercise 7.4.6 *Consider the zero-sum game between Calum and Rowena with the associated matrix showing Rowena's winnings.*

	C_1	C_2	C_3
R_1	4	-2	-5
R_2	-2	4	3
R_3	-3	6	2
R_4	3	-8	-6

$25 \leq n \leq 44$, extend two fingers and guess two fingers and, if $45 \leq n \leq 59$, extend three fingers and guess 1.

Explain why Rowena will never choose R_4 and use this fact to reduce the game to one involving a 3×3 matrix.

By repeated arguments of this type, reduce the game to one involving a 2×2 matrix and solve it. What is the solution for the original game?

Use Lemma 7.3.6 to check your answer.

Exercise 7.4.7 *The theory of two-person zero-sum games was developed partly to deal with games of bluff like poker. Here is a simple example.*

By mistake, you find yourself lured into the following game. You shuffle a pack of cards and deal one card to 'Bluffer' Bingham. Bingham examines the card and then either 'passes' or 'bets'. If he passes, he pays you €1. If he bets, you must decide either to 'fold' or 'call'. If you fold, you pay Bingham €1. If you call, Bingham shows you the card. If it is red he pays you €2. If it is black you pay him €2.

Bingham decides on the strategy 'if the card is black, bet with probability b and, if the card is red, bet with probability p'. You decide that (if Bingham bets) you should call with probability q. What value of b should Bingham choose? What value of p should he choose and what value of q should you choose? What are Bingham's expected winnings per game?

Until very recently, mathematicians had to restrict themselves to toy models of games like poker. However, a new generation of poker players have used the power of modern computers to study various aspects of the real game (or at least create the impression in the minds of their opponents that they have done so).

Exercise 7.4.8 *In the game of* Colonel Blotto, *the two players A and B are each assigned m regiments which they secretly place on n battlefields. If A places a_j regiments on the jth battlefield and B places b_j regiments on the jth battlefield, then A wins the jth battle if $a_j > b_j$, B wins the jth battle if $b_j > a_j$ and the battle is a draw if $a_j = b_j$. The numbers a_j and b_j must be non-negative integers. The player who wins the most battles wins the game. If the players win the same number of battles the result is a draw.*

(i) Why is the game uninteresting if $n = 1$ or $n = 2$?

(ii) Show that, if $n \geq m$, there is a strategy for A which guarantees at least a draw.

(iii) Show that, if $m > n \geq 3$, there is no strategy for A which guarantees at least a draw.

[The name comes from two puzzles by Caliban (we shall see another of his puzzles on page 252) published in [50]. If m is reasonably large compared to n (for example if $n = 10$, $m = 100$) we get an interesting playable game which, so far as I know, has resisted analysis.]

7.5 Can we generalise further?

So far we have only studied games between two people in which one side's gains are the other side's losses. What happens if gains and losses do not balance?

Let us look at the simplest case in which, as before, each of the participants, Rowena and Calum has two options. Let us suppose that Rowena gains r_{ij} if she chooses row i and Calum chooses column j and that Calum gains c_{ij} under the same circumstances. We represent the game by the following diagram.

$$
\begin{array}{c|c}
(r_{11}, c_{11}) & (r_{12}, c_{12}) \\
\hline
(r_{21}, c_{21}) & (r_{22}, c_{22})
\end{array}
$$

If $r_{ij} = -c_{ij}$, we have a zero-sum game.

Some of the problems involved are indicated by the games known by the names 'Prisoner's Dilemma' and 'Chicken'.

Prisoner's Dilemma The police hold two prisoners in separate cells. They go to each prisoner and make the following speech. 'We can prove that you both burgled Heathcliff Mansions last night and we know, but cannot prove, that you both were behind the Bromley jewel thefts. If neither of you confesses to the jewel thefts, you will get one year's imprisonment. If one of you is prepared to confess, we will let him go free and his partner will get three years in jail. If you both confess, you both will get two years in jail.'

We represent these options in the manner suggested above.

	silence	confession
silence	$(-1, -1)$	$(-3, 0)$
confession	$(0, -3)$	$(-2, -2)$

Suppose first that the prisoners have received a good training in mathematics and logic. Each prisoner reasons as follows: 'I do not know what my partner will do, but he must either keep silent or confess. If he keeps silent, then it is in my interest to confess, since instead of suffering a one year sentence, I shall walk free. If he confesses, then I would be a fool not to confess, since confession shortens my sentence from three years to two. Since whatever he does, I am better off confessing, I should confess'. Thus the 'mathematical prisoners' end up with two years' jail each.

On the other hand, if the prisoners are illogical non-mathematicians they may think as follows: 'I have no idea what is going on, but I never trust the

police. I will keep silent.' The illogical non-mathematical prisoners end up
with one year's jail each. This result disturbs mathematicians.[10]

Chicken   Two youths drive cars very fast directly at one another. At the
last moment they can each choose to swerve left or to drive straight on. If one
drives straight on and the other swerves, then the first driver can call the second
'chicken'. If both swerve, they will keep quiet about the whole thing. If neither swerves, the consequences are unpleasant for both. Readers will probably
differ about the value of these various outcomes but the general pattern is the
following.

	swerve	straight
swerve	$(-1,-1)$	$(-5,5)$
straight	$(5,-5)$	$(-100,-100)$

The following argument is clearly unsatisfactory: 'My opponent and I are
both rational. No rational person would risk a crash, so my opponent will
swerve. Since my opponent will swerve, I, being rational, will drive straight
on.' However, it is less clear what argument will be satisfactory.[11]

I have given the standard description of Prisoner's Dilemma and Chicken,
but the next example suggests that we should consider a slightly different setup.
Suppose that Rowena and Calum play a game of Matching Pennies with the
rule that, if they make the same choice, they both get €1 but, if their choices are
different, they both get nothing. (I generously agree to provide the promised
prize.) The game is represented by the following diagram.

	heads	tails
heads	$(1,1)$	$(0,0)$
tails	$(0,0)$	$(1,1)$

If the two players are not allowed to communicate, this game becomes one
of 'trying to make the same guess as the other person' – a game of more
psychological than mathematical interest.[12]

---

[10] There are other ways of approaching the problem. If the prisoners are philosophers they may
reason as follows: 'The situation is symmetric so, whatever choice one of us makes, the other
will make the same. If we are both going to make the same choice, we are obviously better
off keeping silent'. As it stands, I do not find this argument very convincing, but subtler
variations are possible.

[11] It has been suggested (see, for example, [31]) that, since being rational produces these
problems, you should convince your opponent that you are irrational. As the two cars
approach, you should unscrew the steering wheel and throw it out of the car. Of course there
may be problems if your opponent does not know you have done so (see the film *Dr
Strangelove*) or if your opponent does the same thing at the same time.

[12] Apparently most people choose heads (see [58]).

We therefore drop the secrecy condition and allow the players to talk to one another before the game. There are then two different classes of games – cooperative games, in which the players can be trusted to keep any agreement they come to, and non-cooperative games in which they cannot.

If Prisoner's Dilemma is played cooperatively, it ceases to represent a problem, since the prisoners will agree to keep silent. It is not immediately clear what will happen in the case of Chicken. (Perhaps the participants will reason as follows: 'The only rational agreement is that we both swerve. But if I refuse to come to an agreement then, in the absence of an agreement, my opponent, being rational, will have to swerve and I can drive straight ahead. So I should refuse to make any agreement'.)

If Prisoner's Dilemma and Chicken are played non-cooperatively, then preliminary discussions will be useless and all the difficulties we have pointed out remain. However, even if our game of Matching Pennies is played non-cooperatively, Rowena and Calum will have no difficulty in reaching agreement and will keep that agreement.

What happens if, instead of generalising from two-person zero-sum games to two-person non zero-sum games, we generalise from two-person zero-sum games to n-person zero-sum games?

Suppose, for example, we have three players: Rowena who can choose R_1 or R_2, Calum who can choose C_1 or C_2 and Simon who can choose S_1 or S_2. If Rowena chooses R_i, Calum chooses C_j and Simon chooses S_k then Rowena receives r_{ijk}, Calum c_{ijk} and Simon s_{ijk}. In order to make this a zero-sum game we must have

$$r_{ijk} + c_{ijk} + s_{ijk} = 0.$$

A possible game of this type is given by taking

$$r_{ij1} = r_{ij2} = a_{ij}, \ c_{ij1} = c_{ij2} = b_{ij}, \ s_{ij1} = s_{ij2} = -a_{ij} - b_{ij}$$

(so that Simon's choices do not affect the issue and his only job is to provide the winnings for Rowena and Calum). The game reduces to a game[13] between Rowena and Calum given by the following diagram.

(a_{11}, b_{11})	(a_{12}, b_{12})
(a_{21}, b_{21})	(a_{22}, b_{22})

A little reflection shows that any non-zero-sum game between k players can be considered as a zero-sum game between $k+1$ players. In particular, we see that the study of zero-sum games with three players must be at least as difficult as

---

[13] Which might or might not be a zero-sum game.

the study of games between two players. It is also clear that if we study games between any number of players, there is no point in considering the special case of zero-sum games.

Many player games in which the players can be trusted to keep their word give rise to the possibility of coalitions. Consider the following game between Wynken, Blynken and Nod. Each player votes for one of the other players. If a player receives two or more votes, he pays the other two players 50 gold pieces each. If no player receives two votes, then nothing happens. It is clearly advantageous for Wynken and Blynken to agree to vote for Nod.

However, this is not the end of the matter. If Wynken and Blynken are considering such a coalition, Nod can go to Blynken and offer to pay him a gold piece from his own potential winnings if Blynken agrees to join with him against Wynken. (So, in this arrangement, Blynken gets 51 gold pieces and Nod get 49.) Faced with this possibility, Wynken offers to pay Blynken two gold pieces to form a coalition against Nod ...

I have not given a complete catalogue of the problems that can arise in many player games. (Note, for instance, that the examples I have given retain a high degree of symmetry between the players.) However, I hope that I have shown the reader that we cannot hope to 'solve' all many-player games. Can we say anything at all about such games? In the next chapter I hope to convince the reader that the answer is yes.

7.6 A noisy duel

We conclude the chapter with a game of a rather different type. This section is for amusement only and will not be referred to again.

Consider a paintball duel between Ferocious Fred and Gorgeous George. Each is armed with a paintball gun which will fire only once. If a participant is hit by a paintball, he is out of the duel. The sole object of each participant is to paintball the other.[14] When the two duellists are far apart, their chance of hitting each other is essentially zero, but they are certain to hit at point blank range. They start far apart, and slowly walk towards each other. When should George open fire on Fred and when should Fred open fire on George? Note that, if one party fires and misses before the second fires, then the second party will walk up to him and fire point blank.

---

[14] Thus, if they fire simultaneously, they do not care whether or not they themselves are hit but only whether the other player is hit.

Let us produce an idealised mathematical model of this already rather idealised duel. Suppose that, if x is the distance between the participants, then $f(x)$ is the probability that Fred will hit George from that distance and $g(x)$ is the probability that George will hit Fred. We assume that f and g are continuous and strictly decreasing.[15] We take $f(1) = g(1) = 0$, $f(0) = g(0) = 1$ and assume the two participants start 1 unit apart. If neither Fred nor George fire before the distance closes to x and George fires at x but Fred does not, then, with probability $g(x)$, he hits Fred and is not paintballed but, with probability $1 - g(x)$, he misses Fred and will certainly be paintballed. If neither Fred nor George fire before the distance closes to x and Fred fires at x but George does not, then George will be paintballed with probability $f(x)$. If Fred and George fire simultaneously at x, then Fred will be paintballed with probability $g(x)$ and George with probability $f(x)$.

Fred decides to choose a y with $1 \geq y \geq 0$. He will not fire until $x = y$ and, if George fires earlier and misses, he will not fire until $x = 0$. Similarly, George chooses a z with $1 \geq z \geq 0$. He will not fire until $x = z$ and, if Fred fires earlier and misses, he will not fire until $x = 0$. Thus[16]

$$\text{Pr(George paintballed)} = \begin{cases} \text{Pr(Fred hits at } x = y) = f(y) & \text{if } y \geq z \\ \text{Pr(George misses at } x = z) = 1 - g(z) & \text{if } z > y. \end{cases}$$

As usual, let us suppose that each man fears that the other is cleverer. A little fiddling around shows that an important role is played by the unique solution (call it x_0) of the equation $f(x) = 1 - g(x)$.

Exercise 7.6.1 *Write $h(x) = f(x) + g(x)$. Depending on your background knowledge either explain (if you have not done rigorous analysis) or prove (if you have) the following results.*

(i) h is a strictly decreasing continuous function.
(ii) $h(0) = 2$ and $h(1) = 0$.
(iii) The equation $h(x) = 1$ has a unique solution x_0 with $0 < x_0 < 1$.

Exercise 7.6.2 *(i) Suppose that George has chosen $y > x_0$ and Fred knows the value of y. Show that Fred should choose $z < y$ (so that Fred will either be hit when $x = y$ or will not be hit and will fire at $x = 0$) and that, with this choice,*

$$\text{Pr(George paintballed)} > f(x_0).$$

---

[15] If $y > x$, then $f(y) < f(x)$ and $g(y) < g(x)$.
[16] At the battle of Fontenoy the British officer commanding courteously requested 'Gentlemen of the French Guard fire first'. With equal courtesy, his opposite number suggested that the British fire first. Since both sides were armed with muskets, the side which took the initiative would have to stand fire while reloading.

(ii) Suppose that George has chosen $y < x_0$ and Fred knows the value of y. Show that Fred should take $z = y$ and that, with this choice,

$$\text{Pr}(\textit{George paintballed}) > f(x_0).$$

(iii) Suppose that George has chosen $y = x_0$ and Fred knows this. Show that Fred should take any value of z with $z \le x_0$ (so either George decides to fire at x_0 or he decides not to fire at x_0 and, if unhit, to fire when $x = 0$) and that with these choices

$$\text{Pr}(\textit{George paintballed}) = f(x_0).$$

(iv) Advise Fred. Advise George.

Thus, if Fred and George have a reasonable regard for each other's intellectual abilities, they should fire at the distance x_0 where the sum of the two probabilities of hitting $f(x_0) + g(x_0) = 1$.

Exercise 7.6.3 *Show that if the two opponents do not fire simultaneously exactly one of them will be paintballed.*

Write $p = f(x_0)$, $q = g(x_0)$. If both duellists fire at $x = x_0$, what is the probability that both will be paintballed? What is the probability that neither will be paintballed? What is the expected number of paintballed duellists?

What can you say about the expected number of paintballed duellists if they fire simultaneously when $x > x_0$ and if they fire simultaneously when $x < x_0$?

The result is rather pretty, but the reader should be aware that the conditions of the problem have been very carefully chosen so as to ensure that a solution exists. It is easy to construct similar-looking problems for which there is no best strategy.

Mathematicians call duels like this, in which you know if your opponent has fired, 'noisy duels'. If you do not know when your opponent has fired, they refer to a 'quiet duel'.

We conclude with some exercises on related themes. Like the rest of this section, they are not meant to be taken too seriously.

Exercise 7.6.4 *Consider two ice-cream sellers on a beach and suppose that customers will head towards the nearest ice-cream seller. If we take the beach to be the interval*

$$[0, 1] = \{t \,:\, 0 \le t \le 1\},$$

the first ice-cream seller to be at x and the second at y with $0 \le x,\, y \le 1$, convince yourself that it is reasonable to suppose that the value of the first

seller's trade is proportional to $f(x, y)$ *and that of the second to* $1 - f(x, y)$
where

$$f(x, y) = \begin{cases} x + (y - x)/2 = (x + y)/2 & \text{if } x \le y, \\ (1 - x) + (x - y)/2 = 1 - (x + y)/2 & \text{if } y \le x. \end{cases}$$

Show, by arguments similar to those that we employed for the noisy duel, that, if the two sellers have to choose x *and* y *in advance without knowing the other's choice, they should take* $x = y = 1/2$.

Suppose instead that the two sellers do not have to choose x *and* y *in advance but start at different points on the beach and move alternately so as to increase their trade. Discuss informally what will happen.*

Exercise 7.6.5 *Two bus companies run one bus every hour from A to B. The first bus leaves at* u *minutes past the hour and the second at* v *minutes past the hour. Write* $x = u/60$, $y = v/60$. *Convince yourself it is reasonable to suppose that the number of people carried by the first company is proportional to* $f(x, y)$ *and that of the second to* $1 - f(x, y)$ *where*

$$f(x, y) = \begin{cases} (1 - y) + x = 1 - (y - x) & \text{if } x < y, \\ 1/2 & \text{if } x = y, \\ x - y & \text{if } x > y. \end{cases}$$

Suppose that the departures must take place on the minute.[17] *Suppose first that the companies must choose their times in ignorance of each other's choices. The first company decides to choose* $x = X$ *where* X *is some random variable. Explain why, if it chooses*

$$\Pr(X = r/60) = 1/60$$

for each integer r *with* $0 \le r \le 59$, *then the expected proportion of the total passengers that it carries will be* $1/2$, *regardless of what the second company does. Explain why (assuming that both companies make intelligent choices) the first company cannot do better than this.*

Now suppose that the companies are allowed to change their departure times in alternate weeks. What will happen? (Notice that although the ice-cream sellers of the previous exercise eventually settle down, the bus companies never do.[18]*)*

---

[17] If you know about what are called probability densities you can easily drop this condition.
[18] In practice, they seem to settle on leaving at times which are so close that passengers view them as identical.

Our final exercise wanders still further from our main themes but may be found enjoyable.

Exercise 7.6.6 *Two travellers and a horse set off on a journey of x leagues. Each traveller can walk at u leagues per hour and ride at v leagues per hour where v > u. Left to itself, the noble steed will stay where it is. Find the quickest way for the travellers to complete the journey if the first traveller rides the horse a certain distance and then completes the journey on foot whilst the second traveller walks until he reaches the horse and then rides the horse for the rest of the way.*

Show that the travellers cannot do better by adopting some more complicated scheme of walking and riding.

Now suppose that the conditions are the same except that, if so instructed, the noble and sagacious steed will walk at w leagues per hour back the way it has come. Find the quickest way for the travellers to complete the journey and show that it is indeed the quickest.

8
Points of agreement

8.1 An evening out

Alice and Bob (an old married couple) are discussing what to do this evening. If they cannot agree, they will stay at home. If they go out, they could go to the cinema, go to a concert, visit friends or have a slap-up meal. The value they assign to each of these possibilities is given by the following table.

	home	cinema	concert	friends	meal
Alice	4	6	8	10	2
Bob	3	5	5	−1	8

What should they do?

The reader's first instinct may be to say that Bob should take note of how much Alice wants to visit friends and modify his evident aversion to this choice. But suppose that each of them says 'Darling, I just want to do what you want to do'. We then get the following table:

	home	cinema	concert	friends	meal
Alice	3	5	5	−1	8
Bob	4	6	8	10	2

and the problem is no easier. We therefore assume that the value that each participant assigns to the various outcomes includes consideration of the feelings of the other. ('I really dislike dining out, but I know Bob likes it and I like to see him happy'.)

Generalising slightly, we suppose that Alice and Bob have $n + 1$ possible options $C_0, C_1, \ldots, C_n$, that Alice values C_j at a_j and Bob values C_j at b_j. If they cannot reach agreement, they will settle on C_0.

In keeping with the spirit of this book, they decide to choose the outcome at random with choice C_j having probability p_j. Their problem is now to choose a

$$\mathbf{p} = (p_0, p_1, p_2, \ldots, p_n).$$

Since the possible values for $\mathbf{p}$ include $p_i = 1$, $p_j = 0$ for $j \neq i$, (that is to say, just choosing C_i) they have increased the range of possibilities they can choose from, which must be a good thing.[1] The outcome will have an expected value

$$x = a_0 p_0 + a_1 p_1 + \cdots + a_{n-1} p_{n-1} + a_n p_n$$

to Alice and an expected value

$$y = b_0 p_0 + b_1 p_1 + \cdots + b_{n-1} p_{n-1} + b_n p_n$$

to Bob. Is there any way of deciding the 'best $\mathbf{p}$'?

While an undergraduate, the mathematician John Nash took an elective course in 'international economics'. This exposure to economic ideas led him to the arguments which follow.

There is another way of looking at the problem facing Alice and Bob. Consider the set of pairs of expected outcomes

$$\tilde{K} = \left\{ (x, y) : x = \sum_{j=0}^{n} a_j p_j, \, y = \sum_{j=0}^{n} b_j p_j \right.$$

$$\left. \text{with } \sum_{j=0}^{n} p_j = 1 \text{ and } p_k \geq 0 \text{ for all } k \right\}.$$

We then ask what is the 'best possible outcome' $(x, y) \in K$.

We can generalise this question somewhat.

Definition 8.1.1 *A subset K of $\mathbb{R}^n$ is called* convex *if, whenever $\mathbf{u}$, $\mathbf{v} \in K$ and t is a real number with $1 \geq t \geq 0$, we have*

$$t\mathbf{u} + (1 - t)\mathbf{v} \in K,$$

that is to say,

$$\left(tu_1 + (1 - t)v_1, tu_2 + (1 - t)v_2, \ldots, tu_n + (1 - t)v_n \right) \in K.$$

Geometrically speaking, a set is convex if every chord joining two points in the set lies entirely within the set.

---

[1] We return to this idea when we discuss Rule 2 on page 224.

Exercise 8.1.2 *Suppose that* p_j, $q_j \geq 0$ *and* $\sum_{j=0}^{n} p_j = \sum_{j=0}^{n} q_j = 1$. *Show that, if* $1 \geq t \geq 0$, *then*

$$tp_j + (1-t)q_j \geq 0 \quad and \quad \sum_{j=0}^{n}(tp_j + (1-t)q_j) = 1.$$

Deduce that the set $\tilde{K}$ *of expected outcomes for Alice and Bob is convex.*

Exercise 8.1.3 *Suppose that* $f : \mathbb{R} \to \mathbb{R}$ *is a smooth concave function (see Definition 3.5.1). Show that*

$$\{(x, y) \in \mathbb{R}^2 : f(x) \geq y\}$$

is convex.

We can now formulate our problem still more generally. Suppose K is a convex subset of $\mathbb{R}^2$ representing the possible outcomes of an agreement between Alice and Bob. If $(x, y) \in K$, then x represents the value of that outcome to Alice and y the value of that outcome to Bob. If Alice and Bob cannot come to an agreement, then the outcome will be some fixed $(x_0, y_0) \in K$. We call (x_0, y_0) the *status quo point*.

Is there any reasonable set of rules for deciding what would constitute a best choice $(x^*, y^*) \in K$? Will such a best choice be unique (if it exists)? Will such a best choice always exist?

It is a truly remarkable fact that such a reasonable set of rules exist. Our first rule is due to Pareto.[2]

Rule 1 (Pareto optimality) Let K be a set of options in $\mathbb{R}^2$. If (x_1, y_1), (x_2, y_2) are distinct points of K and $x_1 \geq x_2$ and $y_1 > y_2$ or $x_1 > x_2$ and $y_1 \geq y_2$, then (x_1, y_1) is preferred to (x_2, y_2).

This is a very natural rule. If both Alice and Bob do at least as well and one of them does better by choosing point A over point B, then they should surely choose point A. (In the example given, they will prefer going to a concert to going to the cinema.)

Lemma 8.1.4 *Consider a set of options*

$$K = \{(x, y) : x + y \leq 1\}.$$

---

[2] Pareto was an unconventional Italian economist and sociologist. Once, when he was presenting a paper, a distinguished German professor objected that 'There are no laws of economics'. The next day Pareto (a rather shabby dresser) approached him in the street in the character of a beggar and asked 'Can you tell me where I can find a restaurant where I can eat for free?'. 'My dear man,' the professor replied, 'there are no such restaurants but there is a place round the corner where you can eat cheaply'. 'Ah,' replied Pareto, revealing his face, 'so there are laws of economics after all'. In later life, Pareto came to believe that people do not act for rational reasons but afterwards seek rational justifications for their acts.

The set of best choices under Rule 1 is

$$E = \{(x, y) : x + y = 1\}.$$

Proof If $(x, y) \in K$ and $(x, y) \notin E$, then $x + y < 1$ and $(x, 1 - x)$ is to be preferred to (x, y). If $(x_1, y_1) \in E$ and $(x_2, y_2) \in K$, then, if $x_1 < x_2$, we have $y_1 > y_2$ and, if $y_1 < y_2$, we have $x_1 > x_2$. Thus Rule 1 allows all points in E and only points in E as best choices.   ■

The second rule is a very natural expression of the notion of fairness.

Rule 2 (Fairness) Let K be a set of options which is symmetric between Alice and Bob in the sense that $(x, y) \in K$ implies $(y, x) \in K$. If the status quo point (x_0, y_0) is symmetric between Alice and Bob, in the sense that $x_0 = y_0$, any best choice must also be symmetric and so of the form (x, x).

Thus if all the conditions of the problem are symmetric between Alice and Bob any best choice must also be symmetric.

Lemma 8.1.5 *Consider a set of options*

$$K = \{(x, y) : x + y \le 1\}$$

with status quo point $(0, 0)$. *Under Rule 1 and Rule 2 there is a unique best choice* $(1/2, 1/2)$.

Proof Since K and $(0, 0)$ are symmetric we can apply Rule 2. Observe that the only symmetric point in the set E of Lemma 8.1.4 is $(1/2, 1/2)$.   ■

We have only found the best choice for one set and one status quo point. The next rule, of a type which we met before in the context of Arrow's theorem (Section 6.2), allows us to extend this result considerably.

Rule 3 (Indifference to rejected alternatives[3]) Let K and K' be a sets of options with $(x_0, y_0) \in K' \subseteq K$. Suppose that $(x^*, y^*) \in K$ is a best choice for K with status quo point (x_0, y_0). Then, if $(x^*, y^*) \in K'$, it will be a best choice for K' with status quo point (x_0, y_0).

This says that, if Alice and Bob choose (x^*, y^*) when faced with a given set of possibilities, they will still choose (x^*, y^*) from a smaller set of possibilities. The next lemma is immediate.

Lemma 8.1.6 *Consider a set of options K with*

$$K \subseteq \{(x, y) : x + y \le 1\}$$

and status quo point $(0, 0)$. *If* $(1/2, 1/2) \in K$, *then, under Rules 1 to 3, it is the unique best choice.*

---

[3] This is clearly related to, but not the same as, indifference to irrelevant alternatives.

We now turn aside to state and prove a lemma whose use will only become apparent later in the argument.

Lemma 8.1.7 *Suppose that K is a convex set such that $(1/2, 1/2) \in K$ and $xy \leq 1/4$ for all $(x, y) \in K$ with $x, y \geq 0$. Then*

$$K \subseteq \{(x, y) : x + y \leq 1\}.$$

Proof (This is a very clever idea, but one that mathematicians use so often that it has become routine.) Suppose that $(x, y) \in K$ and $x, y \geq 0$. If $1 \geq t \geq 0$, we know by convexity that

$$\left(\tfrac{1}{2} + t(x - \tfrac{1}{2}), \tfrac{1}{2} + t(y - \tfrac{1}{2})\right) = \left((1 - t)\tfrac{1}{2} + tx, (1 - t)\tfrac{1}{2} + ty\right)$$
$$= (1 - t)(\tfrac{1}{2}, \tfrac{1}{2}) + t(x, y) \in K.$$

It follows from the hypotheses that

$$\tfrac{1}{4} \geq \left(\tfrac{1}{2} + t(x - \tfrac{1}{2})\right) \times \left(\tfrac{1}{2} + t(y - \tfrac{1}{2})\right) = \tfrac{1}{4} + t(x + y - 1) + t^2\left((x - \tfrac{1}{2})^2 + (y - \tfrac{1}{2})^2\right).$$

Subtracting $1/4$ from both sides and dividing by t, we get

$$0 \geq (x + y - 1) + t\left((x - \tfrac{1}{2})^2 + (y - \tfrac{1}{2})^2\right)$$

for all t with $1 \geq t > 0$.

By choosing t very small, we can make $t\left((x - \tfrac{1}{2})^2 + (y - \tfrac{1}{2})^2\right)$ as small as we please and so we must have

$$0 \geq x + y - 1.$$

In other words, we have $x + y \leq 1$, as stated. ∎

Exercise 8.1.8 *Draw sketches illustrating this lemma with various choices of K. Observe that, if K is the disc $\{(x, y) : x^2 + y^2 \leq 1/2\}$, then K and the hyperbola $xy = 1/4$ share a common tangent and that this will always happen when 'K is nice and smooth'. (You are not asked to produce rigorous proofs or even to make very exact statements.)*

Combining Lemma 8.1.7 with Lemma 8.1.6, we obtain the following result whose use is again not immediately apparent.

Lemma 8.1.9 *Consider a convex set of options K with status quo point $(0, 0)$. If $xy \leq 1/4$ for all $(x, y) \in K$ and $(1/2, 1/2) \in K$, then, under Rules 1 to 3, $(1/2, 1/2)$ is the unique best choice.*

The next rule we appeal to also corresponds to a notion of fairness.

Rule 4 (Scale invariance) Suppose that (x^*, y^*) is a best choice for some set K with status quo point (x_0, y_0). If $a, b > 0$ and

$$K_{a,b} = \{(ax, by) : (x, y) \in K\},$$

then (ax^*, by^*) is a best choice for $K_{a,b}$ with status quo point (ax_0, by_0).

Observe that, although Bob may be able to assign 'happiness units' to various outcomes, he has no way of comparing his happiness units with those of Alice. Thus decisions should remain the same, whatever the rate of exchange between Bob's units and Alice's. Different businessmen may put a different value on the possession of one unit of currency, but they will come to the same commercial decisions.

Exercise 8.1.10 *(i) Suppose that Rule 4 holds. Let K be a set of options and let $a, b > 0$. Show that, if (ax^*, by^*) is a best choice for $K_{a,b}$ with status quo point (ax_0, by_0), then (x^*, y^*) is a best choice for K with status quo point (x_0, y_0).*

(ii) Suppose that L is a convex set, and $a, b > 0$. Show that

$$L_{a,b} = \{(ax, by) : (x, y) \in L\}$$

is a convex set.

Lemma 8.1.9 now reveals its purpose.

Lemma 8.1.11 *Consider a convex set of options K with status quo point $(0, 0)$. If $(x^*, y^*) \in K$, $x^*, y^* > 0$ and $xy \leq x^* y^*$ for all $(x, y) \in K$ then, under Rules 1 to 4, (x^*, y^*) is the unique best choice.*

Proof Let $a = 2x^*, b = 2y^*$ and

$$L = K_{a^{-1}, b^{-1}} \{(a^{-1} x, b^{-1} y) : (x, y) \in K\}.$$

We observe that L is a convex set, that $(1/2, 1/2) \in L$ and that $xy \leq 1/4$ for all $(x, y) \in L$. Thus $(1/2, 1/2)$ is the unique best choice for L with status quo point $(0, 0)$.

It follows by Rule 4 that (x^*, y^*) is the unique best choice for $K = L_{a,b}$. ∎

We complete our set of rules with a rule that resembles Rule 4, but which, I think, requires less justification.

Rule 5 (Translation invariance) Suppose that (x^*, y^*) is a best choice for some set K with status quo point (x_0, y_0). If $(u, v) \in \mathbb{R}^2$ and

$$K_{(u,v)} = \{(u + x, v + y) : (x, y) \in K\}$$

then $(u + x^*, v + y^*)$ is a best choice for $K_{(u,v)}$ with given status quo point $(u + x_0, v + y_0)$.

This rule reflects the view that happiness is like potential energy: you can always add a constant to everything without affecting the nature of the choice.

Exercise 8.1.12 *Suppose that K is a convex set in $\mathbb{R}^2$ and $(u, v) \in \mathbb{R}^2$. Show that*

$$K_{(u,v)} = \{(u + x, v + y) : (x, y) \in K\}$$

is a convex set.

Theorem 8.1.13 *Consider a convex set of options K with status quo point (x_0, y_0). If $(x^*, y^*) \in K$, $x^* \geq x_0$, $y^* \geq y_0$ and*

$$(x - x_0)(y - y_0) \leq (x^* - x_0)(y^* - y_0)$$

for all $(x, y) \in K$ with $x - x_0$, $y - y_0 \geq 0$, then, under Rules 1 to 5, (x^*, y^*) is the unique best choice.

Proof Let

$$L = K_{(-x_0,-y_0)} = \{(x - x_0, y - y_0) : (x, y) \in K\}.$$

We observe that L is a convex set, that $(x^* - x_0, y^* - y_0) \in L$ and that $xy \leq (x^* - x_0)(y^* - y_0)$ for all $(x, y) \in L$ with x, $y \geq 0$. Thus $(x^* - x_0, y^* - y_0) \in L$ is the unique best choice for L with status quo point $(0, 0)$.

It follows, by Rule 5, that (x^*, y^*) is the unique best choice for $K = L_{(u,v)}$. ■

Exercise 8.1.14 *Perhaps the best way of reviewing the discussion is to extend it to three people Alice, Bob and Caroline. Restate Rules 1 to 5 so that they apply to $\mathbb{R}^3$. (Thus, for example, when extending Rule 2, we say that a set K in $\mathbb{R}^3$ is symmetric if, whenever $(x, y, z) \in K$, it follows that (x, z, y), (y, x, z), $(z, y, x) \in K$.) State and prove a result corresponding to Theorem 8.1.13.*

Write down the appropriate result for n people.

8.2 Technical points

We have found a criterion for a unique best choice, but we have not shown that a best choice always exists. The reader should do Exercise 8.2.1 and then as much of the rest of the section as she feels appropriate.

Exercise 8.2.1 *By making appropriate sketches convince yourself that if our set of options is*

$$\tilde{K} = \left\{ (x, y) : x = \sum_{j=0}^{n} a_j p_j, \ y = \sum_{j=0}^{n} b_j p_j \right.$$

$$\left. \text{with } \sum_{j=0}^{n} p_j = 1 \text{ and } p_k \geq 0 \text{ for all } k \right\}$$

and the status quo point (x_0, y_0) lies in $\tilde{K}$, then, if there is a point $(x_1, y_1) \in \tilde{K}$ with $x_1 > x_0$ and $y_1 > y_0$, it follows that there exists a $(x^*, y^*) \in \tilde{K}$ (which may be (x_0, y_0) itself) such that

$$(x - x_0)(y - y_0) \leq (x^* - x_0)(y^* - y_0)$$

for all $(x, y) \in L$ with $x - x_0, \ y - y_0 \geq 0$. The point (x^*, y^*) is the unique best choice.

The next exercise deals with what can happen in a rather special case when there is no point (x_1, y_1) of the type required in Exercise 8.2.1.

Exercise 8.2.2 *(i) We use the notation of Exercise 8.2.1. Suppose that*

$$(a_0, b_0) = (0,0), \ (a_1, b_1) = (0, -1), \ (a_2, b_2) = (1,0) \text{ and } (a_3, b_3) = (1, -1).$$

Sketch $\tilde{K}$. If the status quo point is $(0, 0)$, use Rule 1 (the Pareto principle) to find the best choice.

(ii) Suppose K is a convex set containing $(0, 0)$ but such that if $x, y > 0$ then $(x, y) \notin K$. Show that at least one of the following two things must be true.

(a) If $(x, y) \in K$ and $x \geq 0$, then $y = 0$.

(b) If $(x, y) \in K$ and $y \geq 0$, then $x = 0$.

Give examples to show that (a) may be false, that (b) may be false or that both (a) and (b) may be true simultaneously.

(iii) Suppose that $\tilde{K}$ and (x_0, y_0) are as in Exercise 8.2.1. Explain informally why, if there is no point (x_1, y_1) of the type specified in that exercise, there is, nonetheless, a unique best choice (x^*, y^*).

The reader may ask if we can replace $\tilde{K}$ in Exercise 8.2.1 by any convex set. The following exercise shows that the answer is no.

Exercise 8.2.3 *(i) Show that, if a and b are real,*

$$a^2 + b^2 \geq 2ab.$$

(ii) We work in $\mathbb{R}^2$ *as usual. Show that the sets*

$$D = \{(x, y) : x^2 + y^2 < 2\} \quad and \quad \bar{D} = \{(x, y) : x^2 + y^2 \leq 2\}$$

are convex.

(iii) Write

$$H_k = \{(x, y) \in \mathbb{R}^2 : xy = k, \, x, \, y \geq 0\}.$$

Show that, if $0 < k < 1$, *then* $H_k \cap D \neq \varnothing$, *but, if* $k \geq 1$, *then* $H_k \cap D = \varnothing$. *Deduce that there does not exist an* $(x^*, y^*) \in D$ *with* $x^*, y^* \geq 0$ *such that* $x^* y^* \geq xy$ *for all* $(x, y) \in D$ *with* $x, y \geq 0$.

(iv) Show, however, that there does exist an $(x^*, y^*) \in \bar{D}$ *with* $x^*, y^* \geq 0$ *such that* $x^* y^* \geq xy$ *for all* $(x, y) \in \bar{D}$ *with* $x, y \geq 0$.

In some sense the difference between $\bar{D}$ and D is that '$\bar{D}$ contains its boundary' but D does not. In advanced analysis, the required notion is formalised by talking about 'closed sets'.

If the reader knows and cares enough about such things, reading the following lemma and doing Exercise 8.2.5 should put her mind at ease.

Lemma 8.2.4 *Suppose that* K *is a closed convex set in* $\mathbb{R}^2$ *and* $(x_0, y_0) \in K$. *Suppose that*

(i) there exists an M *such that, whenever* $(x, y) \in K$, *we have* $x, y \leq M$,
(ii) there exists an $(x_1, y_1) \in K$ *with* $x_1 > x_0$, $y_1 > y_0$.
Then there exists a unique $(x^*, y^*) \in K$ *such that*

$$(x - x_0)(y - y_0) \leq (x^* - x_0)(y^* - y_0)$$

whenever $(x, y) \in K$ *and* $x \geq x_0$, $y \geq y_0$.

Exercise 8.2.5 *The object of this exercise is to prove Lemma 8.2.4.*
(i) Explain why we may take $(x_0, y_0) = (0, 0)$.
(ii) Prove uniqueness. (A diagram may help you see what is going on.)
(iii) Explain why the set

$$E = \{xy : (x, y) \in K \text{ and } x, \, y \geq 0\}$$

is bounded and non-empty. Deduce that[4] E *has a supremum* k.
(iv) Explain why we can find a sequence $(x_n, y_n) \in K$ *with* $x_n, \, y_n \geq 0$ *such that* $x_n y_n \to k$.
(v) By careful use of the Bolzano–Weierstrass theorem, show that there exists a sequence $n(j) \to \infty$ *and* $(x^*, y^*) \in \mathbb{R}^2$ *such that* $x_{n(j)} \to x^*$ *and* $y_{n(j)} \to y^*$.

---

[4] If this makes no sense to you, do not worry but proceed no further with the exercise.

(vi) Show that $(x^*, y^*) \in K$, x^*, $y^* > 0$ *and* $xy \leq x^*y^*$ *whenever* $(x, y) \in K$ *and* x, $y \geq 0$.

8.3 What about reality?

So far as the author is concerned, the ideas behind Theorem 8.1.13 are so pretty that it does not matter whether they have any connection with reality or not. 'People say that life is the thing, but I prefer reading'.[5] However, it is natural to ask whether anything we have said applies to the real world.

The first thing to say is that bargaining is so common and covers so many disparate situations, from the negotiation of the Nazi–Soviet Pact to the setting up of the International Postal Union and from an airline buying a new fleet of aircraft to Alice and Bob deciding on how to spend an evening out, that no one theory should be expected to cover them all. Rather, when considering a particular situation, we should ask which aspects are covered by our theory and which are not.

If we look at Alice and Bob, it seems to me that the chief weakness of our account is the assumption that the participants can actually assign 'happiness units' to the various outcomes. When looking at Arrow's theorem, we remarked that it is hard for human beings to order their preferences in a consistent manner, and even if I know that I prefer eclairs to apple tart and apple tart to ice-cream, I am a long way from assigning 7 'deliciousness units' to eclairs, 6 units to apple tart and 3 to ice-cream. It is possible that the discussion between Alice and Bob is less a negotiation than a learning process in which, not only does Alice discover Bob's preferences, but Bob discovers his own preferences. 'Before we started, I did not realise how much I preferred a nice quiet meal at a restaurant to an evening listening to our friends droning on about their holidays'.

The reader may also object that Alice and Bob are not concerned with 'maximising the expected value of the outcome' but just want a night out. However, even the best relationship may come under strain if one side thinks that the other always gets their way, and using the 'Nash solution' should reassure Alice and Bob that this is not happening.

In larger scale negotiations, it is often much easier to put a monetary value on the outcome. Consider two rival firms A and B engaged in an advertising war. So long as the war continues, the additional costs of advertising mean that the larger firm A loses 3 million pounds a year and the smaller firm B loses 1

[5] Logan Pearsall Smith, *Afterthoughts*.

million pounds a year. If they can agree to cease hostilities, then A will make 8 million a year and B will make 1 million a year. How much should A pay B per year to achieve this end?[6]

From the point of view discussed here, the two firms can make a total of 9 million a year, which they can share as x million to A and y million to B and (if they wish) burn z million (provided $z \geq 0$). Thus

$$x + y + z = 9 \quad \text{and} \quad z \geq 0.$$

They can make any choice of $(x, y) \in K$ with

$$K = \{(x, y) : x + y \leq 9\}$$

and they have a status quo point $(-3, -1)$.

Exercise 8.3.1 *Show that A should pay B 9/2 million pounds a year.*

Of course, there is no absolute reason why the two firms should adopt the 'Nash solution' any more than that countries should choose to make their borders run along rivers. However, just as it is easier to draw a boundary along a natural feature than along an arbitrary line, so, if the two firms are reasonably happy with the Nash rules, they may find it easier to accept the Nash solution than to pick some other arbitrary division.

Players in some of the games we have discussed can also use these ideas to improve the outcome.

Exercise 8.3.2 *(i) In the heat of the moment Jules and Jim have engaged to play the game of Chicken described on page 214. The value they assign to the various outcomes are given below.*

	swerve	*straight*
swerve	$(-1, -1)$	$(-5, 5)$
straight	$(5, -5)$	$(-100, -100)$

Suppose that Jules and Jim both decide that they will drive straight with probability q and swerve otherwise. Show that the best expected outcome for both occurs when $q = 1/101$.

(ii) In the cold light of morning they have second thoughts. They discuss things over the phone and agree on what they will do. They decide to use the random number generator on a calculator and to adopt one of the following courses of action. With probability p_1 they will both swerve, with probability

---

[6] The reader may feel that it would be very difficult for rival firms to come to an agreement in this way. In fact, it appears to be so easy that most countries have strict laws against such behaviour.

p_2 *they will both drive straight, with probability p_3 Jules will swerve and Jim will drive straight and with probability p_4 Jim will swerve and Jules will drive straight. Suppose that they agree that, regardless of the result of part (i), the status quo point is 'both swerve'. By drawing a diagram, or otherwise, find their best course of action (in the sense of Nash). Show that they will reach the same agreement if they take status quo point to be given by the best strategy of the type given in (i).*

(iii) Suppose that the value Jules and Jim attach to the various outcomes is given by the following table.

	swerve	straight
swerve	$(-1,-1)$	$(a,5)$
straight	$(5,a)$	$(-100,-100)$

Discuss, using diagrams rather than calculation, how the agreement will vary with the value of a.

Exercise 8.3.3 *Suppose that, in the Prisoner's Dilemma described on page 213, the prisoners are allowed to confer, that they trust each other to keep their word and have decided to follow the same kind of system as that just described for Chicken. They take the status quo point to be 'both confess'. Show that their best course of action (in the sense of Nash) is to keep silent.*

In order to agree on the kind of bargains that we have considered in this chapter, each side must know the value the other side attaches to each particular outcome. There is a possibility that they may decide to lie about these values. However, there are many negotiations where the value of the various outcomes is so clear that lying is useless.

It is also true that a particular bargain cannot be considered in isolation. When the Canadian and US governments negotiate, they know that the two countries will have to negotiate many times in the future about many issues. Although underhand tactics might bring immediate advantage, it is likely to be in the interests of the two countries that all negotiations be conducted in good faith.

Often, the real difficulty in applying the Nash method lies in the choice of status quo point. I have chosen examples in which this choice is straight-forward, but this is not always the case, for example, in wage negotiations between employers and workers. The appropriate 'status quo point' may corre-spond not to ordinary working but to a strike or lockout. In such circumstances it may be appropriate to use the other common name for the 'status quo point' and call it the 'threat point'.

Consider, in particular, the non-zero sum games that we looked at in the preceding chapter. By cooperating among themselves, the players may gain a larger total than if they do not cooperate, but, unless they can agree on how to share that total, they cannot cooperate. A natural solution is for the players to agree on how they would act if they did not cooperate. The result provides the threat point (i.e. status quo point) for the application of the Nash method.

We have already looked at the game of Chicken and the Prisoner's Dilemma in this way. In both cases there were natural threat points and the idea worked well. However, there are many non-zero sum games for which there is no universally agreed solution to their non-cooperative form and for which the method of the previous paragraph fails.

In the remainder of this chapter we look at non-cooperative non-zero sum games. We shall see that, although we cannot say everything about how to play every such game, we can say something. The ideas are, once again, due to Nash. We require a piece of mathematics which, at first sight, has nothing to do with the matter in hand.

8.4 Fixed points

Fixed point theorems are among the jewels of mathematics. Here is one. Recall that

$$[-1, 1] = \{x \in \mathbb{R} : -1 \le x \le 1\}.$$

Theorem 8.4.1 *If $f : [-1, 1] \to [-1, 1]$ is continuous, then there exists a $x_0 \in [-1, 1]$ with $f(x_0) = x_0$.*

In other words, a continuous function of the interval $[-1, 1]$ into itself has a fixed point. If we take a piece of elastic cord covering a metre rule and stretch it, compress it, lay it backwards and forwards and so on, then, if the result still lies above the metre rule, one point of the cord will be in the place where it starts.

Proof Suppose, if possible, that f has no fixed point. Then $t - f(t) \ne 0$ for all $t \in [-1, 1]$ and so

$$F(t) = \frac{t - f(t)}{|t - f(t)|}$$

is a well-defined continuous function on $[0, 1]$.

We now observe that

$$|F(t)| = \frac{|t - f(t)|}{|t - f(t)|} = 1$$

so F can only take the values 1 and -1. Thus F must be constant. However, $f(-1) \neq -1$ and $f(-1) \geq -1$ so $f(-1) > -1$ and $F(-1) = -1$. Similarly, $F(1) = 1$ and we have a contradiction.

Thus f must have a fixed point. ∎

To extend this result to two dimensions we need the following result.

Theorem 8.4.2 *Let*

$$\bar{D} = \{(x, y) \in \mathbb{R}^2 : x^2 + y^2 \leq 1\} \text{ and } \partial D = \{(x, y) \in \mathbb{R}^2 : x^2 + y^2 = 1\}.$$

We cannot find a continuous function $F : \bar{D} \to \partial D$ *with* $F(x, y) = (x, y)$ *for all* $(x, y) \in \partial D$.

Thus there is no continuous way of mapping the unit disc $\bar{D}$ to its boundary ∂D, keeping the boundary points fixed.[7]

If the reader thinks of a piece of rubber stretched over a ring, it is fairly obvious that there is no way of pulling all the rubber back to the ring without the rubber tearing. However, although it is easy to supply any additional technical detail required for most of the proofs in this book, this is not the case here. All the known proofs of this theorem and its generalisation to higher dimensions require a mixture of new techniques and sheer cleverness. Proofs vary in the proportion of technique and cleverness but all belong to advanced undergraduate or beginning graduate courses.

Theorem 8.4.3 *If* $\bar{D}$ *is the unit disc in* $\mathbb{R}^2$ *and* $f : \bar{D} \to \bar{D}$ *is continuous, then there is an* $(x_0, y_0) \in \bar{D}$ *such that* $f(x_0, y_0) = (x_0, y_0)$.

Proof We follow a similar path to that we used in proving Theorem 8.4.1. Suppose that $f : \bar{D} \to \bar{D}$ is a continuous function with no fixed point, that is to say that $f(\mathbf{x}) \neq \mathbf{x}$ for all $\mathbf{x} \in \bar{D}$.

We can now define a function $F : \bar{D} \to \partial D$ by the recipe 'starting at $f(\mathbf{x})$, draw a line through $\mathbf{x}$ and continue it until it cuts the boundary ∂D at $F(\mathbf{x})$'.

Observe that, if we make a small change in $\mathbf{x}$, there will only be a small change in $f(\mathbf{x})$ (since f is continuous) and so only a small change in $F(\mathbf{x})$. Thus F is continuous.[8] By construction, $F(\mathbf{x}) = \mathbf{x}$ for all $\mathbf{x} \in \partial D$ and we have a contradiction with the result of Theorem 8.4.2.

Thus no f of the type described in the first paragraph can exist and the result follows. ∎

---

[7] We need some such condition to prevent the use of the continuous function $F(x, y) = (0, -1)$ which takes every point of $\bar{D}$ to the same boundary point.

[8] If the reader is so inclined, it is easy, but a little tedious, to write this in terms of epsilons and deltas.

It is not hard to guess the appropriate extensions of our results to higher dimensions.

Theorem 8.4.4 *We work in* $\mathbb{R}^3$. *Let*

$$\bar{B} = \{(x, y, z) \in \mathbb{R}^3 : x^2 + y^2 + z^2 \leq 1\}.$$

If $f : \bar{B} \to \bar{B}$ *is continuous, then there exists an* $\mathbf{x} \in \bar{B}$ *such that* $f(\mathbf{x}) = \mathbf{x}$.

Exercise 8.4.5 *(i) State the appropriate extension of Theorem 8.4.2 to three dimensions and use it to prove Theorem 8.4.4.*

(ii) State the appropriate generalisations of Theorems 8.4.2 and 8.4.3 to n *dimensions. We note that arguments about rubber toys, which seem convincing in two and three dimensions, begin to lose their persuasive power in higher dimensions. However, the results remain true. The generalisation of Theorem 8.4.3 to higher dimensions is known as Brouwer's fixed point theorem.*

We have proved a fixed point result for maps from the disc to the disc, but the method of proof suggests that the result will hold for any set which 'can obtained from the disc by a reasonable amount of stretching and compression'. Fortunately, this rather vague statement can be put into precise terms.

Lemma 8.4.6 *Let* $\bar{D}$ *be the unit disc in* $\mathbb{R}^2$. *Suppose that* E *is a set in* $\mathbb{R}^2$ *such that there exist continuous functions* $h_1 : \bar{D} \to E$ *and* $h_2 : E \to \bar{D}$ *with the properties that*

$$h_2\big(h_1(\mathbf{x})\big) = \mathbf{x} \text{ for all } \mathbf{x} \in \bar{D}$$
$$h_1\big(h_2(\mathbf{u})\big) = \mathbf{u} \text{ for all } \mathbf{u} \in E.$$

Then, if $g : E \to E$ *is continuous, there is an* $\mathbf{u}_0 \in E$ *such that* $g(\mathbf{u}_0) = \mathbf{u}_0$.

(The conditions on E can be stated more concisely by saying that $h_1 : \bar{D} \to E$ is a bijective continuous function with continuous inverse.)

Proof If we set

$$f(\mathbf{x}) = h_2\Big(g\big(h_1(\mathbf{x})\big)\Big),$$

then f is a continuous function from $\bar{D}$ to $\bar{D}$ and so has a fixed point $\mathbf{x}_0$. Set $\mathbf{u}_0 = h_2(\mathbf{x}_0)$. Then

$$g(\mathbf{u}_0) = g\big(h_2(\mathbf{x}_0)\big) = h_2\bigg(h_1\Big(h_2\big(g(h_1(\mathbf{x}_0))\big)\Big)\bigg) = h_2\big(f(\mathbf{x}_0)\big) = \mathbf{u}_0.$$

∎

Exercise 8.4.7 *Assuming the result for the ball, state and prove the result corresponding to Lemma 8.4.6 in* $\mathbb{R}^3$. *What is the generalisation to* $\mathbb{R}^n$?

8.5 Nash equilibrium

Suppose Albert and Bertha sit down to play an $n \times m$ game. Albert may choose
to play one of $A_1, A_2, \ldots, A_n$ and Bertha to play one of $B_1, B_2, \ldots, B_m$.
If Albert plays A_i and Bertha B_j, then Albert receives a_{ij} and Bertha b_{ij}.
Albert decides to play A_i with probability p_i and Bertha decides to play B
with probability q_j so that Albert's expected winnings are

$$\alpha(\mathbf{p}, \mathbf{q}) = \sum_{i=1}^{n} \sum_{j=1}^{m} a_{ij} p_i q_j$$

and Bertha's expected winnings are

$$\beta(\mathbf{p}, \mathbf{q}) = \sum_{i=1}^{n} \sum_{j=1}^{m} b_{ij} p_i q_j.$$

(We say that Albert chooses strategy $\mathbf{p}$ andBertha chooses strategy $\mathbf{q}$.)

We proved that, if $n = m = 2$ and $a_{ij} = -b_{ij}$, that is to say, if we have a
zero-sum 2×2 game, then there is a pair of strategies $\mathbf{p}^*, \mathbf{q}^*$ such that

$$\alpha(\mathbf{p}^*, \mathbf{q}^*) \geq \alpha(\mathbf{p}, \mathbf{q}^*) \text{ for all } \mathbf{p},$$
$$\beta(\mathbf{p}^*, \mathbf{q}^*) \geq \beta(\mathbf{p}^*, \mathbf{q}) \text{ for all } \mathbf{q}.$$

Thus, if Bertha knows for certain that Albert will choose $\mathbf{p}^*$, she has no reason
to change her choice from $\mathbf{q}^*$ and, if Albert knows for certain that Bertha will
choose $\mathbf{q}^*$, he has no reason to change his choice from $\mathbf{p}^*$.

Nash showed that this remains true for any n and any m and whether the
game is zero-sum or not.

Theorem 8.5.1 *In the situation described at the beginning of this section, there
always exists a pair of strategies* $\mathbf{p}^*, \mathbf{q}^*$ *such that*

$$\alpha(\mathbf{p}^*, \mathbf{q}^*) \geq \alpha(\mathbf{p}, \mathbf{q}^*) \text{ for all } \mathbf{p}$$
$$\beta(\mathbf{p}^*, \mathbf{q}^*) \geq \beta(\mathbf{p}^*, \mathbf{q}) \text{ for all } \mathbf{q}.$$

Proof Consider the set E in $\mathbb{R}^{n+m}$ consisting of all possible strategies.
We have

$$E = \left\{ (p_1, p_2, \ldots, p_n, q_1, q_2, \ldots, q_m) : \right.$$

$$\left. \sum_{i=1}^{n} p_i = 1, \sum_{j=1}^{n} q_j = 1, p_r, q_s \geq 0 \right\}.$$

A little reflection[9] shows that E is a $n + m - 2$ dimensional parallelepiped and we can apply the multidimensional version of Lemma 8.4.6 discussed in Exercise 8.4.7. The next part of the proof deals with the construction of a suitable g.

Let $u_r(\mathbf{p}, \mathbf{q})$ be the amount (if positive) that Albert can gain by playing A_r (that is to say, playing A_r with probability 1) rather than $\mathbf{p}$ if Bertha plays $\mathbf{q}$. If Albert will not gain anything by this, then we set $u_r(\mathbf{p}, \mathbf{q}) = 0$. Thus

$$u_r(\mathbf{p}, \mathbf{q}) = \max\left(0, \alpha(\mathbf{p}^{[r]}, \mathbf{q}) - \alpha(\mathbf{p}, \mathbf{q})\right)$$

where $p_k^{[r]} = 0$ for $k \neq r$ and $p_r^{[r]} = 1$. Writing this out in detail, we have

$$u_r(\mathbf{p}, \mathbf{q}) = \max\left(0, \sum_{j=1}^m a_{rj} q_j - \sum_{i=1}^n \sum_{j=1}^m a_{ij} p_i q_j\right).$$

In the same way, we define

$$v_s(\mathbf{p}, \mathbf{q}) = \max\left\{0, \beta(\mathbf{p}, \mathbf{q}^{[s]}) - \beta(\mathbf{p}, \mathbf{q})\right\}$$

$$= \max\left(0, \sum_{i=1}^n b_{is} p_i - \sum_{i=1}^n \sum_{j=1}^m b_{ij} p_i q_j\right),$$

the available gain to Bertha if she plays B_s instead of $\mathbf{q}$ when Albert plays $\mathbf{p}$.

We now set $g(\mathbf{p}, \mathbf{q}) = (\mathbf{p}', \mathbf{q}')$ where

$$p_r' = \frac{p_r + u_r(\mathbf{p}, \mathbf{q})}{1 + \sum_{i=1}^n u_i(\mathbf{p}, \mathbf{q})} \text{ for } r = 1, 2, \ldots, n,$$

$$q_s' = \frac{q_s + v_s(\mathbf{p}, \mathbf{q})}{1 + \sum_{j=1}^m v_j(\mathbf{p}, \mathbf{q})} \text{ for } s = 1, 2, \ldots, m.$$

We check that $p_r', q_s' \geq 0$ and

$$\sum_{r=1}^n p_r' = \sum_{s=1}^m q_s' = 1,$$

so g does, indeed, take points in E to points in E. Since g is continuous, it has a fixed point $(\mathbf{p}^*, \mathbf{q}^*)$. The rest of the proof consists in the careful examination of the properties of this point.

Suppose, if possible, that

$$\alpha(\mathbf{p}^{[r]}, \mathbf{q}^*) > \alpha(\mathbf{p}^*, \mathbf{q}^*) \text{ whenever } p_r^* > 0.$$

---

[9] Or take the author's word for it.

Then

$$p_r^* \alpha(\mathbf{p}^{[r]}, \mathbf{q}^*) \geq p_r^* \alpha(\mathbf{p}^*, \mathbf{q}^*) \text{ for all } r$$

and

$$p_{r_0}^* \alpha(\mathbf{p}^{[r_0]}, \mathbf{q}^*) > p_{r_0}^* \alpha(\mathbf{p}^*, \mathbf{q}^*) \text{ for some } r_0$$

so

$$\alpha(\mathbf{p}^*, \mathbf{q}^*) = \sum_{r=1}^n p_r^* \alpha(\mathbf{p}^{[r]}, \mathbf{q}^*) > \sum_{r=1}^n p_r^* \alpha(\mathbf{p}^*, \mathbf{q}^*) = \alpha(\mathbf{p}^*, \mathbf{q}^*)$$

which is absurd. Thus our original assumption must be wrong and there must be some r_1 with

$$\alpha(\mathbf{p}^{[r_1]}, \mathbf{q}^*) \leq \alpha(\mathbf{p}^*, \mathbf{q}^*) \quad \text{and} \quad p_{r_1}^* > 0.$$

Without loss of generality and to fix ideas, we suppose $r_1 = 1$.

We now know that

$$\alpha(\mathbf{p}^{[1]}, \mathbf{q}^*) \leq \alpha(\mathbf{p}^*, \mathbf{q}^*) \quad \text{and} \quad p_1^* > 0$$

and so

$$u_1(\mathbf{p}^*, \mathbf{q}^*) = \max \left\{ 0, \alpha(\mathbf{p}^{[1]}, \mathbf{q}^*) - \alpha(\mathbf{p}^*, \mathbf{q}^*) \right\} = 0.$$

Thus, by the definition of g,

$$p_1^* = \frac{p_1^* + u_1(\mathbf{p}^*, \mathbf{q}^*)}{1 + \sum_{i=1}^n u_i(\mathbf{p}^*, \mathbf{q}^*)} = \frac{p_1^*}{1 + \sum_{i=1}^n u_i(\mathbf{p}^*, \mathbf{q}^*)}.$$

Since $p_1^* > 0$, it follows that

$$\sum_{i=1}^n u_i(\mathbf{p}^*, \mathbf{q}^*) = 0$$

and so $u_i(\mathbf{p}^*, \mathbf{q}^*) = 0$ for all i.

Thus

$$\alpha(\mathbf{p}^{[i]}, \mathbf{q}^*) \leq \alpha(\mathbf{p}^*, \mathbf{q}^*)$$

for all i and so

$$\alpha(\mathbf{p}, \mathbf{q}^*) = \sum_{i=1}^n p_i \alpha(\mathbf{p}^{[i]}, \mathbf{q}^*) \leq \sum_{i=1}^n p_i \alpha(\mathbf{p}^*, \mathbf{q}^*) = \alpha(\mathbf{p}^*, \mathbf{q}^*)$$

for all possible choices of $\mathbf{p}$.

The same argument shows that

$$\alpha(\mathbf{p}^*, \mathbf{q}) \leq \alpha(\mathbf{p}^*, \mathbf{q}^*)$$

for all possible choices of $\mathbf{q}$, so we are done.   ∎

The argument just given is one of the most difficult in the book, but is not quite as difficult as it may seem to the reader. The use of fixed point theorems to establish the existence of solutions goes back a long way in mathematics: von Neumann used a fixed point theorem in his original investigation of two player zero-sum games. If a mathematician decides to try using a fixed point theorem, then there will only be a few plausible candidates for such a theorem and she can investigate each in turn. If we want to use the Brouwer theorem we must find an appropriate E and g.

Our choice of E is about the simplest we could make and our choice of g is just as natural.[10] We now examine the resulting fixed point $(\mathbf{p}^*, \mathbf{q}^*)$ and find it rather harder than we expected to show that it has the right properties. A little playing around reveals that the root of the problem is that some of the p_i^* may be zero[11] but that our proof will go through if we can show that there exists an i for which $p_i^* \neq 0$ and, simultaneously, $u_i(\mathbf{p}^*, \mathbf{q}^*) = 0$. The whole matter reduces to asking the right question, selecting the right tool and using a little perseverance and ingenuity.[12]

There are two standard methods to help one understand proofs like the above. The first is to prove a special case with fewer notational difficulties.

Exercise 8.5.2 *Prove Theorem 8.5.1 in the special case of a 2 × 2 game.*

The second is to extend it to a more general case.

Exercise 8.5.3 *Suppose that we have 3 players A, B and C and that, if A makes her ith choice, B her jth choice and C her kth choice, the outcome is worth a_{ijk} to A, b_{ijk} to B and c_{ijk} to C. Thus, if A, B and C make their choices with probabilities corresponding to the vectors $\mathbf{p}$, $\mathbf{q}$ and $\mathbf{r}$, the expected values of the outcomes to the various players are*

$$\alpha(\mathbf{p}, \mathbf{q}, \mathbf{r}) = \sum_{i=1}^{n}\sum_{j=1}^{m}\sum_{k=1}^{l} a_{ijk} p_i q_j r_k,$$

$$\beta(\mathbf{p}, \mathbf{q}, \mathbf{r}) = \sum_{i=1}^{n}\sum_{j=1}^{m}\sum_{k=1}^{l} b_{ijk} p_i q_j r_k,$$

$$\gamma(\mathbf{p}, \mathbf{q}, \mathbf{r}) = \sum_{i=1}^{n}\sum_{j=1}^{m}\sum_{k=1}^{l} c_{ijk} p_i q_j r_k.$$

[10] Moreover, other choices of g will work.
[11] Recall that this actually happens in in our discussion of Morra in Section 7.4.
[12] And all you need to be rich is to have a lot of money.

Show that there exist $\mathbf{p}^*$, $\mathbf{q}^*$ *and* $\mathbf{r}^*$ *such that*

$$\alpha(\mathbf{p}^*, \mathbf{q}^*, \mathbf{r}^*) \geq \alpha(\mathbf{p}, \mathbf{q}^*, \mathbf{r}^*) \text{ for all } \mathbf{p},$$
$$\beta(\mathbf{p}^*, \mathbf{q}^*, \mathbf{r}^*) \geq \beta(\mathbf{p}^*, \mathbf{q}, \mathbf{r}^*) \text{ for all } \mathbf{q},$$
$$\gamma(\mathbf{p}^*, \mathbf{q}^*, \mathbf{r}^*) \geq \gamma(\mathbf{p}^*, \mathbf{q}^*, \mathbf{r}) \text{ for all } \mathbf{r}.$$

Exercise 8.5.4 *Convince yourself that the extension to n players is just a case of choosing an appropriate notation. If you think it will be useful to you, find such a notation and carry out the proof.*

Now consider the special case of a two-person zero-sum $n \times m$ game. In this case, $a_{ij} = -b_{ij}$ and the conclusion of Theorem 8.5.1 can be rewritten as follows.

In the situation described at the beginning of this section, there always exists a pair of strategies $\mathbf{p}^*, \mathbf{q}^*$ such that

$$\sum_{i=1}^{n} \sum_{j=1}^{m} a_{ij} p_i^* q_j^* \geq \sum_{i=1}^{n} \sum_{j=1}^{m} a_{ij} p_i q_j^* \text{ for all } \mathbf{p} \text{ and}$$

$$-\sum_{i=1}^{n} \sum_{j=1}^{m} a_{ij} p_i^* q_j^* \geq -\sum_{i=1}^{n} \sum_{j=1}^{m} a_{ij} p_i^* q_j \text{ for all } \mathbf{q}.$$

If we set $v = \sum_{i=1}^{n} \sum_{j=1}^{m} a_{ij} p_i^* q_j^*$, these inequalities can be rewritten as

$$v \geq \sum_{i=1}^{n} \sum_{j=1}^{m} a_{ij} p_i q_j^* \text{ for all } \mathbf{p} \text{ and } \sum_{i=1}^{n} \sum_{j=1}^{m} a_{ij} p_i^* q_j \geq v \text{ for all } \mathbf{q}.$$

Thus, by Lemma 7.3.6, all two player zero-sum games are soluble. (This fulfils the promise made on page 211.)

8.6 Hawks, doves and others

We have seen that there always exists a selection of strategies for the n players of the games considered here such that, if $n - 1$ of the players maintain their strategies unchanged, the remaining player gains no advantage by changing strategy. We call the joint strategy a Nash equilibrium point.

If the n players meet before the game to coordinate their strategies then, if they do not trust each other to keep agreements, in order for any choice to be useful, it must be a Nash equilibrium point.

There is another way to look at the matter which we might call the 'no regrets argument'. Suppose that, after the game, it turns out the players have used strategies which do not constitute a Nash equilibrium point. Then, in retrospect, at least one player could have done better by using a different strategy. It is clearly irrational to play a strategy that one regrets afterwards. The only agreement possible for the players which none of them will regret afterwards is the choice of a Nash equilibrium point.

In view of this, it might be thought that, once we know that every game has an equilibrium point, all our problems are resolved. Unfortunately this is not the case. Let me repeat one of our leitmotifs.

The statement of a mathematical theorem is like a legal contract. It does not say what we think it means, it says what it means and only that.

The first point to note is that Nash's theorem says that there exists a collection of strategies such that it is in the interest of no *single* player to change their strategy. It says nothing about what happens when two or more players change their strategies. Consider a game played by 3 players. Each player writes down a letter A or B. An umpire then looks at the results and sees which letter was chosen by the majority. The umpire then divides €6 equally among all those who chose that letter. Those who chose the other letter get nothing. The joint strategy 'everybody chooses A' is a Nash equilibrium point, since if any *single* player decides to choose B, they will be worse off. However if two players write B and the remaining player sticks to the original joint strategy, the two players who change will be better off. Any consideration of many player games has to consider the possibility of coalitions.

The second point, which is still more important, is that Nash equilibrium points need not be unique even for two player games. Let us consider the game 'Traffic'. The two players must decide whether to drive on the left hand side or the right hand side of the road.[13] Here is a possible set of payoffs.

	left	right
left	$(0, 0)$	$(-1, -1)$
right	$(-1, -1)$	$(0, 0)$

It is clear that 'both drive on the left' and 'both drive on the right' are Nash equilibrium points and, after some reflection, we see that 'each driver chooses drive on the left with probability $1/2$' is also a Nash equilibrium (if only one driver changes her strategy the probability of a collision remains the same whatever she does).

---

[13] An Englishman is asked why he has never visited France. 'I know that they drive on the right there so I tried it one day in London. Never again!'

Let us consider a related but very slightly more complicated set of payoffs:

	left	right
left	$(1, 1)$	$(-1, -1)$
right	$(-1, -1)$	$(0, 0)$

and try to find all the Nash equilibrium points.

Suppose that Albert is driving one car and Bertha the other. If Albert chooses left with probability p and right with probability $1 - p$ while Bertha chooses left with probability q and right with probability $1 - q$, then the expected value of the game to Albert is

$$\alpha(p, q) = pq - p(1 - q) - q(1 - p) = 3pq - p - q = (3q - 1)p - q.$$

Thus

if $q > 1/3$, then $\alpha(1, q) > \alpha(p, q)$ for all $1 > p \geq 0$,

if $q < 1/3$, then $\alpha(0, q) > \alpha(p, q)$ for all $1 \geq p > 0$.

In addition, we observe that $\alpha(1/3, 1/3) = \alpha(p, 1/3)$ for all p. Since similar results apply to Bertha with left and right interchanged, we see that there are exactly three Nash equilibrium points: $(p, q) = (0, 0)$ with expected value to each player of 0, $(p, q) = (1, 1)$ with expected value to each the players of 1 and $(p, q) = (1/3, 1/3)$ with expected value to the two players of

$$\left(\frac{1}{3}\right)^2 - 2 \times \frac{1}{3} \times \frac{2}{3} = -\frac{1}{3}.$$

Observe that, if the players are playing the strategy 'both drive on the right', then, although both would prefer 'both drive on the left', any unilateral decision by one of the drivers to change strategy will make both drivers worse off.

Exercise 8.6.1 *Find the Nash equilibrium points for the game*

	left	right
left	(a, a)	$(-1, -1)$
right	$(-1, -1)$	$(0, 0)$

for all values of a. Write down the value of the game to the players for each of the points. What happens when $a < -1$?

In the example just given, both players prefer one Nash equilibrium point to the others. However, it is easy to modify our example to prevent this. We now consider the following payoffs (for Albert driving row).

	left	right
left	$(1, 0)$	$(-1, -1)$
right	$(-1, -1)$	$(0, 1)$

Almost exactly the same arguments as before show that there are exactly three Nash equilibrium points: $(p, q) = (1, 0)$ with expected value to Albert of 1 and to Bertha of 0, $(p, q) = (0, 1)$ with expected value to Albert of 0 and to Bertha of 1 and $(p, q) = (2/3, 1/3)$ with expected value to the two players of

$$\frac{1}{3} \times \frac{2}{3} - \left(\frac{1}{3}\right)^2 - \left(\frac{2}{3}\right)^2 = -\frac{1}{3}.$$

Exercise 8.6.2 *Find the Nash equilibrium points for the game*

	left	*right*
left	$(a, 0)$	$(-1, -1)$
right	$(-1, -1)$	$(0, a)$

for all values of a. Write down the value of the game to the players for each of the points. What happens when $a < -1$?

Exercise 8.6.3 *It should now be apparent to the reader that it is relatively easy to find the Nash equilibrium points of a general 2×2 game. Use this idea to prove algebraically (that is without using a fixed point theorem) that every 2×2 game does, indeed, have a Nash equilibrium point.*

By looking at $2 \times 2 \times 2$ three-person games, convince yourself that simple algebra is unlikely to give a proof of the existence of Nash equilibria in general many-person games.

Our next game is particularly relevant to problems of bargaining. The two players may choose to play hawk or dove. The idea is that a hawk will always fight and a dove will never fight. When a hawk meets a dove, the dove flees and the hawk gains the prize V. When two doves meet, they divide the prize with each gaining $V/2$. When two hawks meet, they fight and end up with $(V - D)/2$ each (here D represents the damage incurred in fighting). Thus the payoffs are as follows.

	hawk	dove
hawk	$((V - D)/2, (V - D)/2)$	$(V, 0)$
dove	$(0, V)$	$(V/2, V/2)$

We suppose $V, D > 0$.

Suppose that Albert and Bertha engage in a role playing game with this structure. If Albert chooses hawk with probability p and dove with probability

$1 - p$ while Bertha chooses hawk with probability q and dove with probability $1 - q$, then the expected value of the game to Albert is

$$\alpha(p, q) = \frac{V - D}{2}pq + Vp(1-q) + \frac{V}{2}(1-p)(1-q) = \frac{V - Dq}{2}p + \frac{V}{2}(1-q).$$

Thus

if $V > Dq$, then $\alpha(1, q) > \alpha(p, q)$ for all $1 > p \geq 0$,

if $V < Dq$, then $\alpha(0, q) > \alpha(p, q)$ for all $1 \geq p > 0$.

In addition, we observe that, if $V = Dq$, then $\alpha(V/D, V/D) = \alpha(p, V/D)$ for all p. Similar results apply for Bertha.

If $V > D$, then $V > Dq$ for all $1 \geq q \geq 0$. Thus there is only one Nash equilibrium point $(p, q) = (1, 1)$. (If the rewards of being a hawk exceed the rewards of being a dove in every case, then one should always be hawk.) The expected value to each player is $(V - D)/2$.

If $V < D$, then there are exactly three Nash equilibrium points. Two of the points are given by $(p, q) = (1, 0)$ and $(p, q) = (0, 1)$, so one player plays hawk and the other dove. The expected value to the hawk is V and the expected value to the dove 0. The third point is given by $(p, q) = (V/D, V/D)$ with expected value to each player $V(D - V)/(2D)$.

Exercise 8.6.4 *(i) Verify the statements just made.*

(ii) Suppose that $D > V$ and the players choose $(p, q) = (V/D, V/D)$. What happens to their behaviour and to the expected value of the outcome to each player as D gets large? Why should you expect this?

What happens to their behaviour and to the expected value of the outcome to each player when D is close to V? Why should you expect this?

(iii) Find the Nash equilibrium points when $V = D$.

Now suppose that we take the biological analogy further and consider a collection of 'real' hawks and doves. From time to time, two birds will clash over some resource. The expected value of the outcome to each bird is given in 'fitness points' by the table above. The more fitness points a bird has, the more descendants it is likely to have. (Thus a bird with 5 fitness points is likely to have more descendants than a bird with 0 fitness points and a bird with 0 fitness points is likely to have more descendants than a bird with -5 fitness points.)

If we introduce a few hawks into a large flock of doves, then any particular bird is far more likely to meet a dove than a hawk. Since hawks do better than doves in an encounter with a dove, the average hawk is likely to accumulate more fitness points than the average dove and so the number of hawks is likely to increase from generation to generation.

Suppose now that $D > V$ and we introduce a few doves into a large flock of hawks. Under these circumstances, any particular bird is far more likely to meet a hawk than a dove. Since doves do better than hawks in an encounter with a hawk, the average dove is likely to accumulate more fitness points than the average hawk and so the number of doves is likely to increase from generation to generation.

Looking more carefully at our argument, we see that if the proportion of hawks in a large flock is p, then (since the probability that a particular bird meets a hawk in any particular encounter is p) the expected value of an encounter will be

$$H(p) = p\frac{V - D}{2} + (1 - p)V$$

for a hawk and

$$D(p) = (1 - p)\frac{V}{2}$$

for a dove. Since

$$H(p) - D(p) = \frac{V - Dp}{2},$$

we see that, if $p < V/D$, the average hawk is likely to accumulate more fitness points than the average dove and so the proportion of hawks is likely to increase from this generation to the next, but, if $p > V/D$, the average dove is likely to accumulate more fitness points than the average hawk and so the proportion of hawks is likely to decrease from this generation to the next.

Putting this information together, it seems plausible that, if $V/D > 1$, a mixed flock of hawks and doves will end up as a flock of hawks but, if $V/D < 1$, the proportion of hawks will settle down to about V/D of the flock. The expected value of an encounter will be

$$\frac{D - V}{D} \times \frac{V}{2}$$

for both doves and hawks.[14]

Exercise 8.6.5 *(i) Suppose that we have flock of lefters and righters and the fitness points resulting from an encounter is that which we gave for the game of traffic.*

---

[14] Books could be and have been written on the ways in which our model fails to match reality. I shall confine myself to a footnote. In nature, competition for resources is usually most intense between closely related species. Since it is important to the hawks that the doves recognise them for what they are, the hawks may have to develop some method of signalling their hawkishness by, for example, aggressive displays. This adds to the cost of being a hawk.

	lefter	righter
lefter	$(0, 0)$	$(-1, -1)$
righter	$(-1, -1)$	$(0, 0)$

Show, using the type of argument used above, that, if the proportion p of left-ers exceeds $1/2$ in one generation, the proportion of lefters is likely to increase from this generation to the next. Show that if $p < 1/2$, the proportion of lefters is likely to decrease from this generation to the next. Conclude that we must expect mixed flocks to be replaced by flocks consisting entirely of lefters or entirely of righters. (Note that the laws of mechanics do not exclude the possibility that a pencil can be balanced on its point for a week but that the balance can be upset by a random puff of air.)

(ii) Extend your result to the situations described in Exercise 8.6.1.

If we set $V = 4$ and $D = 8$ for our game of Hawks and Doves, then we obtain the following table.

	hawk	dove
hawk	$(-2, -2)$	$(4, 0)$
dove	$(0, 4)$	$(2, 2)$

We now suppose that each bird patrols its own territory occasionally wandering off into some other bird's territory, so that all conflicts take place between an 'owner' and an 'intruder'. We introduce a third type of bird 'the bourgeois'.[15] This bird is no better at fighting than the hawk and receives no greater value from a prize than either the hawk or dove. However, bourgeois birds have a sense of territory. Outside their own territory they behave like doves (so they flee hawks and cooperate with doves) but within their own territory they behave like hawks (so they fight hawks and frighten off doves). If two bourgeois birds meet, the intruder behaves like a dove and flees, and the owner of the territory collects the prize.

If we assume that each bird is equally likely to be owner or intruder, we obtain the following table.[16]

	hawk	dove	bourgeois
hawk	$(-2, -2)$	$(4, 0)$	$(1, -1)$
dove	$(0, 4)$	$(2, 2)$	$(1, 3)$
bourgeois	$(-1, 1)$	$(3, 1)$	$(2, 2)$

If we introduce a few hawks and doves into a large flock of bourgeois, then any particular bird is far more likely to meet a bourgeois than a hawk. Since

---

[15] This is the standard name.

[16] The reader can, of course, ignore our preliminary discussion and just take the table as given.

bourgeois do better than hawks in an encounter with a bourgeois, the average hawk is likely to accumulate fewer fitness points than the average bourgeois and so the number of hawks is likely to decrease from generation to generation. The same is true of the doves.

Thus, although the release of a few doves into a flock of hawks or a few hawks into a flock of doves is likely to lead to a mixed flock, this is not true if we release a few hawks and doves into a flock of bourgeois.

Exercise 8.6.6 *What will happen if we release a few bourgeois into a flock of doves? What will happen if we release a few bourgeois into a flock of hawks?*

(If the reader asks what happens when a flock contains many doves, hawks and bourgeois she will find that stronger assumptions need to be made about the relationship between breeding and fitness points than our simple rule that a bird with more fitness points will, on average, outbreed one with fewer.)

This simple model suggests why territorial behaviour is so widespread among animals.

There are two standard objections to this kind of argument. The first is that the same ideas can be conveyed by words alone without using algebra. In the present case, when both author and reader are happy with a bit of simple algebra, this provokes the reply 'So what?'.

The second objection is that armchair theorising is useless without corroborative evidence. This is very strong objection and the reader who wants a proper discussion is directed to [41][17] where the ratio of mathematics to observational biology is the more appropriate 1 : 10.

We have seen that a population of hawks and doves will settle down to a mixed flock. This will happen even though the average value of encounters within a flock of doves will be greater than that within a mixed flock so, taken as a whole, a flock of doves will do better than a mixed flock.

This corresponds to the fact that in a game like Hawks and Doves the participants would do better not to chose a Nash equilibrium point, but it will then be to the immediate advantage of at least one player to break the agreement.

In commercial life, this problem is avoided by the use of contracts. By adding extra penalties for certain behaviour, we can make our desired outcome a Nash equilibrium. Simple-minded cynics say that the law exists to protect the rich from the poor, but much of the law is concerned with the enforcement of contracts and protects the rich from each other.

---

[17] Note that, whereas we have merely shown that certain types of behaviour are advantageous, [41] deals with the more difficult problem of how such behaviour could have arisen.

The use of some superior power to enforce cooperation in non-zero sum games does not resolve all problems. We have already talked about the problem of sharing the surplus created by cooperation. Enforced cooperation carries its own costs. It is easier to drop litter than to dispose of it tidily and, if everybody drops litter, we are all worse off. However, most people would object to policemen at the end of each street to enforce anti-littering laws.

In dealings between nations, there is no superior power and the problems raised by Prisoner's Dilemma and Chicken remain unresolved.[18] McNamara, the US Defence Secretary at the time of the Cuban crisis, was of the opinion that all the participants behaved in a perfectly rational manner and only good luck prevented a full scale nuclear war.[19]

---

[18] The reader who wishes to learn more will find the mathematical side of things well explained in [61] and the human side well illustrated in [53].

[19] 'I want to say, and this is very important: at the end we lucked out. It was luck that prevented nuclear war. We came that close to nuclear war at the end. Rational individuals: Kennedy was rational; Khrushchev was rational; Castro was rational. Rational individuals came that close to total destruction of their societies'. [Filmed in *The Fog of War*.]

9

Long duels

9.1 *A, B* and *C*

The following tale entitled *A, B, and C. The Human Element in Mathematics* is taken from Stephen Leacock's *Literary Lapses.*

The student of arithmetic who has mastered the first four rules of his art, and successfully striven with money sums and fractions, finds himself confronted by an unbroken expanse of questions known as problems. These are short stories of adventure and industry with the end omitted, and though betraying a strong family resemblance, are not without a certain element of romance.

The characters in the plot of a problem are three people called *A, B* and *C*. The form of the question is generally of this sort:– '*A, B* and *C* do a certain piece of work. *A* can do as much work in one hour as *B* in two, or *C* in four. Find how long they work at it.'

Or thus:–'*A, B* and *C* are employed to dig a ditch. *A* can dig as much in one hour as *B* can dig in two, and *B* can dig twice as fast as *C*. Find how long, etc., etc.'

Or after this wise:–'*A* lays a wager that he can walk faster than *B* or *C*. *A* can walk half as fast again as *B*, and *C* is only an indifferent walker. Find how far, and so forth.'

The occupations of *A, B* and *C* are many and varied. In the older arithmetics they contented themselves with doing 'a certain piece of work.' This statement of the case, however, was found too sly and mysterious, or possibly lacking in romantic charm. It became the fashion to define the job more clearly and to set them at walking matches, ditch digging, regattas, and piling cord wood. At times, they became commercial and entered into partnership, having with their old mystery, a 'certain' capital. Above all they revel in motion. When they tire of walking matches, – *A* rides on horseback, or borrows a bicycle and competes with his weaker minded associates on foot. Now they race on locomotives; now they row; or again they become historical and engage stage coaches; or at times they are aquatic and swim. If their occupation is actual work they prefer to pump water into cisterns, two of which leak through holes in the bottom and one of which is water-tight. *A*, of course, has the good one; he also takes the bicycle, and the best locomotive, and the right of swimming with the current. Whatever they do they put money on it, being all three sports. *A* always wins.

249

In the early chapters of the arithmetic, their identity is concealed under the names John, William and Henry, and they wrangle over the division of marbles. In algebra they are often called X, Y, Z. But these are only their Christian names, and they are really the same people.

Now to one who has followed the history of these men through countless pages of problems, watched them in their leisure hours dallying with cord wood, and seen their panting sides heave in the full frenzy of filling a cistern with a leak in it, they become something more than mere symbols. They appear as creatures of flesh and blood, living men with their own passions, ambitions, and aspirations like the rest of us. Let us view them in turn. A is a full-blooded blustering fellow, of energetic temperament, hot headed and strong willed. It is he who proposes everything, challenges B to work, makes the bets and bends the others to his will. He is a man of great physical strength and phenomenal endurance. He has been known to walk forty-eight hours at a stretch, and to pump ninety-six. His life is arduous and full of peril. A mistake in the working of a sum may keep him digging a fortnight without sleep. A repeating decimal in the answer might kill him.

B is a quiet easy going fellow, afraid of A and bullied by him, but very gentle and brotherly to little C, the weakling. He is quite in A's power, having lost all his money in bets.

Poor C is an undersized, frail man, with a plaintive face. Constant walking, digging and pumping has broken his health and ruined his nervous system. His joyless life has driven him to drink and smoke more than is good for him, and his hand often shakes as he digs ditches. He has not the strength to work as the others can, in fact, as Hamlin Smith has said, 'A can do more work in one hour than C in four.'

The first time that ever I saw these men was one evening after a regatta. They had all been rowing in it, and it had transpired that A could row as much in one hour as B in two, or C in four. B and C had come in dead fagged and C was coughing badly. 'Never mind, old fellow,' I heard B say, 'I'll fix you up on the sofa and get you some hot tea.' Just then A came blustering in and shouted, 'I say, you fellows, Hamlin Smith has shown me three cisterns in his garden and he says we can pump them until tomorrow night. I bet I can beat you both. Come on. You can pump in your rowing things you know. Your cistern leaks a little I think, C.' I heard B growl that it was a dirty shame and that C was used up now, but they went, and presently I could tell from the sound of the water that A was pumping four times as fast as C.

For years after that I used to see them constantly about town and always busy. I never heard of any of them eating or sleeping. Then owing to a long absence from home, I lost sight of them. On my return I was surprised to no longer find A, B and C at their accustomed tasks; on enquiry I heard that work in this line was now done by N, M, and O, and that some people were employing for algebraical jobs four foreigners called Alpha, Beta, Gamma and Delta.

Now it chanced one day that I stumbled upon old D, in the little garden in front of his cottage, hoeing in the sun. D is an aged labouring man who used occasionally to be called in to help A, B and C. 'Did I know 'em, Sir?' he answered, 'why, I knowed 'em ever since they was little fellows in brackets. Master A, he were a fine lad, Sir, though I always said, give me master B for kind heartedness like. Many's the job as we've been on together, Sir, though I never did

no racing nor ought of that, but just the plain labour, as you might say. I'm getting a
bit too old and stiff for it now-a-days, Sir – just scratch about in the garden here
and grow a bit of a logarithm, or raise a common denominator or two. But
Mr Euclid he use me still for them propositions, he do.'

From the garrulous old man I learned the melancholy end of my former
acquaintances. Soon after I left town, he told me, *C* had been taken ill. It seems that
A and *B* had been rowing on the river for a wager, and *C* had been running on the
bank and then sat in a draft. Of course the bank had refused the draft and *C* was
taken ill. *A* and *B* came home and found *C* lying helpless in bed. *A* shook him
roughly and said, 'Get up, *C*, we're going to pile wood.' *C* looked so worn and
pitiful that *B* said, 'Look here, *A*, I won't stand this, he isn't fit to pile wood
to-night.' *C* smiled feebly and said, 'Perhaps I might pile a little if I sat up in bed.'
Then *B* thoroughly alarmed said, 'See here, *A*, I'm going to fetch a doctor; he's
dying.' *A* flared up and answered, 'you've no money to fetch a doctor.' 'I'll reduce
him to his lowest terms,' *B* said firmly, 'that'll fetch him.' *C*'s life might even then
have been saved but they made a mistake about the medicine. It stood at the head of
the bed on a bracket, and the nurse accidentally removed it from the bracket
without changing the sign. After the fatal blunder *C* seems to have sunk rapidly. On
the evening of the next day as the shadows deepened in the little room, it was clear
to all that the end was near. I think that even *A* was affected at the last as he stood
with bowed head, aimlessly offering to bet with the doctor on *C*'s laboured
breathing. '*A*,' whispered *C*, 'I think I'm going fast.' 'How fast do you think you'll
go, old man,' murmured *A*. 'I don't know,' said *C*, 'but I'm going at any rate.'

The end came soon after that. *C* rallied for a moment and asked for a certain
piece of work that he had left downstairs. *A* put it in his arms and he expired. As
his soul sped heavenward, *A* watched its flight with melancholy admiration. *B*
burst into a passionate flood of tears and sobbed, 'Put away his little cistern and the
rowing clothes he used to wear, I feel as if I could hardly ever dig again.'

The funeral was plain and unostentatious. It differed in nothing from the
ordinary, except that out of deference to sporting men and mathematicians, *A*
engaged two hearses. Both vehicles started at the same time, *B* driving the one
which bore the sable parallelepiped containing the last remains of his ill-fated
friend. *A* on the box of the empty hearse generously consented to a handicap of a
hundred yards, but arrived first at the cemetery by driving four times as fast as *B*.
(Find the distance to the cemetery.) As the sarcophagus was lowered, the grave was
surrounded by the broken figures of the first book of Euclid.

It was noticed that after the death of *C*, *A* became a changed man. He lost
interest in racing with *B*, and dug but languidly. He finally gave up his work and
settled down to live on the interest of his bets.

B never recovered from the shock of *C*'s death; his grief preyed upon his
intellect and it became deranged. He grew moody and spoke only in
monosyllables. His disease became rapidly aggravated, and he presently spoke
only in words whose spelling was regular and which presented no difficulty to the
beginner. Realising his precarious condition he voluntarily submitted to be
incarcerated in an asylum, where he abjured mathematics and devoted himself to
writing the History of the Swiss Family Robinson in words of one syllable.

Exercise 9.1.1 *Show that the distance to the cemetery is at least* $133\frac{1}{3}$ *yards.*

Exercise 9.1.2 *C owns a tortoise and A owns a hare. The two pets decide to race to the vegetable patch, a distance X kilometres from the starting post, and back. The tortoise sets off immediately, at a steady speed v kilometres per hour. The hare goes to sleep for half per hour and then sets off at a steady speed V kilometres per hour. The hare overtakes the tortoise half a kilometre from the starting post, and continues on to the vegetable patch, where she has another half an hour's sleep before setting off for the return journey at her previous pace. One and a quarter kilometres from the vegetable patch, she passes the tortoise still plodding gallantly and steadily towards the vegetable patch. Show that*

$$V = \frac{10}{4X - 9}$$

and find v in terms of X. Find X if the hare arrives back at the starting point one and a half hours after the start of the race.

How long does it take the tortoise to reach the vegetable patch?

9.2 The three-sided duel

I first came across this question in a book [51] by the noted puzzler 'Caliban' but it may well be older.

Example 9.2.1 *A, B and C decide to hold a 3-cornered paintball duel. A hits every target he aims at, B hits any target he aims at with probability b and C hits any target he aims at with probability c. As might be expected, $1 > b > c > 0$. The rules of the contest are as follows.*

C fires first, B fires second (unless he has been hit, in which he case he drops out), A third (unless he has been hit, in which case he drops out), then C (unless he has been hit), then B (unless he has been hit) and so on.

What should C do?

Since, at some point, the three-sided duel will become a two-sided duel, it makes sense to start with this.

Lemma 9.2.2 *Suppose that P and Q fire alternately at one another until one is hit. Suppose further that P has probability p of hitting Q with each shot and Q has probability q of hitting P. If P fires first, the probability that she remains unhit is*

$$\frac{p}{p+q-pq}.$$

We give two proofs. The first is straight-forward and the second less so.

First proof Observe that

$$\Pr(P \text{ hits}) = p,$$
$$\Pr(P \text{ misses}, Q \text{ misses}, P \text{ hits}) = (1-p)(1-q)p,$$
$$\Pr(P \text{ misses}, Q \text{ misses}, P \text{ misses}, Q \text{ misses}, P \text{ hits})$$
$$= (1-p)(1-q)(1-p)(1-q)p = (1-p)^2(1-q)^2 p$$

and so on. Thus, summing an infinite geometric series,

$$\Pr(P \text{ unhit}) = \Pr(P \text{ hits}) + \Pr(P \text{ misses}, Q \text{ misses}, P \text{ hits}) + \cdots$$
$$= p + (1-p)(1-q)p + (1-p)^2(1-q)^2 p + \cdots$$
$$= p\Big(1 + \big((1-p)(1-q)\big) + \big((1-p)(1-q)\big)^2 + \cdots\Big)$$
$$= \frac{p}{1-(1-p)(1-q)} = \frac{p}{p+q-pq}$$

as required. ∎

Second proof Let p_0 be the probability that P wins (that is to say, remains unhit) if she fires first and let q_0 be the probability that Q wins if she fires first. Then

$$p_0 = \Pr(P \text{ wins starting first})$$
$$= \Pr(P \text{ hits first time}) + \Pr(P \text{ misses first time but wins})$$
$$= p + (1-p)\Pr(P \text{ wins if } Q \text{ fires first})$$
$$= p + (1-p)\big(1 - \Pr(Q \text{ wins if } Q \text{ fires first})\big)$$
$$= p + (1-p)(1-q_0).$$

Reversing the roles of P and Q gives

$$q_0 = q + (1-q)(1-p_0)$$

and so, combining our two results,

$$p_0 = p + (1-p)(1-q_0)$$
$$= p + (1-p)\big(1 - q - (1-q)(1-p_0)\big)$$
$$= p + (1-p)(1-q)p_0.$$

Thus $(1-(1-p)(1-q))p_0 = p$ and the result follows. ∎

We now look at the various situations that our duellists may face.

(1) Suppose A, B and C are still standing when it is A's turn to fire. If A hits B, then C will fire at A and hit him with probability c. If A survives, he will certainly get C on his next shot, so his probability of emerging unscathed if he fires at C is $1 - c$. Similarly, if A hits C, his probability of emerging unscathed is $1 - b$. A should therefore do the obvious thing and fire at his most dangerous opponent B.

(2) Suppose A, B and C are still standing when it is B's turn to fire. If B does not hit A, then (1) tells us that A will use his turn to eliminate B. Thus B must fire at A.

We can now turn to the results of C's first shot.

(3) If C hits B, then A will dispose of C with his first shot, so C will certainly lose.

(4) If C hits A, then the result is a duel between C and B in which C and B fire alternately and B has first shot. The probability that C will win is

$$p_A = 1 - \Pr(B \text{ wins}) = 1 - \frac{b}{b + c - bc} = \frac{c(1 - b)}{b + c - bc}.$$

(5) If C misses both, then, by (2), we know that B will fire at A. If B hits A, then the result is a duel between C and B in which C and B fire alternately and C has first shot. The probability that C will win is then

$$1 - \Pr(B \text{ wins}) = 1 - \frac{b}{b + c - bc} = \frac{c(1 - b)}{b + c - bc}.$$

If B misses A, then, by (1), we know that A will fire at B. C now has one shot at A. With probability c, he hits A and wins the match. If he misses A then he must lose. The probability that C wins the match if his first shot goes wide is thus

$$
\Pr(B \text{ hits } A)\frac{c(1 - b)}{b + c - bc} + \Pr(B \text{ misses } A)c
$$
$$
= \frac{bc(1 - b)}{b + c - bc} + (1 - b)c = c(1 - b)\frac{2b + c - bc}{b + c - bc}.
$$

Thus, if $2b + c - bc > 1$, C is better off if he misses both A and B and should therefore make sure to miss. If $2b + c - bc < 1$, C should aim for A. If $2b + c - bc = 1$, he can do either.

Exercise 9.2.3 (i) *Show that, if $b \leq 1/3$, C should always try to hit with his first shot. Show that, if $b \geq 1/2$, C should always shoot wide with his first shot.*

(ii) Give a simple non-algebraic argument to show that, if C misses on his first shot, he has probability at least c of winning the duel.

(iii) Show that we can choose b and c so that the probability of A winning is close to 1. Show that we can choose b and c so that the probability of B winning is close to 1. Show that we can choose b and c so that the probability of C winning is close to 1.

Exercise 9.2.4 *(i) Consider the situation described in Lemma 9.2.2. Show that, if p and q are small, the probability that P wins is approximately*

$$\frac{p}{p+q},$$

whether she shoots first or second. Explain, without algebra, why it makes little difference whether she shoots first or second.

(ii) Suppose that A, B and C are very poor shots. The probability that A hits his target is a, the probability that B hits is b and the probability that C hits is c with $a > b > c > 0$. They engage in a three-cornered duel using the rules already given. What tactics should they adopt? Show that the probability that A wins is roughly

$$\frac{a^2}{(a+b+c)(a+c)}$$

and find the approximate probabilities that B and C win.

(iii) Show that (for appropriate choices of a, b and c) A's probability of winning can be anywhere between 1 and 1/6. Obtain the corresponding results for B and C. Give an informal verbal explanation for these results.[1]

We gave two proofs of Lemma 9.2.2. They are rather less different than they look at first sight. Let us consider two ways to sum a geometric series.

Lemma 9.2.5 *If $|r| < 1$, then*

$$1 + r + r^2 + \cdots = \frac{1}{1-r}.$$

First proof Observe that

$$(1 + r + r^2 + \cdots + r^n)(1 - r)$$
$$= (1 + r + r^2 + \cdots + r^n) - (r + r^2 + r^3 + \cdots + r^{n+1})$$
$$= 1 - r^{n+1}.$$

---

[1] Kilgour and Brams give an entertaining discussion of more general three-sided duels (or 'truels') in [34]. They conclude that 'optimal play is very sensitive to slight changes in the rules'.

Thus

$$1 + r + r^2 + \cdots + r^n = \frac{1 - r^{n+1}}{1 - r}$$

and, allowing $n \to \infty$, we obtain

$$1 + r + r^2 + \cdots = \frac{1}{1 - r}$$

as required. ∎

Second proof Let $S = 1 + r + r^2 + \cdots + r^n + \cdots$. Then

$$rS = r + r^2 + \cdots + r^n + \cdots = S - 1$$

and so $1 = S - rS = (1 - r)S$ whence

$$S = \frac{1}{1 - r}$$

as required. ∎

The weakness of the second method, is that it *assumes* the existence of the sum S. If S does not exist, the argument leads to statements like

$$1 + 2 + 2^2 + 2^3 + \cdots = \frac{1}{1 - 2} = -1.$$

However, in more complicated situations than those discussed in this section (for example, the HHH game discussed in Section 9.4), the reader will find that it is much easier to apply the second method than the first. It also provides a much clearer picture of the underlying probabilistic process. In the questions we shall consider and most of probability theory it is clear that an appropriate answer exists and we only need to find it.

Exercise 9.2.6 *A and B start from towns* 60 *miles apart and cycle towards each other. A travels at a miles per hour and B at b miles per hour. A friendly fly starts from the tip of A's nose and flies at c miles per hour towards B. (We have $c > a, b$.) When it reaches B it turns round and flies back to A, on reaching A it turns back to B and so on. How far will it have flown when A and B meet? Here is one approach.*

(i) Suppose the cyclists are x miles apart. If the fly starts at A, calculate $d_A(x)$ the distance it will have flown when it first gets to B. Calculate $D_A(x)$ the distance apart the cyclists will be when the fly reaches B.

(iii) Let $S_A(x)$ be the total distance the fly will go if the cyclists start a distance x apart and the fly starts at A. Let $S_B(x)$ be the total distance the fly will go if the cyclists start a distance x apart and the fly starts at B.

Write down an equation connecting $S_A(x)$, $d_A(x)$ and $S_B(D_A(x))$ and an equation connecting $S_B(x)$, $d_B(x)$ and $S_A(D_B(x))$ (where $d_B(x)$ and $D_B(x)$ have the appropriate meaning). Hence find $S_A(x)$.

(iv) Suppose the cyclists are x miles apart. If the fly starts at A, calculate the distance it will have flown when it first returns to A and how far apart the two cyclists are at this first return. Calculate $S_A(x)$ by summing a geometric series,

(v) If you have not already spotted the trick, find the time it takes for the two cyclists to meet and hence compute $S_A(x)$ directly.

[This is a traditional brain teaser with $x = 20$ and $a = b = 10$. The story goes that someone told the problem to von Neumann who instantly gave the correct answer. 'Oh you must have heard the trick before!' 'What trick? All I did was sum an infinite series'.[2]]

Exercise 9.2.7 *The rules of the dice game 'Craps' are as follows. On each throw you throw two dice. If the first throw is 7 or 11, then you win, and if it is 2, 3 or 12, then you lose. If your first throw is none of these, then you throw repeatedly until you again score the same as your first throw, in which case you win, or you throw a 7, in which case you lose. Find the odds against you. [This is a long question but worth doing since it shows that whoever drew up the rules was mathematically very astute.]*

9.3 One-person duels

In this section we deal with problems similar to, but simpler than, the three- and two-person duels of the previous sections.

From time to time, the popular scientific press carries reports of some scientist or engineer who has constructed a coin-tossing device so perfect that it always throws heads. Presumably, even under normal conditions, a coin will have slight bias toward heads or tails. Is there any way to use a coin with probability p of heads and $1 - p$ of tails to imitate a fair coin?

Lemma 9.3.1 *Suppose that I have a coin which shows heads with probability p and tails with probability $1 - p$ where $0 < p < 1$. I play the following game. In each round I throw the coin twice. If it first comes down heads and then tails, I record 'left' and stop the game. If it first comes down tails and then heads, I record 'right' and stop the game. If neither event occurs, then I move on to the next round.*

---

[2] Moral. Do not waste time being clever, if you do not need to be.

The probability that I record left is $1/2$ *and the probability that I record right is* $1/2$.

We give three proofs.

First proof If we toss our coin twice, the probability w that it comes down first heads and then tails is given by $w = p(1 - p)$ and the probability that it comes down first tails and then heads is $(1 - p)p = w$. The probability that neither occurs is $1 - 2w$. Thus the probability that r rounds produce no result and I write 'left' in the $r + 1$th round is $(1 - 2w)^r w$.

Summing a geometric series, we see that

Pr(I write left) = Pr(left in 1st round) + Pr(left in 2nd round) + $\cdots$

$$= w + (1 - 2w)w + (1 - 2w)^2 w + \cdots = \frac{w}{1 - (1 - 2w)} = \frac{1}{2}.$$

The same argument works for right. ∎

Second proof Let l be the probability that I write left. With probability $w = p(1 - p)$, I will write left in the first round. With probability $1 - 2w$ I will proceed to a further round, in which case the probability that I will finally write left is again l. Thus

$l = $ Pr(I write left)

$= $ Pr(I write left in 1st round)

$\quad + $ Pr(there is a second round and I end up writing left)

$= w + (1 - 2w)l,$

so $w = 2wl$ and $l = 1/2$. ∎

Third proof Since our procedure is completely symmetric between left and right, the two probabilities must be equal. ∎

How long does the procedure take? We may be lucky and get a decision in the first round or it may take us many rounds. In order to investigate, we need a preliminary lemma.

Lemma 9.3.2 *If x is a real number with $x \neq 1$, then*

$$1 + 2x + 3x^2 + \cdots + nx^{n-1} = \frac{1 - x^{n+1}}{(1 - x)^2} - \frac{(n + 1)x^n}{1 - x}.$$

Proof This follows on differentiating both sides of the equality

$$1 + x + x^2 + \cdots + x^n = \frac{1 - x^{n+1}}{1 - x}.$$

∎

Lemma 9.3.3 (*i*) *Suppose that A is firing at a target. The probability of his hitting the target with any one shot is a. If he fails to hit the target in n shots he stops.*

Under these conditions, the expected number e_n of shots that he fires is

$$\frac{1 - (1 - a)^{n+1}}{a} - (1 - a)^n.$$

(*ii*) *We have*

$$e_n(a) \to a^{-1}$$

as $n \to \infty$.

Proof The probability that *A* misses with his first *r* shots and hits with $r + 1$th is $(1 - a)^r a$. The probability that he misses with all *n* shots is $(1 - a)^n$. Thus

$$e_n = \sum_{r=1}^{n} r \, \text{Pr(hits on } r\text{th try)} + n \, \text{Pr(misses } n \text{ tries)}$$

$$= \left(a + 2a(1 - a) + 3a(1 - a)^2 + \cdots + na(1 - a)^{n-1} \right) + n(1 - a)^n$$

$$= a \left(1 + 2(1 - a) + 3(1 - a)^2 + \cdots + n(1 - a)^{n-1} \right) + n(1 - a)^n$$

$$= a \left(\frac{1 - (1 - a)^{n+1}}{(1 - (1 - a))^2} - \frac{(n + 1)(1 - a)^n}{1 - (1 - a)} \right) + n(1 - a)^n$$

$$= \frac{1 - (1 - a)^{n+1}}{a} - (n + 1)(1 - a)^n + n(1 - a)^n$$

$$= \frac{1 - (1 - a)^{n+1}}{a} - a(1 - a)^n$$

as stated.

(ii) Immediate. ∎

It seems reasonable to interpret this result as saying that if *A* fires at a target until he hits it, the expected number of shots will be $1/a$ and we shall adopt this interpretation. In particular the expected number of throws to decide left or right in Lemma 9.3.1 is $p^{-1}(1 - p)^{-1}$. Thus, for example, if we know that $1/4 \le p \le 3/4$, we can deduce that the expected number of throws we shall need is no greater than $16/3$.

However, before committing ourselves to this path, we should note that matters need not be as straightforward as in the last example.

Exercise 9.3.4 [St Petersburg paradox][3] *I offer to play the following game with you. I toss a fair coin repeatedly until it comes up heads or until tails have*

---

[3] So called because the first printed version was published by Daniel Bernoulli in the *Commentaries of the Imperial Academy of Science of Saint Petersburg.*

come up n times. If heads comes up first on the rth go, I pay you 2^r roubles. If heads do not come up in the first n goes, I pay you nothing. Show that you have expected winnings $u_n = n$ roubles.

We observe that $u_n \to \infty$ and the same argument as above suggests that, if I throw a fair coin repeatedly and pay you 2^r roubles if heads comes up first on the rth go, then your expected winnings are infinite. While admitting that many clever people have found this conclusion troublesome for a variety of reasons, I think that all this 'paradox' does is remind us that the expected winnings from a game of unlimited duration may be infinite.

The next example is more worrying.

Exercise 9.3.5 [**Double or quits**] *I owe you 1 rouble, so I offer to play you 'double or quits'. I lose and now owe you 2 roubles, so I offer to play you 'double or quits'. I lose and now owe you 4 roubles, so I offer to play you 'double or quits' . . .*

This game comes in two versions.

(i) I offer to play the following game with you. I toss a fair coin repeatedly until it comes up heads or until tails have come up n times. If heads comes up first on the rth throw, you pay me 1 rouble. If heads do not come up in the first n throws, I pay you $2^n - 1$ roubles. Show that you have expected winnings $v_n = 0$.

(ii) I offer to play the following game with you. I toss a fair coin repeatedly until it comes up heads or until tails have come up n times. If heads comes up first on the rth throw, you pay me 1 rouble. If heads do not come up in the first n throws, I pay you nothing. Show that you have expected winnings $w_n = -1 + 2^{-n}$.

We observe that $v_n \to 0$ but $w_n \to -1$ as $n \to \infty$. What value (if either) should we assign to the infinite game? In this case the answer is clear. If we play a game of double or quits of unlimited duration which ends at the first head with you paying me 1 rouble, then (with probability 1) you will pay me 1 rouble. The expected value to you is -1 rouble. We shall revisit this example when we talk about the martingale system on page 290.

Fortunately, troublesome games like these reveal themselves by the fact that the stakes increase very rapidly as the game proceeds.

Exercise 9.3.6 *In Exercises 9.3.4 and 9.3.5 we used a fair coin. Investigate what happens if all the conditions are unchanged, except that the coin has probability $1 - p$ of coming down heads.*

After this detour, we return to Lemma 9.3.3 and use our second method to do the calculation.

Second proof of Lemma 9.3.3 Let l be the expected number of shots required. Observe that, with probability a, A will hit the target first time and require exactly one shot. With probability $1 - a$, he will miss so the expected number of shots he now needs is l and the expected total number of shots needed (including the first miss) is $l + 1$. Thus

$$l = \mathbb{E}(\text{number of shots needed})$$
$$= a \times 1 + (1 - a) \times (l + 1) = 1 + (1 - a)l.$$

Thus $al = 1$ and the result follows. ∎

We have already observed that the weakness of this proof lies in the fact that we must know in advance that l exists. We add that existence of l is to be interpreted in the strong sense 'l exists and is finite', since, if we allow for the possibility that l is infinite, we run into problems with 'equations' of the form '$\infty = \infty$'.

We shall see a third proof of Lemma 9.3.3 in Exercise 9.6.4 (ii).

Exercise 9.3.7 *I arrive home from a feast and attempt to open my front door with one of the n keys in my pocket. (You may assume that exactly one key will open the door and that if I use it I will be successful.) Find the expected number a_n of tries that I will need if I take the keys at random from my pocket but drop any key that fails onto the ground. Find the expected number b_n of tries that I will need if I take the keys at random from my pocket and immediately put back in my pocket any key that fails. Find the expected number c_n of tries that I will need if I take the keys at random from my pocket and put back in my pocket any key that fails in such a way that my next try is taken at random from all my keys with the exception of the one last tried.*
Show that $a_n/b_n \to 1/2$ and $b_n/c_n \to 1$ as $n \to \infty$.

We have shown how to obtain a fair coin from an unfair coin. Can we obtain an unfair coin from a fair one? More precisely, can we obtain the equivalent of a coin which comes down heads with probability p by a procedure involving only a fair coin? If you do the following exercise you should get a good idea of how to do it.

Exercise 9.3.8 *(i) You toss a fair coin twice. Can you find events of probability 1/4, 1/2 and 3/4?*
(ii) You toss a fair coin three times. What probabilities are associated with possible events?
(iii) If r is an integer with $1 \le r \le 2^n - 1$, show that you can imitate the toss of a coin which has probability $r/2^n$ of coming down heads by using n consecutive tosses of a fair coin. What can you say if $r = 0$ or $r = 2^n$?

(If you cannot do this exercise do not worry, but return to it after the discussion that follows.)

Our attack on the general case depends on the following fact which the reader may consider obvious.

Lemma 9.3.9 *If r and n are integers with $n \geq 1$ and $0 \leq r \leq 2^n - 1$, then we can find unique $e_j \in \{0, 1\}$ such that*

$$r = e_1 2^{n-1} + e_2 2^{n-2} + e_3 2^{n-3} + \cdots + e_{n-1} 2 + e_n.$$

Sketch proof If r is odd then $e_n = 1$. If r is even, then $e_n = 0$. Now consider $(r - e_n)/2$ and repeat the argument with e_{n-1}. ∎

Exercise 9.3.10 *If you are not satisfied with the sketch, write out a complete proof using induction.*

In what follows, we discuss the case of a general p. However, the reader may find things a little easier if she excludes the case when $p = r2^{-n}$ for some positive integers r and n. More generally, until she gets the general idea of what is going on, she should ignore the details and once she understands what is going on she will not need them.

Lemma 9.3.11 *Suppose that I toss a fair coin n times and take $X_j = 1$ if the coin comes down heads on the jth throw and $X_j = 0$ if it comes down tails.*

(i) If r is an integer with $0 \leq r \leq 2^n - 1$, then

$$\Pr(X_1 2^{-1} + X_2 2^{-2} + X_3 2^{-3} + \cdots + X_{n-1} 2^{-n+1} + X_n 2^{-n} = r2^{-n}) = 2^{-n}.$$

(ii) If k is an integer with $0 \leq k \leq 2^n - 1$ and $(k-1)2^{-n} \leq p < k2^{-n}$, then

$$\Pr(X_1 2^{-1} + X_2 2^{-2} + X_3 2^{-3} + \cdots + X_{n-1} 2^{-n+1} + X_n 2^{-n} \leq p) = k2^{-n},$$
$$\Pr(X_1 2^{-1} + X_2 2^{-2} + X_3 2^{-3} + \ldots + X_{n-1} 2^{-n+1} + X_n 2^{-n} > p) = (2^n - k)2^{-n}.$$

(iii) If p is a real number with $0 \leq p < 1$,

$$p \leq \Pr(X_1 2^{-1} + X_2 2^{-2} + X_3 2^{-3} + \cdots + X_n 2^{-n} \leq p) \leq p + 2^{-n},$$
$$1 - p - 2^{-n} \leq \Pr(X_1 2^{-1} + X_2 2^{-2} + X_3 2^{-3} + \cdots + X_n 2^{-n} > p) \leq 1 - p,$$
$$\Pr(p - 2^{-n} < X_1 2^{-1} + X_2 2^{-2} + X_3 2^{-3} + \cdots + X_n 2^{-n} \leq p) = 2^{-n}.$$

Proof (i) Write

$$r = e_1 2^{n-1} + e_2 2^{n-2} + e_3 2^{n-3} + \cdots + e_{n-1} 2 + e_n$$

as in Lemma 9.3.9. Then

$$\Pr(X_1 2^{-1} + X_2 2^{-2} + X_3 2^{-3} + \cdots + X_{n-1} 2^{-n+1} + X_n 2^{-n} = r2^{-n})$$
$$= \Pr(X_j = e_j \text{ for } 1 \leq j \leq n) = 2^{-n}.$$

(ii) Observe that

$$\Pr(X_1 2^{-1} + X_2 2^{-2} + X_3 2^{-3} + \cdots + X_{n-1} 2^{-n+1} + X_n 2^{-n} \le p)$$

$$= \Pr(X_1 2^{-1} + X_2 2^{-2} + \cdots + X_n 2^{-n} = r 2^{-n} \text{ for some } 0 \le r \le k-1)$$

$$= \sum_{r=0}^{k-1} \Pr(X_1 2^{-1} + X_2 2^{-2} + \cdots + X_{n-1} 2^{-n+1} + X_n 2^{-n} = r 2^{-n}) = k 2^{-n}.$$

The second inequality is left as an exercise for the reader.

(iii) Let k be the integer with $(k-1)2^{-n} \le p < k 2^{-n}$. Then

$$p \le k 2^{-n} < p + 2^{-n},$$

so, using (ii), we have

$$p \le \Pr(X_1 2^{-1} + X_2 2^{-2} + X_3 2^{-3} + \cdots + X_n 2^{-n} \le p) \le p + 2^{-n}.$$

The remaining inequalities are again left as an exercise. ∎

Lemma 9.3.12 *Suppose that I toss a fair coin n times and take $X_j = 1$, if the coin comes down heads on the jth throw, and $X_j = 0$, if it comes down tails.*

(i) If

$$X_1 2^{-1} + X_2 2^{-2} + X_3 2^{-3} + \cdots + X_r 2^{-r} \le p - 2^{-r}$$

for some $r \le n$, then

$$X_1 2^{-1} + X_2 2^{-2} + X_3 2^{-3} + \cdots + X_n 2^{-n} \le p - 2^{-n}.$$

(ii) If

$$X_1 2^{-1} + X_2 2^{-2} + X_3 2^{-3} + \cdots + X_r 2^{-r} > p$$

for some $r \le n$, then

$$X_1 2^{-1} + X_2 2^{-2} + X_3 2^{-3} + \cdots + X_n 2^{-n} > p.$$

Proof (i) If

$$X_1 2^{-1} + X_2 2^{-2} + X_3 2^{-3} + \cdots + X_r 2^{-r} \le p - 2^{-r},$$

then

$$X_1 2^{-1} + X_2 2^{-2} + X_3 2^{-3} + \cdots + X_n 2^{-n}$$

$$\le X_1 2^{-1} + X_2 2^{-2} + X_3 2^{-3} + \cdots + X_r 2^{-r} + 2^{-r-1} + \cdots + 2^{-n}$$

$$\le p - 2^{-r} + 2^{-r-1} + \cdots + 2^{-n} = p - 2^{-n}$$

(ii) Obvious. ∎

Using Lemmas 9.3.11 and 9.3.12, we get our key result.

Lemma 9.3.13 *Suppose that $0 < p < 1$. I play the following game. In each round, I throw a fair coin. If it comes down heads in the jth round, I set $X_j = 1$. If it comes down tails in the jth round, I set $X_j = 0$.*
   If, in the rth round

$$X_1 2^{-1} + X_2 2^{-2} + X_3 2^{-3} + \cdots + X_r 2^{-r} \le p - 2^{-r},$$

I record 'left' and stop the game. If, in the rth round

$$X_1 2^{-1} + X_2 2^{-2} + X_3 2^{-3} + \cdots + X_r 2^{-r} > p,$$

I record 'right' and stop the game. If neither event occurs, then I move on to the next round.
   Under these conditions, the probability that I record left is p and the probability that I record right is $1 - p$.

Proof Using Lemmas 9.3.12 (i) and 9.3.11 (iii), we see that the probability l_n that I write left in the nth round or earlier satisfies

$$p - 2^{-n} \le l_{-n} \le p.$$

Similarly, the probability r_n that I write right in the nth round or earlier satisfies

$$1 - p - 2^{-n} \le r_{-n} \le 1 - p$$

and the probability that I have written neither is exactly 2^{-n}.
   We observe that $l_n \to p$ and $r_n \to 1 - p$ as $n \to \infty$. ∎

Exercise 9.3.14 *Choose a value of p and carry out the suggested procedure several times.[4] Try and find a value of p (with $0 < p < 1$) for which the average number of tosses before a decision is reached is particularly long. Does experiment confirm your view?*

If the reader has done Exercise 9.3.14, she should be primed for the next exercise.

Exercise 9.3.15 *We use the set up of Lemma 9.3.13. Suppose that no decision has been reached after n rounds.*
   (i) Explain why

$$X_1 2^{-1} + X_2 2^{-2} \cdots + X_n 2^{-n} \le p < X_1 2^{-1} + X_2 2^{-2} + \cdots + X_n 2^{-n} + 2^{-n}.$$

   (ii) Explain why exactly one of the following two statements must be true

---

[4] If you feel that mathematicians should not condescend to toss coins, use the random number generator of your calculator or a table of random numbers. Odd digits correspond to heads, even to tails.

(A) If $X_{n+1} = 0$, then there is a decision in the $n + 1$th round, but, if $X_{n+1} = 1$, then there is no decision.

(B) If $X_{n+1} = 1$, then there is a decision in the $n + 1$th round, but, if $X_{n+1} = 0$, then there is no decision.

(iii) Deduce that the probability of a decision in the $n + 1$th round is $1/2$.

Combining Exercise 9.3.15 with Lemma 9.3.3 we obtain a rather unexpected result.

Lemma 9.3.16 *The expected number of tosses required by the procedure outlined in Lemma 9.3.13 is 2 whatever the value of p (with $0 < p < 1$).*

The ideas of this section can also be used to give an estimate for the expected number of proposals required by the marriage algorithm of Section 6.1 in the case that the preferences of the individuals involved are entirely random.

We start with a traditional problem.

Example 9.3.17 *The manufacturers of Kangaroo Cereal, 'the cereal that makes you jump for joy', decide to boost their sales by including a small plastic bust of one of the major Australian lyric poets in each of its packets. The company is far too public spirited to create an artificial shortage of a particular poet, so each packet is equally likely to contain any poet. Show that the expected number e of packets that you will need to buy to get the full set of n poets satisfies the inequality*

$$n \log n \le e \le n(1 + \log n).$$

Solution Let e_r be the expected number of packets you will need to buy before you get a new poet if you already have r poets. Since the probability of getting a new poet with any particular packet is $(n - r)/n$, Lemma 9.3.3 tells us that $e_r = n/(n - r)$. The total expected number of packets is thus

$$\sum_{r=0}^{n-1} e_r = \sum_{r=0}^{n-1} \frac{n}{n-r} = \sum_{r=1}^{n} \frac{n}{r} = n \sum_{r=1}^{n} \frac{1}{r}.$$

We now use Exercise 5.2.3 (iv). ∎

Exercise 9.3.18 *Swapping busts with other cereal buyers will usually reduce the number of packets you have to buy. Fix some $\epsilon > 0$ and suppose that N families club together to buy $(1 + \epsilon)Nn$ packets of the cereal described in Example 9.3.17 (so that each family only pays the cost of $(1 + \epsilon)n$ packets). Show, using Tchebychev's inequality (see, for example, Lemma 2.5.7), or otherwise, that there exists an $N_0(\epsilon)$ such that, if $N \ge N_0(\epsilon)$,*

$$\Pr(\textit{they obtain fewer than } N \textit{ busts of the } j\textit{th poet}) < \epsilon/n.$$

Deduce that, if $N \geq N_0(\epsilon)$, they will have at least N full sets of poets with probability at least $1 - \epsilon$.

Example 9.3.17 is the key to the proof of Theorem 9.3.19.

Theorem 9.3.19 *Consider the Gale–Shapley marriage algorithm of Section 6.1. Suppose that the preferences of the gentlemen involved are entirely random (so that each order of preference is equally likely) and independent of the preferences of the other gentlemen. Then the expected number of proposals is no greater that $n(1 + \log n)$.*

Proof Observe first that, when a gentleman makes a proposal, he is equally likely to make his proposal to any of the ladies he has not yet proposed to.

Let f_r be the expected number of proposals that will be made before a proposal is made to an unaffianced lady if there are already r affianced ladies. If there are r affianced ladies, then every proposal, will be made by someone who has not proposed to any of the $n - r$ unaffianced ladies. By the previous paragraph, the probability that he will propose to an unaffianced lady is

$$\frac{n - r}{\text{number of ladies he has not proposed to}} \geq \frac{n - r}{n}.$$

Since the probability of a successful outcome is at least as great in the case of the ladies as in the case of the cereals, we must have $f_r \leq e_r$, where e_r is the corresponding expectation in the cereal packet problem.

It follows that

$$f_r \leq \frac{n - r}{r}$$

and the total expected number of proposals is

$$\sum_{r=0}^{n-1} f_r \leq \sum_{r=0}^{n-1} \frac{n}{n - r} = n \sum_{r=1}^{n} \frac{1}{r} \leq n(1 + \log n),$$

using Exercise 5.2.3 (iv).   ∎

Exercise 9.3.20 *(i) Explain why the marriage algorithm must terminate after at most n^2 proposals.*

(ii) Suppose that all the gentlemen have the same order of preference for the ladies. Show that the marriage algorithm requires $n(n + 1)/2$ proposals.

9.4 HHH

Suppose that we are idly tossing pennies giving a sequence like

$$HHTHT\ldots$$

(heads, heads, tails, heads, tails ...). It is natural to bet on whether the sequence HTT will occur before the sequence HHH or not. Thus, if we get the sequence,

$$HHTHTHHT\mathbf{HTT}TTHHT\ldots$$

the person who has chosen HTT will win.

I ask you to choose a three letter sequence, after which I choose another three letter sequence. Does it matter what choice you make? To see that it does matter, suppose that you choose HHH. I will then choose THH.

Lemma 9.4.1 *The only way the first occurrence of HHH can precede the first occurrence of THH in a sequence of Ts and Hs is for it to be the first three letters.*

Proof Suppose the first occurrence of HHH starts at the rth letter with $r \geq 2$. The $r - 1$th letter cannot be H (since then there would be an occurrence of HHH starting at the $r - 1$th letter) so it must be T and the sequence THH starting at at the $r - 1$th letter precedes the first occurrence of HHH. ∎

Thus, if you choose HHH and I choose THH, the probability that you will win is the probability that the first three throws are heads, that is to say, $1/8$.

The situation may be made clearer by Figure 9.1, which shows how we can get from triples at the rth, $r + 1$th and $r + 2$th place in our sequence to triples at the $r + 1$th, $r + 2$th and $r + 3$th place. (Note that the only 'free choice' is that for the $r + 3$th place. The $r + 1$th and $r + 2$th places are fixed.)

We may think of HHH as being perfectly blockaded by THH, since, in order to get to HHH from anywhere except HHH itself, we have to pass through THH. If we look for other blockades, we see that only HHH and

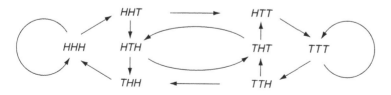

Figure 9.1. Triplet transitions.

TTT have perfect blockades, but there are other 'partial blockades' that look promising. For example, if you choose HTH, it seems worth my while to try HHT.

To see whether my guess works, we need to sit down and calculate. Let p_{XYZ}, be the probability that, starting from the triple XYZ, the sequence goes through HHT before it goes through HTH. Since the probability of each initial triple is $1/8$, we have

$$\text{Pr(I win)} = \frac{1}{8}(p_{HHH}+p_{HHT}+p_{HTH}+p_{HTT}+p_{THH}+p_{THT}+p_{TTH}+p_{TTT}).$$

Now we need to find the p_{XYZ}.

If we are at HHT, then I have won and if we are at HTH, then you have won. Thus

$$p_{HHT} = 1, \quad p_{HTH} = 0.$$

If we are at HTT, then with probability $1/2$, the next triple will be TTH and, with probability $1/2$, the next triple will be TTT. Thus

$$p_{HTT} = \frac{1}{2}p_{TTH} + \frac{1}{2}p_{TTT}.$$

Similar arguments show that

$$p_{TTT} = \frac{1}{2}p_{TTH} + \frac{1}{2}p_{TTT}, \quad p_{TTH} = \frac{1}{2}p_{THH} + \frac{1}{2}p_{THT}$$
$$p_{THT} = \frac{1}{2}p_{HTH} + \frac{1}{2}p_{HTT}, \quad p_{THH} = \frac{1}{2}p_{HHH} + \frac{1}{2}p_{HHT}$$
$$p_{HHH} = \frac{1}{2}p_{HHH} + \frac{1}{2}p_{HHT}.$$

It is easy to solve these 8 simultaneous equations as they stand, but we can reduce them to 4 by remarking that it is only the last two places of the initial triple that matter and so it is reasonable to look at

$$p_{YZ} = p_{HYZ} + p_{TYZ}.$$

We now get

$$\text{Pr(I win)} = \frac{1}{4}(p_{HH} + p_{HT} + p_{TH} + p_{TT})$$

and

$$p_{HH} = \frac{1}{2}(p_{HH} + 1), \quad p_{HT} = \frac{1}{2}(0 + p_{TT})$$
$$p_{TH} = \frac{1}{2}(p_{HH} + p_{HT}), \quad p_{TT} = \frac{1}{2}(p_{TH} + p_{TT}).$$

Exercise 9.4.2 *Explain the formulae just given in terms of probabilities.*

A first glance at our four equations reveals that

$$p_{HH} = 1, \quad p_{HT} = \frac{1}{2}p_{TT}, \quad p_{TH} = \frac{1}{2}(p_{HH} + p_{HT}), \quad p_{TT} = p_{TH}$$

and substitution in the third equation gives $\frac{1}{2}p_{TT} = \frac{1}{2}(1 + \frac{1}{2}p_{TT})$, so

$$p_{TT} = \frac{2}{3}, \quad p_{HH} = 1, \quad p_{HT} = \frac{1}{3}, \quad p_{TH} = \frac{2}{3}$$

and Pr(I win) $= \frac{2}{3}$.

Exercise 9.4.3 *(i) Show that, if you choose HTT and I choose HHT, then I have probability 2/3 of winning.*

(ii) Show that, if you choose HHT and I choose THH, then I have probability 3/4 of winning.

(iii) Use symmetry to give appropriate choices for me when you choose TTT, THT, THH and TTH.

Thus whatever triple you choose, I can choose a triple which gives me at least a 2/3 chance of winning. We thus have another example of non-transitive probabilities to add to those discussed on page 191.

Exercise 9.4.4 *Suppose we change to betting on two letter sequences. Are there any foolish choices for the first player? If one player chooses TH and the other HT how many tosses are required before the outcome of the game is known?*

Exercise 9.4.5 *We have shown that I can always choose a triple which gives me at least a 2/3 chance of winning. Can I do better? The simplest way forward may be to compute all the possibilities. At first sight this seems to involve 56 calculations, but things are not quite as bad as that.*
(i) Explain why

$$\Pr(XYZ \text{ beats } UVW) = 1 - \Pr(UVW \text{ beats } XYZ).$$

(ii) Let us write $\tilde{X} = T$ if $X = H$ and $\tilde{X} = H$ if $X = T$. Explain why

$$\Pr(XYZ \text{ beats } UVW) = \Pr(\tilde{X}\tilde{Y}\tilde{Z} \text{ beats } \tilde{U}\tilde{V}\tilde{W}).$$

Why does

$$\Pr(HTH \text{ beats } THT) = 1/2?$$

(iii) Explain why

$$\Pr(XYH \text{ beats } XYT) = 1/2.$$

Even so, the reader may prefer to take my word for most of the results of Table 9.1 and just check one or two entries.

Table 9.1 suggests another version of the HHH game. Since the second chooser has the advantage in the original game, let us demand that each player secretly writes down their choice and then reveals it to the other player. If the two players choose the same triple, then we have a draw. Otherwise, they play the HHH game and the loser pays the winner 1 unit. The expected value of the game to each side is then given by the Table 9.2.

Looking at Table 9.2, we see that it is always at least as good to choose *HHT* as to choose *HHH* and always at least as good to choose *HHT* as to choose *HTH*. Similarly, it is always at least as good to choose *TTH* as to choose *TTT* and always at least as good to choose *THH* as to choose *THT*. Thus it makes sense to look at the case when both you and I choose from the remaining strategies. We write out the table of expectations for the remaining strategies as Table 9.3.

Table 9.1. *Probability row triplet beats column triplet*

	HHH	HHT	HTH	HTT	THH	THT	TTH	TTT
HHH		1/2	2/5	2/5	1/8	5/12	3/10	1/2
HHT	1/2		2/3	2/3	1/4	5/8	1/2	7/10
HTH	3/5	1/3		1/2	1/2	1/2	3/8	7/12
HTT	3/5	1/3	1/2		1/2	1/2	3/4	7/8
THH	7/8	3/4	1/2	1/2		1/2	1/3	3/5
THT	7/12	3/8	1/2	1/2	1/2		1/3	3/5
TTH	7/10	1/2	5/8	1/4	2/3	2/3		1/2
TTT	1/2	3/10	5/12	1/8	2/5	2/5	1/2	

Table 9.2. *Expected value to row of row triplet against column triplet*

	HHH	HHT	HTH	HTT	THH	THT	TTH	TTT
HHH	0	0	−1/5	−1/5	−3/4	−1/6	−2/5	0
HHT	0	0	1/3	1/3	−1/2	1/4	0	2/5
HTH	1/5	−1/3	0	0	0	0	−1/4	1/6
HTT	1/5	−1/3	0	0	0	0	1/2	3/4
THH	3/4	1/2	0	0	0	0	−1/3	1/5
THT	1/6	−1/4	0	0	0	0	−1/3	1/5
TTH	2/5	0	1/4	−1/2	1/3	1/3	0	0
TTT	0	−2/5	−1/6	−3/4	−1/5	−1/5	0	0

Table 9.3. *Reduced triplet table*

	HHT	HTT	THH	TTH
HHT	0	1/3	−1/2	0
HTT	−1/3	0	0	1/2
THH	1/2	0	0	−1/3
TTH	0	−1/2	1/3	0

Table 9.3 is rather symmetric and has many zeros. After a certain amount of head scratching[5] it may occur to us that HTT 'looks rather better' than HHT and we should investigate strategies involving only HTT and THH. Symmetry suggests choosing each of these triples with probability 1/2.

We have now guessed a good strategy for one player. A guess is not a proof, but, fortunately, Lemma 7.3.6 enables us to check if our guess is correct.

Exercise 9.4.6 *Use Lemma 7.3.6 (and the symmetric nature of the game) to show that an optimal strategy is to choose the triples HHT and TTH each with probability $1/2$ and no other triples.*

If your opponent just plays one triple XYZ, for which of the triples XYZ will you have strictly positive expected winnings?

If your opponent is foolish enough always to choose either HTH or THT and foolish enough to let you know this, how should you change your tactics?

9.5 Tit for tat

In this section[6] we take a final look at the Prisoner's Dilemma described on page 213. We shall make the stakes smaller by supposing that the game is played for money.

The Friends of Italian Opera run an annual 'Prisoner's Dilemma Contest'. In each round each player may press a hidden switch. If they both press their switches, they both get 1 dollar, if neither presses their switches, then they get 2 dollars but if one presses their switch and the other does not, the one who presses the switch gets 3 dollars and the other gets nothing.

	no press	press
no press	(2, 2)	(0, 3)
press	(3, 0)	(1, 1)

---

[5] In the case of the author, rather more than he cares to admit.

[6] The ideas of this section are discussed in much greater detail and with more optimism in Axelrod's *The Evolution of Cooperation* [2].

We assume that the two participants can talk to one another but have no way of making each other keep their word.

Exercise 9.5.1 *Explain why this is the same Prisoner's Dilemma as described on page 213. Explain why 'mathematical reasoning' results in both partici- pants getting 1 dollar while 'reliable cooperation' results in both participants getting 2 dollars*

It has been suggested that, although 'mathematical reasoners' will always press the switch in a single game, they may cooperate if they have to play a long series of games. The shopkeeper may cheat the visitor but will play fair with his regular customers.

However, things are not this simple. Suppose that Little Bonaparte and Spats Columbo sit down to play this game 100 times. Consider the 100th game. Bonaparte knows that, whatever he does, there is no way that Spats can re- taliate since this is the last game. He therefore presses the switch and Spats, an equally acute reasoner, does the same. Now consider the 99th game. Both Bonaparte and Spats have worked out what will happen in the 100th game. Since Bonaparte knows that Spats will press the switch in the 100th game, whatever Bonaparte has done in the 99th game, there is no advantage to Bona- parte from not pressing the switch in the 99th game. Thus Bonaparte will press the switch in the 99th game and Spats will do the same. Now consider the 98th game . . . 'Backward induction' shows that Bonaparte and Spats will both press the switch each time.

Exercise 9.5.2 *The* Traveller's Club *has* 50 *members. Some of the members are gifted raconteurs and some are tedious bores. Naturally, each member knows which class every other member belongs to, but does not know about himself. One afternoon, the club secretary (who is not a member) makes the following announcement: 'At least one of the members is a bore. Anyone who knows that he is a bore must deliver his resignation, privately and in writing, to me on the evening of the day that he becomes certain that he is a bore. Each morning, I will post a list of resignations'. If, on the 49 mornings following, no resignations are announced, what will happen on the 50th morning and why?*

If the players do not know how many games will be played, the situation may be different.

Example 9.5.3 *Tania and Sonia play a series of rounds of the game described at the beginning of the section. At the end of each round they toss a coin to decide whether to play another round and there is a probability p that they will play a further round.*

Tania announces that she will play a 'tit for tat' strategy as follows. In the first round she will not press the switch and in every succeeding round she will play whatever Sonia played in the previous round. Sonia knows that Tania will keep her word. Advise Sonia.

A little thought shows that it may be easier to solve Example 9.5.3 if we combine it with Example 9.5.4.

Example 9.5.4 *Tania and Sonia play a series of matches as described in Example 9.5.3.*

Tania announces that she will play a 'tat for tit' strategy as follows. In the first round she will press the switch and in every succeeding round she will play whatever Sonia played in the previous round. Sonia knows that Tania will keep her word. Advise Sonia.

Suppose that Sonia has decided on her first move against tit for tat and her first move against tat for tit. If Sonia presses the switch in the $n - 1$th round and the experiment continues to the nth round, then Sonia knows that her nth decision will correspond exactly to the decision she would make in her first move against tat for tit. If Sonia does not press the switch in the $n - 1$th round and the match continues to the nth round, then Sonia knows that her nth decision will correspond exactly to the decision she would make in her first move against tit for tat.

Thus Sonia's decisions reduce to choosing 'press' or 'do not press' as first move against tit for tat and choosing 'press' or 'do not press' as first move against tat for tit. A little thought shows that the 4 possible choices can be thought of as 'always press', 'never press', 'always do the same as Tania', 'always do the opposite to Tania'.

Let

$B(I, P) =$ expected winnings against tit for tat of 'always press',
$B(I, N) =$ expected winnings against tit for tat of 'never press',
$B(I, S) =$ expected winnings against tit for tat of 'same as Tania',
$B(I, O) =$ expected winnings against tit for tat of 'opposite of Tania',
$B(A, P) =$ expected winnings against tat for tit of 'always press',
$B(A, N) =$ expected winnings against tat for tit of 'never press',
$B(A, S) =$ expected winnings against tat for tit of 'same as Tania',
$B(A, O) =$ expected winnings against tat for tit of 'opposite of Tania'.

If Sonia plays 'always press', then she will gain 3 in the first round. With probability p she will then have to play a game against tat for tit in which her expected winnings will be $B(A, P)$. Thus

$$B(I, P) = 3 + pB(A, P).$$

Similarly,

$$B(A, P) = 1 + pB(A, P)$$

and so

$$B(A, P) = \frac{1}{1 - p}, \quad B(I, P) = 3 + \frac{p}{1 - p} = \frac{3 - 2p}{1 - p}.$$

In the same way, we have

$$B(I, O) = 3 + pB(A, O), \quad B(A, O) = 0 + pB(I, O) = pB(I, O)$$

so that $B(I, O) = 3 + p^2 B(I, O)$ and

$$B(I, O) = \frac{3}{1 - p^2}, \quad B(A, O) = \frac{3p}{1 - p^2}.$$

Exercise 9.5.5 *Use similar arguments to prove the remaining results in this table:*

$$B(I, P) = 3 + \frac{p}{1 - p}, \qquad B(A, P) = \frac{1}{1 - p},$$

$$B(I, N) = \frac{2}{1 - p}, \qquad B(A, N) = \frac{2p}{1 - p},$$

$$B(I, S) = \frac{2}{1 - p}, \qquad B(A, S) = 0,$$

$$B(I, O) = \frac{3}{1 - p^2}, \qquad B(A, O) = \frac{3p}{1 - p^2}.$$

Exercise 9.5.6 *Obtain the results of the previous exercise by summing appropriate geometric series. (You may well find the geometric series method more illuminating in this particular case.)*

We now observe that

$$B(I, N) - B(I, P) = \frac{2 - p}{1 - p} - 3 = \frac{2p - 1}{1 - p},$$

$$B(I, N) - B(I, O) = \frac{2(1 + p) - 3}{1 - p^2} = \frac{2p - 1}{1 - p^2},$$

$$B(I, P) - B(I, O) = -\frac{3p^2}{1 - p^2} + \frac{p}{1 + p} = \frac{p(1 - 4p)}{1 - p^2}.$$

Thus

$$B(I, N) > B(I, P) \text{ for } p > 1/2, \quad B(I, P) > B(I, N) \text{ for } 1/2 > p,$$
$$B(I, N) > B(I, O) \text{ for } p > 1/2, \quad B(I, O) > B(I, N) \text{ for } 1/2 > p,$$
$$B(I, O) > B(I, P)) \text{ for } p > 1/4, \quad B(I, P) > B(I, O) \text{ for } 1/4 > p.$$

Looking at these results, we advise Sonia to play the strategy 'never press' (or, what turns out to be exactly the same strategy, 'do the same as Tania') whenever $p \geq 1/2$, to follow the strategy 'do the opposite of Tania' when $1/2 \geq p \geq 1/4$ (that is to say, press on the first, do not press on the second, press on the third, ...) and always press when $1/4 \geq p$. (If $p = 1/2$ or $p = 1/4$ there is a free choice between the two recommended strategies.)

Exercise 9.5.7 *Explain to a non-mathematician why we switch strategies as p becomes smaller.*

Exercise 9.5.8 *Suppose that Tania announces that she will follow a tat for tit strategy. Advise Sonia.*

Although this does suggest how 'long run' matches can be more favourable to cooperation than single games, we must be careful not to read too much into a single example.

Exercise 9.5.9 *Suppose that we have the situation discussed in Example 9.5.3, except that the table of outcomes is changed to read as follows.*

	no press	press
no press	(2, 2)	(0, 6)
press	(6, 0)	(1, 1)

Advise Sonia.

Explain to a non-mathematician why, when p is large, and Tania follows the tit for tat strategy Sonia should follow the tactic 'do the opposite of Tania'. Is the 'tit for tat strategy' well named in this case?

Exercise 9.5.10 *This exercise lies outside the main line of argument. However, the reader may wish to make sure that there are no other radically different versions of Exercise 9.5.9.*

Suppose that we have the situation discussed in Example 9.5.3 except that the table of outcomes is changed to read as follows,

	no press	press
no press	(c, c)	(0, d)
press	(d, 0)	(e, e)

with $d > c > e > 0$. Advise Sonia in the case when $2c > d$ and p is large. Advise Sonia in the case when $2c < d$ and p is large.

Even in the particular case which we have concentrated on in this section, it is not clear that we are seeing true cooperation. Observe that, if $p > 1/2$, and Tania declares that she is going to play 'tat for tit', we would still advise Sonia to play the strategy 'never press' although now Sonia will get less than Tania from the game.

There have been several experiments to see what will happen when two people play the Prisoner's Dilemma game many times without being allowed to talk to one another. The early experiments did not show any clear pattern. Presented with a non-zero sum game, people often converted it into a zero-sum game 'I win if I have more points than you' rather than 'My object is to collect as many points as possible irrespective of your score' or wander between these two possible goals.[7] They also tended to think in terms of 'punishing' or 'rewarding' the other players. The fact that the two players could only signal to each other by means of their choices of moves added a further layer of complication.[8]

More clear cut results have emerged when computer programs are allowed to play one another. There are two ways of organising such contests. In one, the various programs each play matches (which we will assume to be of the type discussed earlier with p close to 1) against each other once and the program with most points wins. In the other, we return to the biological metaphor of page 244 and let randomly chosen programs play matches. At some point, low scoring programs are eliminated and extra copies of high scoring programs are added. It turns out that 'tit for tat' is remarkably successful in both sorts of contests.

[7] This difficulty is inherent in the way we see the world. Should we use a *relative* standard and say that X is poor if he earns less than 90% of the population? In that case, nothing can reduce the number of poor people. Or should we use an *absolute* standard and say that someone is poor if they cannot afford certain necessities? In that case we could claim that anyone who always has enough to eat is not poor, since such a person would have been considered well-off in 1500. Neither choice seems entirely satisfactory.

[8] This paragraph has been heavily criticised by two readers. The first objected that the experiments were irrelevant. 'It is as if, having decided that the question "How high is a mouse when it spins?" is not one of applied mathematics you then tried to decide it by using an opinion poll.' The second objected that, perhaps because the early experiments involved mathematicians and economists, they gave untypical results. A great deal of experimental work has been done (see, for example, [14]) and reveals a strong bias towards cooperation. I have left the paragraph as it stands to provide an excuse for this footnote.

Exercise 9.5.11 *Suppose that you have to write a program to compete in a contest in which the success of a program is measured by*

$$number\ of\ wins + number\ of\ draws$$

rather than points scored. What would you do?

It appears that the reasons for the success of 'tit for tat' is similar to the reasons for the success of the 'bourgeois bird' in our discussion of hawks and doves. However, 'tit for tat' does not always win and the actual result depends on the initial mix of programs and the detailed scoring system.

Exercise 9.5.12 *In this question you should use the informal arguments of our discussion of hawks and doves. We assume the reward system set out in the first paragraph of this section (page 271).*

(i) Explain why, if we introduce a few 'always presses switch' programs into a very large flock of 'tit for tats', the 'always presses switch' programs will probably be eliminated.

(ii) Explain why, if we introduce a few 'tit for tats' into a very large flock of 'always presses' programs, the 'tit for tat' programs will probably be eliminated.

(iii) What will probably happen if we introduce a few 'always presses' into a very large flock of 'never presses'? What can you say about the case when we introduce a few 'tit for tats' into a very large flock of 'never presses'?

When practised by human beings, 'tit for tat' suffers from the fact that what one side sees as a 'measured response' may be seen as 'totally disproportionate' by the other.[9]

9.6 Foundational matters

In Section 2.1, we took a fair amount of trouble to set up the rules for a simple theory of probability. However, the kind of readers I expect for this book[10] will probably have noted that the simple theory we set up dealt with finite probability spaces, that is to say, systems in which there are only a finite set of outcomes. The long duels that we looked at in this chapter do not have a finite set of outcomes.

---

[9] A different but related problem arises if our game is played by computers but we introduce a some uncertainty so that, occasionally, the instruction 'do not press' is transmitted as 'press' and vice versa. This is known as the 'problem of the trembling hand'.

[10] A small subset of the readers I would like for this book

In our simplest long duel, A fires at a target. The probability of his hitting the target with any one shot is a and he stops when he first hits the target. Since he may stop after 1, 2, 3, ... shots, there are an infinity of outcomes.

Once we start tossing coins, things get even more complicated, since Cantor's diagonal argument[11] shows that there are uncountably many ways of throwing a sequence of heads and tails.

Exercise 9.6.1 *(For those who know the word uncountable.) Consider the HHH game of Section 9.4. Suppose I have chosen HHH and you have chosen TTT. By considering patterns of throws in which the 2rth and $2r + 1$th throw are HT or TH, show that there uncountably many different sequences for which the game goes on forever.*

I shall give the three different ways of dealing with this difficulty. I hope that the reader will be satisfied with one of them.

Level one Section 2.1 was intended to show that it was possible to set up a complete theory of probability in a very simple case. This book is a 'taster' rather than a textbook, so it makes sense to look at more general questions without setting up the full underlying theory.

Level two If we look at the three duellists of Section 9.2, we are really considering a duel in which A, B and C follow certain strategies and which finishes when only one player is left standing or n shots have been fired. We are interested in

$$p_n(A) = \Pr(A \text{ wins the } n \text{ shot match}),$$
$$p_n(B) = \Pr(B \text{ wins the } n \text{ shot match}),$$
$$p_n(C) = \Pr(C \text{ wins the } n \text{ shot match}),$$
$$q_n = \Pr(n \text{ shot match undecided}).$$

If $q_n \to 0$, $p_n(A) \to p(A)$, $p_n(B) \to p(B)$ and $p_n(C) \to p(C)$ as $n \to \infty$, it is surely reasonable to say that the probability of A, B or C winning the unlimited duel is $p(A)$, $p(B)$ or $P(C)$ respectively.

All the long duels we consider can be dealt with in this way. Sometimes we neglect to show explicitly that

$$\Pr(\text{duel undecided after } n \text{ rounds}) \to 0,$$

but this is usually easy to do using an argument like the following.

Lemma 9.6.2 *A plays a succession of rounds of a game. If he has not stopped in the first $n - 1$ rounds, the probability that he stops after the nth round is at least q, independent of what happened earlier. If $q > 0$,*

[11] If you do not know what the words of the sentence mean, ignore it and all similar sentences.

$$\Pr(A \text{ stops at or before the } n\text{th round}) \to 1.$$

Proof If q_n is the probability that A plays the nth round, then

$$q_{n+1} \le q_n(1 - q)$$

so, by induction, $q_n \le (1 - q)^n$. Thus, if $q > 0$, $q_n \to 0$ as $n \to \infty$. ∎

Exercise 9.6.3 *A fruit machine has three windows which show one of three letters A, B or C. When the window changes, each of the letters has probability 1/3 of appearing, independently of what happens to the other windows. It costs 10 pence to operate the machine which pays out 30 pence if, at the end of a go, all windows show the same letter.*

After the coin is inserted, but before anything else happens, there is a proba- bility 1/2 that 'hold' lights. If 'hold' lights and all the windows show the same letter, the go ends (and the machine pays out). If 'hold' lights and 2 windows show the same letter, only the other window changes at random and the go ends. If 'hold' lights and all the windows show different letters then all three windows change at random.

A (small time) gambler starts playing with the machine and stops at the first time after that when all the windows show different letters. Show that the probability that she plays an nth round is no greater than $(8/9)^{n-1}$ and deduce that the expected number of rounds that she plays and so her expected winnings are finite.

If she starts when all the windows show different letters, show that her expected loss is 5 pence.

Exercise 9.6.4 *There is another method for finding the expected length of duels which does not fit so well with later parts of this book, but which is both simple and useful.*

(i) Let X be the number of rounds before a certain game ends. If it is certain that $X \le N$, show that

$$\mathbb{E}X = \sum_{r=1}^{N} \Pr(X \ge r).$$

Convince yourself that, more generally, provided that $\sum_{r=1}^{\infty} \Pr(X \ge r)$ converges, we have

$$\mathbb{E}X = \sum_{r=1}^{\infty} \Pr(X \ge r).$$

(ii) Use the method of (i) to provide a third proof of Lemma 9.3.3.

(iii) The chairman of the Tripos Reform Procrastination Committee decides to plant an oak forest containing n oaks. Showing the patience and sense of system which make her such a good chairman, she decides to proceed as follows.

In the first year she plants n acorns. Each year thereafter she plants fresh acorns wherever the previously planted acorns have failed to germinate. If each acorn has probability q of failing and $0 < q < 1$, show that the expected number of yearly plantings is

$$\binom{n}{1}\frac{1}{1-q} - \binom{n}{2}\frac{1}{1-q^2} + \binom{n}{3}\frac{1}{1-q^3} - \ldots + (-1)^{n-1}\binom{n}{n}\frac{1}{1-q^n}.$$

In cases like our HHH game we need to modify our argument slightly.

Exercise 9.6.5 *Suppose that we observe two people playing the HHH game, but we only look at the game after the 3rd 6th, 9th toss and so on. Show that, if the game has not been decided after the 3rth toss, then,* whatever the state of the game, *the probability that it will have finished before or on the $3(r+1)$th toss is at least $1/4$. Deduce that the probability that the game is undecided after n tosses tends to zero as $n \to \infty$.*

Level three[12] As we said earlier on page 36, our version of probability on finite probability spaces is easily extended to countably infinite probability spaces

$$\Omega = \{\omega_1, \omega_2, \ldots\}$$

if we assign probabilities to an event $A \subseteq \Omega$ by writing

$$\Pr(A) = \sum_{\omega_j \in A} p(\omega_j).$$

When we consider the HHH game we do, indeed, treat an uncountable probability space. However, in games of this type, Ω has a countable subset

$$\hat{\Omega} = \{\omega_1, \omega_2, \ldots\}.$$

associated with real numbers $p(\omega_j)$ such that

$$p(\omega_j) \geq 0 \text{ for all } j \geq 1 \quad \text{and} \quad \sum_{j=1}^{\infty} p(\omega_j) = 1.$$

We now assign probabilities to every event $A \subseteq \Omega$ by writing

$$\Pr(A) = \sum_{\omega_j \in A} p(\omega_j).$$

---

[12] Ignore this unless you have a reasonable grasp of the notion of countability.

Exercise 9.6.6 *Explain why in the HHH game there are only countably many terminating games.*

Dealing with the kind of probability space described in the previous paragraph is no more difficult than dealing with a countable probability space. However, the structure of the space is very simple

$$\Omega = \hat{\Omega} \cup \hat{\Omega}^c$$

with $\hat{\Omega}$ countable and $\Pr(\hat{\Omega}) = 1$ while $\Pr(\hat{\Omega}^c) = 0$. Speaking informally, we deal only with probability spaces which can be split into a countable part and an uncountable part in such a way that the uncountable part has probability 0. Such spaces are far too special for the needs of modern probability theory.

To show what those needs are, let me quote some results by Borel from the dawn of the modern subject. Borel was able to produce a convincing argument for their truth so, perhaps, the reader may also be able to do so. Even if she does, she will see that she has called on ideas which go far beyond those used in this book.

Theorem 9.6.7 *Suppose we toss a fair coin infinitely often. Write $X_n = 1$ if the nth throw is heads and $X_n = -1$ if the nth throw is tails.*
(i) The probability that

$$\sum_{n=1}^{\infty} \frac{X_n}{n}$$

converges is 1.
(ii) The probability that

$$\sum_{n=1}^{\infty} \frac{X_n}{n^{1/2}}$$

converges is 0.

10

A night at the casino

10.1 How to gamble if you must

If there was one lesson to be drawn from the discussions in the first part of this book, it was never to gamble when the odds are against you. But, sometimes, we have no choice.

Example 10.1.1 *Dubrovsky sits down to a night of gambling with his fellow officers. Each time he stakes u roubles, there is a probability p that he will win and receive back 2u roubles (including his stake), and a probability $1 - p$ that he will lose his stake. At the beginning of the night he has 8000 roubles. If ever he has 256 000 roubles he will marry the beautiful Natasha and retire to his estates in the country. Otherwise he will join a monastery. He decides to follow one of two courses of action.*

(i) To stake 1000 roubles each time until the issue is decided.

(ii) To stake everything each time until the issue is decided.

Advise him (a) if $p = 1/4$ and (b) if $p = 3/4$. What is the probability of a conventional happy ending in each case if he follows your advice?

We know how to calculate the appropriate probabilities if he follows (ii).

Exercise 10.1.2 *Show that the chances of a happy ending if Dubrovsky follows (ii) are p^5. Verify that, if $p = 1/4$ the probability of a happy ending is roughly 0.001 and, if $p = 3/4$, roughly 0.237.*

What happens if he follows (i)? We re-use an idea from the previous chapter and write q_n for the probability of a happy ending if he starts with n thousand roubles. Obviously $q_0 = 0$ and $q_{256} = 1$. Suppose $1 \leq n \leq 255$. In the next round of play, there is a probability $(1 - p)$ that he will lose and be left with $n - 1$ thousand roubles and a probability q_{n-1} of a happy ending. On the other hand, there is a probability p that he will win and be left with $n + 1$ thousand roubles and a probability q_{n+1} of a happy ending. Putting this together, we see that

$$q_n = (1 - p)q_{n-1} + pq_{n+1}$$

or, rearranging,

$$pq_{n+1} - q_n + (1 - p)q_{n-1} = 0. \qquad \bigstar$$

How do we solve equations like $\bigstar$? I shall give a systematic treatment of such *difference equations* in Section 10.3, but an obvious way to start is to look at a simpler equation.

Exercise 10.1.3 *Show that if $t \neq 0$ and*

$$q_{n+1} - tq_n = 0$$

for all n then

$$q_n = q_0 t^n.$$

It is natural to look for solutions to $\bigstar$ of the form $q_n = t^n$ with $m \neq 0$. If we do so, we obtain

$$pt^{n+1} - t^n + (1 - p)t^{n-1} = 0,$$

whence

$$pt^2 - t + (1 - p) = 0,$$

Simple factorisation[1] (or the quadratic formula) gives

$$\big(pt - (1 - p)\big)(t - 1) = 0$$

so

$$t = 1 \quad \text{or} \quad t = \frac{p}{1 - p}.$$

It is natural to try

$$q_n = A + B \left(\frac{1 - p}{p} \right)^n$$

as a solution to $\bigstar$.

Exercise 10.1.4 *(i) Verify that*

$$q_n = A + B \left(\frac{p}{1 - p} \right)^n$$

---

[1] Aided by the knowledge that, in probability questions, 1 frequently arises as a root.

is a solution of ★ *for all choices of A and B.*

  (ii) Find A and B so that $q_0 = 0$ *and* $q_{256} = 1$.

Exercise 10.1.4 shows that

$$q_n = \frac{1 - \left(\frac{1-p}{p}\right)^n}{1 - \left(\frac{1-p}{p}\right)^{256}}$$

satisfies all the conditions we have placed on q_n and it is not hard to persuade ourselves that this is the only possible solution. (See Exercise 10.3.5 for details.)

If $p = 1/4$, we get

$$q_8 = \frac{3^8 - 1}{3^{256} - 1} < 3^{-248},$$

so the chance of a happy ending is negligible. If $p = 3/4$, we get

$$q_8 = \frac{1 - 3^{-8}}{1 - 3^{-256}} \approx 1 - 3^{-8}$$

and the probability of an unhappy ending is less than $1/5000$.

This result accords with common sense. If we have positive expectation in each bet, we want to make many small bets so that we cannot be wiped out by a short run of bad luck. If we have negative expectation, our only hope of winning is to have a run of good luck, so we should make our bets as large as possible in order to make the most of any such run.[2] This common sense argument is reinforced by 'laws of large numbers' like Theorem 2.5.13 which tells us that we are almost certain to lose if we make a large number of bets with negative expectation.

More generally, if you must win a certain fixed sum in a series of identical unfavourable games, the example above suggests that you should always bet the minimum required to reach the required outcome or your entire fortune if you cannot reach the outcome in one game (this strategy is called *bold play*). It is not immediately clear how to turn this suggestion into a theorem, but Dubins and Savage showed in their book *How To Gamble If You Must* [17] that this advice is correct in quite general circumstances. However, they also point out that, under certain circumstances, the strategy suggested may not be the only one.

---

[2] The Kelly bettor bets because she *can* and she wishes to maximise her expected fortune over the long term. She will not bet in unfavourable situations and her bets may be quite large if the situation is very favourable. We bet because we *must* and our only goal is to maximise our chance of obtaining a fixed sum.

Exercise 10.1.5 *(i) I am in Las Vegas and, as a result of events which I do not intend to relate, find that I must pay some rather sinister individuals 2560 dollars by day break. The events have left me with 1440 dollars in my pockets. I decide to play* even *or* odd *at roulette, so I have a probability* $1 - p$ *of losing my stake in any game and a probability p of having twice my stake returned. Naturally* $p < 1/2$. *Show that the following two strategies have identical probabilities of working.*

(a) I bet the minimum required to reach the required 2560 dollars (so my first bet is 1120 dollars) or my entire fortune if I cannot reach the outcome in one game (so if I lose the first bet, I stake my remaining 320 dollars on the second game).

(b) I split my capital into a back pocket of 1280 dollars and a purse of 160 dollars. I then try to reach 1280 dollars using my purse alone (so my first bet is 160 dollars, then, if I win, my second bet is 320 and so on). If I lose my purse money, I now bet my entire back pocket money.

(ii) Suppose that I actually have 2010 dollars. Can you suggest there distinct strategies which are as good as bold play?
[*Hint: observe that* $770/1280 = 1440/2560$.]

Exercise 10.1.6 *Suppose you play a game in which for a stake of 1 dollars you receive* p^{-1} *dollars back with probability p and nothing with probability* $1 - p$. *Suppose you start with k dollars and need to win l dollars with* $l > k$. *(Thus you leave the game either with l dollars or with nothing.) Explain why* whatever reasonable strategy you adopt[3] *you have probability* k/l *of leaving with your desired fortune. (This book does not supply the equipment needed for a watertight proof but you should be able to give a simple argument which caries conviction.)*

Exercise 10.1.7 *We have advised those who gamble because they must to gamble as slowly as possible if the odds are favourable. However, this advice is only useful if the gambler has unlimited time.*

Consider someone who is retiring in n years time. If her fortune on retirement is 1 or more, she will be happy, otherwise she will be unhappy. If at the beginning of the rth year before retirement she has a fortune f she may gamble an amount g with $0 \leq g \leq f$ $[1 \leq r \leq n]$. *With probability p she will win and her fortune will be* $f + g$ *and with probability* $1 - p$ *she will lose and her fortune will be* $f - g$.

---

[3] No strategy which involves betting arbitrarily small sums is reasonable. No strategy which allows your total fortune to exceed *l* at any time is reasonable.

(i) What should she do if $p < 1/2$?

(ii) From now on we assume $p \geq 1/2$. Let f be her fortune at the end of the rth year before retirement and let $p_{r-1}(f)$ be the probability that, if she uses the best tactics, she retires with a fortune of 1 or greater. Explain why

$$p_0(f_0) = \begin{cases} 1 = p_0(1) & \text{if } f \geq 1, \\ 0 = p_0(0) & \text{if } f < 1. \end{cases}$$

Show, by induction, or otherwise that

$$p_r(f) = p_r(k2^{-r}) \text{ if } k2^{-r} \leq t < (k+1)2^{-r}$$

for integers k with $0 \leq k \leq 2^r$.

(iii) Let $p(k, r)$ be the probability that, if she has a fortune $k2^{-r}$ at the end of the rth year before retirement and uses the best tactics, she retires with a fortune of 1 or greater. Explain how, knowing $p(m, r)$ for all $0 \leq m \leq 2^n$, she can find the best tactics if she has a fortune $k2^{-r-1}$ at the end of the $r + 1$th year before retirement and compute $p(k, r + 1)$.

(iv) Suppose p and F are fixed with $1 > p > 1/2$ and $1 > F > 0$. Suppose the subject of the question has a fortune F at the beginning of the nth year. How and why will her behaviour differ between the cases n large and n small. (You are only asked for a brief plausible answer.)

10.2 Boldness be my friend

We have seen that, if it is necessary to win a certain sum in a series of identical unfavourable games, it seems best to adopt a bold strategy. Casinos offer a variety of games. Which should we choose?

Exercise 10.2.1 *A modified roulette wheel has the integers r with $0 \leq r \leq 32$. The probability that any particular r 'wins' is a particular spin is 1/33 independent of anything that may have gone before. The casino offers the following bets.*

(a) You may bet on any single number r. If it loses, you lose your stake, if it wins, the house returns 32 times your stake.

(b) You may bet on 'odd'. If the number which wins is even you lose your stake, if it is odd the house returns 2 times your stake.

Show that the expected return on each bet is 32/33 times your initial stake.

Suppose that you have a fortune of 100 *dollars and that you need* 3200. *What is the probability of leaving with the required sum if you decide to use the bold strategy with bets of type (b)? What is the probability of leaving with the required sum if you decide to use a bold strategy with bets of type (a) and bet your entire* 100 *dollars on a single number?*

Exercise 10.2.1 suggests that, in the circumstances discussed, given two bets with identical negative expectation, we should choose the one with greatest scatter. Since we are trying to mitigate the effect of Theorem 2.5.13 this suggests (but, of course, does not prove) that we should choose bets with large variance. The next exercise shows that this can only be considered a rule of thumb.

Exercise 10.2.2 *(i) You need to win* $\$2^8$ *starting with one dollar. You are offered the choice between game A which, if you bet $\$k$, returns $\$2^8k$ with probability* 2^{-10} *(but nothing otherwise), game B which, if you bet $\$k$, returns $\$2^2k$ with probability* 2^{-3} *(but nothing otherwise) and game C which, if you bet $\$k$, returns $\$2^{80}k$ with probability* 2^{-120} *(but nothing otherwise). Show that game B has the highest expected value for a given bet and game C has the highest variance, but, if you adopt the bold play strategy, you will prefer game A.*

(ii) Game D combines the virtues of games B and C. If you bet $\$k$ it returns $\$2^2k$ with probability 2^{-3} *and $\$2^{80}k$ with probability* 2^{-120} *(but nothing otherwise). Show that game D has higher mean and higher variance than game A but give a reasonably convincing argument (you are not asked for anything resembling a proof) that, if you need to win* $\$2^8$ *starting with one dollar, you should still prefer game A.*

An interesting illustration of the virtues of high variance in unfavourable games is given by a fictitious casino which is prepared to offer games in which, with probability $1 - p$, you lose your stake and, with probability p, you win back kp^{-1} times your stake. The value of k is fixed by the casino with $k < 1$, but you may choose p. (In other words the casino offers a 'fair game' at any odds you choose but retains a fixed proportion $1 - k$ of your winnings 'to cover expenses'.)

Exercise 10.2.3 *Let X be the amount you win in one round of such a game when you wager* 1 *unit. Show that* $\mathbb{E}X = k$ *for all p with* $0 < p \leq 1$. *Compute* var X *and show that* var $X \to 0$ *as* $p \to 1$ *and* var $X \to \infty$ *as* $p \to 0$.

Suppose that you enter such a casino with a fortune $0 < \mathcal{F} < 1$ and you wish to maximise the probability that you leave the casino with fortune 1. A simple bold strategy is to request a game with $p = k\mathcal{F}$ and bet your entire fortune.

Exercise 10.2.4 *Explain why the probability of success with this strategy is at most k, independent of the value of $\mathcal{F}$. Is the strategy reasonable when $\mathcal{F}$ is close to unity?*[4]

Remembering our prejudice in favour of games with large variance, you might be inclined to try the following rather different strategy. Split your fortune into n equal parts and place each sum $\mathcal{F}/n$ in a separate purse. Bet the contents of the first purse at odds which will enable you to walk out of the casino if you win. If your first bet comes off, leave. If you lose your first bet, bet the contents of the second purse at odds which will enable you to walk out of the casino if you win. If your second bet comes off, leave. If you lose your second bet, bet the contents of the third purse at odds which will enable you to walk out of the casino if you win and so on.

Exercise 10.2.5 *(i) Show that, if you have to bet the contents of your rth purse, you will choose*

$$p = p_r = \frac{\mathcal{F}k}{n(1 - \mathcal{F}) + r\mathcal{F}}$$

for this bet.

(ii) Show that the probability q_n of failure with this tactic satisfies the equation

$$\log q_n = \sum_{r=1}^{n} \log(1 - p_r).$$

(iii) Show that $0 < p_r \le \mathcal{F}(1 - \mathcal{F})^{-1}/n$ for all r with $1 \le r \le n$.

We know, from Exercise A.9 or elsewhere, that

$$\log(1 - x) \approx -x$$

---

[4] See Damon Runyon's *All Horseplayers die Broke*.

when x is small. Thus, using the standard approximation to the integral of a well-behaved function by an appropriate sum, we obtain

$$\log q_n = \sum_{r=1}^{n} \log(1 - p_r) \approx - \sum_{r=1}^{n} p_r$$

$$= - \sum_{r=1}^{n} \frac{\mathcal{F}k}{n(1 - \mathcal{F}) + r\mathcal{F}} = -\mathcal{F}k \sum_{r=1}^{n} \frac{1}{n} \times \frac{1}{(1 - \mathcal{F}) + \frac{r}{n}\mathcal{F}}$$

$$\approx -\mathcal{F}k \int_{0}^{1} \frac{dx}{(1 - \mathcal{F}) + x\mathcal{F}} = -k[\log\left((1 - \mathcal{F}) + x\mathcal{F}\right)]_0^1$$

$$= -k \log(1 - \mathcal{F})$$

when n is large.

Thus, when n is large, we have,

$$\Pr(\text{success}) \approx 1 - (1 - \mathcal{F})^k$$

which is obviously better than our simple bold strategy when $\mathcal{F}$ is close to 1.

Exercise 10.2.6 *(Only do this if you want to make the preceding argument rigorous.)*
(i) *Let* $g(x) = x + x^2 + \log(1 - x)$. *Show that* $g'(x) \le 0$ *for* $1/2 \ge x \ge 0$ *and deduce that*

$$x + x^2 \ge -\log(1 - x)$$

for $0 \le x \le 1/2$. *Show also that*

$$-\log(1 - x) \ge x$$

for $0 < x < 1$.
(ii) *Use (i) and an appropriate theorem on integrals (to be quoted precisely) to show that, if we define* q_n *as above, then, as* $n \to \infty$,

$$q_n \to (1 - \mathcal{F})^k.$$

Exercise 10.2.7 (i) *Let* $1 > k > 0$ *and* $g(x) = 1 - (1 - x)^k - kx$. *Show that* $g'(x) > 0$ *for all* $0 < x < 1$ *and deduce that* $g(x) > g(0) = 0$ *for all* $1 > x > 0$.

Deduce that our division strategy (with n sufficiently large) is always better than our simple bold strategy.[5]

---

[5] It can be shown that as $n \to \infty$ the division strategy approaches the upper bound on what is possible with any strategy (see [17]), but the arguments here do not show this.

(ii) Let $h(x) = 1 - (1 - x)^k$. Explain why

$$\frac{1 - (1 - x)^k}{kx} = \frac{1}{k} \times \frac{h(x) - h(0)}{x} \to \frac{h'(0)}{k} = 1$$

and deduce that our division strategy is not much better better than the simple bold strategy when $\mathcal{F}$ is small.

The suggestion 'presented with two unfavourable games with the same expectation, go for the one with the higher variance' is less useful than it seems. Unlike their clients, casinos are risk averse (see page 105) and demand higher rewards (greater expectations) for more risky (higher variance) bets.

Exercise 10.2.8 *Suppose that the casino of Exercise 10.2.1 also offers a more general version of option (b) in which with probability $1 - p$ you lose your stake and with probability p you win back twice your stake. For what values of p will you prefer this game to (a)?*

What is the expected value to the casino of the outcome if you place 1 dollar on game (a)? What is the expected value to the casino of the outcome if you place 1 dollar on the game described in this exercise at the smallest value of p for which you do not prefer game (a)?

To some extent, what is good for us is bad for the casino, so we may expect casinos to dislike games in which stakes can be redoubled many times or which have very high variance. However, we delay discussion of the casino's point of view until Section 10.4.

It is worth making a practical point. If we wish to compare the advantages of two bets we need to know their expected mean. We also need to have some measure of there 'riskiness' and, although (as we saw in Exercise 10.2.2) the variance is not a perfect measure of riskiness, it is often quite a good one. However, when comparing bets it is better to use standard deviation σ rather than the variance σ^2. Observe that, if we increase the sum bet by a factor of 10 so that the expected mean increases by a factor of 10, the standard deviation also increases by a factor of 10 but the variance increases by a factor of 100.

We now look at a strategy which, instead of seeking to get a certain sum out of the casino and then leave, seeks to draw a steady income from the casino. This is the martingale or 'double or quits strategy'. Suppose that I start with a capital of 2 560 000 dollars. At the beginning of each month I go to my local casino and stake 10 000 dollars on a bet which returns nothing with probability p and returns twice my stake with probability $(1 - p)$. If I win, I live off the 10 000 dollars I have added to my fortune for the month and return the next month. If I lose, then I stake 20 000 dollars. If I win, I have again added 10 000 dollars to my fortune and I live off this until the next month. If I lose for the second time, then I stake 40 000 dollars and so on. If I ever lose my 2 560 000 dollars, there is nothing I can do.

In this way, unless I suffer the unlikely misfortune of losing 8 times in a row, I can live happily for many months secure in my possession of an 'infallible gambling system'.

Exercise 10.2.9 *Verify the statements just made. Show that, if* $p = 1/2$, *the probability* q *that I will lose in a particular month is* $1/256$.

Observe that the result of the last sentence is predictable, since, if $p = 1/2$, the game is fair so the expected value of the game to me is 0. Thus

$$2\,560\,000 = \text{amount I enter casino with}$$
$$= \mathbb{E}(\text{amount I leave casino with})$$
$$= q \times 0 + (1 - q)(2\,560\,000 + 10\,000)$$

and $2\,560\,000q = 10\,000$.

So far so good. However, even if the game is fair, my luck cannot last forever. The probability that I do not lose my entire fortune in the first 256 months is

$$\left(1 - \frac{1}{256}\right)^{256} \approx e^{-1} \approx 0.37$$

so, if I rely on my infallible system to keep me going for my lifetime, I am likely to have a very nasty shock.

The problem with martingale systems is shown vividly if we calculate my expected return. This will be

$$10\,000 \times \mathbb{E}(\text{expected number of months to bankruptcy}).$$

If $p = k/2$, then the probability of bankruptcy in a particular month is $p^8 = 2^{-8}k^8$ and so, by Lemma 9.3.3, the expected number of times I enter the casino is $1/p^8 = 2^8k^8$. Thus the expected return from an investment of 2 560 000 is

$$2\,560\,000 \times k^8.$$

Taking a typical value of $k = .95$ we get an expected return of about 1 698 356 dollars.

Exercise 10.2.10 *(i) What is the expected return if* $k = 1$? *Why should we expect this.*

(ii) What is the return if k *takes the extremely favourable value* $k = 0.99$?

Exercise 10.2.11 [Nick Leeson's game][6] *If the casino allows unlimited credit, things look rather different. Suppose that, if you announce a bet of c dollars, the casino will pay you uc dollars with probability p but otherwise you lose c dollars. If you start with a (a may be negative since the casino allows you run up unlimited debts), give a strategy which ends with you owning b using the least expected number of bets. Show that your strategy does indeed use the least expected number of bets. (Naturally p > 0, u > 0, b > a and you must choose c > 0.) Show that the expected number of bets with this strategy does not depend on a, b or u.*

In some sense, every bet we make is taxed by the casino. The bold strategy attempts to reduce our exposure to this tax by making our stay in the casino as short as possible. In the martingale strategy we allow the casino to tax largish sums of money repeatedly.

The fact that (under quite wide conditions) no strategy will give either side a strictly positive expectation in a series of fair games is the basis for a very powerful mathematical theory. Mathematicians have taken over the name of an unwise gambling system and call this theory 'Martingale Theory'.[7]

10.3 Difference and differential equations

This section consists of a series of exercises comparing linear second order differential equations with linear second order difference equations (both having constant coefficients) and showing how to solve them in some standard situations. Although these results form a useful background to the next section, they are not essential.

Exercise 10.3.1 *Suppose that a, b and c are constants and $f : \mathbb{R} \to \mathbb{R}$ is continuous. If*

$$au''(x) + bu'(x) + cu(x) = f(x)$$

and

$$av''(x) + bv'(x) + cv(x) = 0,$$

[6] The name comes from a trader who demonstrated the difference between *unlimited* credit and *apparently unlimited* credit by losing so much money that he destroyed the bank he worked for.
[7] Mathematicians believe that the word martingale comes from a system of straps intended to keep a horse's head at its proper level. The *Oxford English Dictionary* provides the alternative theory 'that the inhabitants of Martigues, a remote town, were eccentric and naive; hence ... the application to an apparently foolish system of gambling'.

show that, if k is constant and $w = kv + u$, then

$$aw''(x) + bw'(x) + cw(x) = f(x).$$

State and prove a similar result for the equation $au'(x) + bu(x) = f(x)$.

Exercise 10.3.2 *(i) Let u be twice differentiable and let us write $Iv(x) = v(x)$, $Dv(x) = v'(x)$ and $D^2v(x) = v''(x)$. Show that*

$$(D - aI)(D - bI)u = (D^2 - (a+b)D + abI)u.$$

(ii) Suppose that

$$u'(x) - au(x) = f(x).$$

Show that

$$\frac{d}{dx}(e^{-ax}u(x)) = e^{-ax}f(x)$$

and deduce that

$$u(x) = u(0)e^{ax} + e^{ax}\int_0^x e^{-as}f(s)\,ds.$$

(iii) By using (i) and applying (ii), twice show that, if $a \neq b$ and

$$u''(x) - (a+b)u'(x) + abu(x) = 0, \qquad\qquad \bigstar$$

then

$$u(x) = Ae^{ax} + Be^{bx}$$

for some constants A and B. Show further that, if u_0 and u_1 are specified, equation $\bigstar$ has exactly one solution with $u(0) = u_0$ and $u'(0) = u_1$.
(iv) Show that if

$$u''(x) - 2au'(x) + a^2u(x) = 0$$

then

$$u(x) = (Ax + B)e^{ax}$$

for some constants A and B. Show further that, if u_0 and u_1 are specified, the equation has exactly one solution with $u(0) = u_0$ and $u'(0) = u_1$.
(v) Suppose that f is continuous and u_0 and u_1 are specified. Show, using (ii), that the equation

$$u''(x) - (a+b)u'(x) + abu(x) = f(x)$$

has exactly one solution with $u(0) = u_0$ and $u'(0) = u_1$.

(vi) Suppose that

$$u''(x) - (a + b)u'(x) + abu(x) = f(x)$$

has a solution v. Show that the general solution of the equation is

$$u(x) = Ae^{ax} + Be^{bx} + v(x),$$

if $a \neq b$ and

$$u(x) = (Ax + B)e^{ax} + v(x)$$

if $a = b$. (We call $v(x)$ a particular solution and $Ae^{ax} + Be^{bx}$ or $(Ax + B)e^{ax}$ complementary solutions.)

Exercise 10.3.3 *Use the results of Exercise 10.3.2 (in particular part (ii) rather than guesswork) to find the general solutions of the following equations.*
(i) $u'(x) - au(x) = Ke^{cx}$ when $c \neq a$.
(ii) $u'(x) - au(x) = Ke^{ax}$.
(iii) $u''(x) - (a + b)u'(x) + abu(x) = Ke^{cx}$ when $c \neq a, b$.
(iv) $u''(x) - (a + b)u'(x) + abu(x) = Ke^{cx}$ when $c = a \neq b$.
(v) $u''(x) - (a + b)u'(x) + abu(x) = Ke^{cx}$ when $a = b = c$.

Our treatment of difference equations parallels our treatment of differential equations but there are several significant changes.

Exercise 10.3.4 *(i) Suppose that $a \neq 0$ and y_n is a sequence. By using induction, show that the system of equations*

$$u_{n+1} - au_n = y_n$$

for all $n \in \mathbb{Z}$ has exactly one solution with $u_0 = \tilde{u}_0$.
(ii) Suppose that $b \neq 0$ and z_n is a sequence. By using induction, show that the system of equations

$$u_{n+2} + au_{n+1} + bu_n = z_n$$

for all $n \in \mathbb{Z}$ has exactly one solution with $u_0 = \tilde{u}_0$, $u_1 = \tilde{u}_1$.

Exercise 10.3.5 *Suppose a, b and c are constants and y_n is a sequence. If*

$$au_{n+2} + bu_{n+1} + cu_n = y_n$$

and

$$av_{n+2} + bv_{n+1} + cv_n = 0,$$

show that, if k is constant and $w_n = kv_n + u_n$, then

$$aw_{n+2} + bw_{n+1} + cw_n = y_n.$$

State and prove a similar result for the equation $au_{n+1} + bu_n = y_n$.

Exercise 10.3.6 *(i) Let u_n be a sequence. Suppose that we write $Iu_n = u_n$, $Eu_n = u_{n+1}$ and $E^2 u_n = u_{n+2}$. Show that*

$$(E - aI)(E - bI)u_n = (E^2 - (a + b)E + abI)u_n.$$

(ii) Suppose that $a \neq 0$ and y_n is a sequence. Show that if

$$(E - aI)u_n = w_n$$

for all n and $v_n = a^{-n} u_n$ then

$$(E - I)v_n = a^{(n+1)} w_n$$

for all n.

(iii) Suppose that y_n is a sequence and

$$(E - I)u_n = y_n$$

for all n. Show that

$$u_n = \begin{cases} u_0 + \sum_{r=0}^{n} y_r & \text{for } n \geq 1, \\ u_0 & \text{for } n = 0, \\ u_0 - \sum_{r=1}^{-n} y_{-r} & \text{for } n \leq -1. \end{cases}$$

Exercise 10.3.7 *Use the results of Exercise 10.3.6 (rather than guess work and verification) to solve the following systems of equations.*

(i) $(E - I)u_n = 0$ for all n, $u_0 = 1$.

(ii) $(E - I)u_n = b^n$ for all n, $u_0 = (b - 1)^{-1}$ where $b \neq 1$.

(iii) $(E - I)u_n = 1$ for all n, $u_0 = 0$.

(iv) $(E - I)u_n = n$ for all n, $u_0 = 0$.

(v) $(E - aI)u_n = 0$ for all n, $u_0 = 1$.

(vi) $(E - aI)u_n = b^n$ for all n, $u_0 = (b - a)^{-1}$ where $b \neq a$.

(vii) $(E - aI)u_n = a^n$ for all n.

(viii) $(E - aI)u_n = na^n$ for all n, $u_0 = 0$.

What are the general solutions of $(E - aI)u_n = 0$ and $(E - aI)u_n = b^n$?

Exercise 10.3.8 *Use the results of the previous exercises to prove the following results.*

(i) Suppose that $a \neq b$ and $a, b \neq 0$. If

$$v_{n+2} - (a + b)v_{n+1} + abv_n = y_n,$$

then the general solution of the system of equations

$$u_{n+2} - (a + b)u_{n+1} + abu_n = y_n$$

is

$$u_n = Aa^n + Bb^n + v_n,$$

where A and B are freely chosen constants.

(ii) If $a \neq 0$

$$v_{n+2} - 2av_{n+1} + a^2 v_n = y_n,$$

then the general solution of the system of equations

$$u_{n+2} - 2au_{n+1} + a^2 u_n = y_n$$

is

$$u_n = (A + Bn)a^n + v_n,$$

where A and B are freely chosen constants.

(We call v_n a particular solution *and the term corresponding to $Aa^n + Bb^n$ or $(An + B)a^n$ a* complementary solution.*)*

Exercise 10.3.9 *In practice, we usually guess the form of the complementary solution and then use a little experimentation to get it exactly. However, the reader may find it interesting to work through the following exercise without using guesswork.*

(i) Suppose that a, b and x are all distinct and non-zero. Use Exercise 10.3.6 to find the general solution of

$$u_{n+2} - (a + b)u_{n+1} + abu_n = Cx^n.$$

(ii) Suppose that a and b are distinct and non-zero. Find the general solution of

$$u_{n+2} - (a + b)u_{n+1} + abu_n = Ca^n.$$

(iii) Find the general solution of

$$u_{n+2} - 2u_{n+1} + u_n = C.$$

Exercise 10.3.10 *Suppose we work over $\mathbb{C}$, that a and b are real, $a \neq 0$ and that $t^2 + bt + c = 0$ has roots α and β. Suppose α and β are not real. Explain why $\beta = \bar{\alpha}$ the conjugate of α. What conditions on A and B will ensure that*

$$A\alpha^n + B\bar{\alpha}^n$$

is real for all n?

Suppose that $c \neq 0$ and $u_{n+2} + bu_{n+1} + cu_n = 0$ for all n and u_0 and u_1 are real. Show that u_n is real for all n both

(i) by induction and

(ii) by computing A and B such that $u_n = A\alpha^n + B\bar{\alpha}^n$.

Exercise 10.3.11 *In this exercise we let*

$$\binom{n}{r} = \frac{n(n-1)\cdots(n-r+1)}{r!}$$

for all integer values of r.

  (i) *Show that*

$$(E-I)\binom{n}{r} = \binom{n}{r-1}$$

for all r ≥ 1.

  (ii)*Write down the general solution of*

$$(E-I)\binom{n}{r} = \binom{n}{r-1}.$$

  (ii) *Show that the general solution of*

$$(E-I)^k u_n = 0$$

is

$$u_n = \sum_{j=0}^{k-1} A_{k-1-j}\binom{n}{j}$$

with A_j arbitrary. Show that the general solution can also be written

$$u_n = \sum_{j=0}^{k-1} B_j n^j$$

with B_j arbitrary.

  (Because of the analogy with differential equations we tend to use the second form, but the first form is likely to be computationally easier to handle.)

  (iii) *Find the general solution of*

$$(E-aI)^k u_n = 0$$

if $a \neq 0$.

  (iv) *Find the general solution of*

$$(E-I)^k u_n = \binom{n}{r}.$$

Exercise 10.3.12 *Solve the difference equation*

$$u_{n+3} + a u_{n+2} + b u_{n+1} + c u_n = 0,$$

with $c \neq 0$ distinguishing the various cases.

Exercise 10.3.13 *A wealthy flea owns a black dog and a white dog. If it is on the black dog, then, just before the end of each minute, it either jumps to the white dog with probability w or stays where it is with probability $1 - w$. If it is on the white dog, then, just before the end of each minute, it either jumps to the black dog with probability b or stays where it is with probability $1 - b$.*

(i) Suppose that $2 > w + b > 0$ and it starts on the white dog. Let us write p_n for the probability that it is on the white dog after n minutes have elapsed. Find a difference equation relating p_n and p_{n-1} and use it to calculate p_n. Show that

$$p_n \to \frac{w}{w + b}$$

as $n \to \infty$.

(ii) Suppose that $2 > w + b > 0$ and it starts on the black dog. Find the probability q_n that it is on the white dog after n minutes have elapsed. What happens as $n \to \infty$?

(iii) Discuss the cases $w + b = 0$ and $w + b = 2$.

10.4 The casino's view

The mathematician understands that everything, from crossing the road to building a new factory, involves a gamble. She knows that we often gamble because we must. The mathematician understands that it is reasonable to gamble voluntarily if the expected outcome of a gamble is positive. However, she finds it hard to sympathise with those who voluntarily make bets with negative expected outcomes.[8]

Because of this rather puritanical attitude, mathematicians tend to view casinos as buildings which people enter carrying sums of cash and leave some hours later carrying rather less. However, people also enter theatres, concert halls and football stadiums carrying sums of cash and leave some hours later carrying rather less. Let us put aside some of our prejudices and try to get a simple idea of the problems of running a casino.

One problem with running a casino, like running any business, is that you may run out of money. You may reply that, so long as the casino makes reasonable bets, the weak law of large numbers (Theorem 2.5.13 and its generalisation in Exercise 2.5.14) ensure that the casino must remain solvent. There

---

[8] The 2002 *World Directory of Mathematicians* contains 57 000 names so there must be many exceptions to this general statement. However, a mathematician who regularly visits casinos may well have the same misgivings as a doctor consulting a homoeopath.

are two objections to this argument. The first is that the weak law of large numbers is a *snapshot*. Theorem 2.5.13 says that (with the notation and under the hypotheses given) if we choose a large number n then, *for that fixed n,*

$$\Pr\left(\left|\frac{X_1 + X_2 + \cdots + X_n}{n} - \mu\right| \geq a\right) \leq \frac{\sigma^2}{na^2}.$$

It *does not say* that there is a high probability that $(X_1 + X_2 + \cdots + X_m)/m$ is close to μ for all sufficiently large m. With more advanced tools, we can state and prove results of this more general form but even these more general results do not necessarily mean what a casino owner would wish them to mean. The second problem is that general laws of large numbers tell us what will happen in the long run without telling us how long that run must be. We are likely to gain more insight by direct computation using simple models than by looking at such general results.

Consider a simple casino in which the players place one bet of unit value at a time. With probability p, the bettor gets 2 units back and, with probability $1 - p$ nothing. All the bets are independent. Suppose that the casino sets itself a target of owning N units and then stops. Let u_n be the probability that the casino avoids bankruptcy (more exactly, never has zero wealth) and attains its goal if it starts with n units. At each bet the casino's wealth decreases by 1 unit with probability p and increases by 1 unit with probability $1 - p$. Thus

$$u_n = pu_{n-1} + (1 - p)u_{n+1}.$$

We have $u_0 = 0$ and $u_N = 1$.

Our standard method of solution tells us to look at the roots of

$$(1 - p)t^2 - t + p = 0$$

and a standard formula tells us that the roots are 1 and $p/(1 - p)$. Thus, if $p \neq 1/2$,

$$u_n = A + B\left(\frac{p}{1 - p}\right)^n.$$

Since $u_0 = 0$, we have $B = -A$, so, using the fact that $u_N = 1$, we have

$$u_n = u_n(N) = \frac{1 - \left(\frac{p}{1-p}\right)^n}{1 - \left(\frac{p}{1-p}\right)^N}.$$

Exercise 10.4.1 *If $p = 1/2$, show that*

$$u_n = u_N(n) = \frac{n}{N}.$$

To see what happens if the casino has no intention of stopping unless it goes bankrupt let $N \to \infty$. If $p > 1/2$, then $p/(1-p) > 1$ and $u_N(n) \to 0$ as $N \to \infty$. Not surprisingly, a casino which makes bets favourable to its customers will eventually go bankrupt with probability 1. Slightly more surprisingly, the same holds if $p = 1/2$. Whatever sum n a fair casino starts with, it will eventually go bankrupt with probability 1. Observe that, although each of its customers may be poorer than the casino, together they have, essentially, infinite wealth.

If $p < 1/2$, then $p/(1-p) < 1$ and

$$u_N(n) \to 1 - \left(\frac{p}{1-p}\right)^n,$$

which may reasonably be interpreted as saying that the probability that the casino goes bankrupt, starting with n units, is $\left(p/(1-p)\right)^n$. Roughly speaking, the long term prospects for the casino are good but it could be ruined by an initial run of bad luck.

Observe that if the casino starts off with n units and its customers bet $1/2$ a unit at a time then the probability of bankruptcy is reduced to $\left(p/(1-p)\right)^{2n}$. A casino which wishes to stay in business will limit the amount wagered on a single bet to a small proportion of its total wealth.

Life is more complicated than this. The casino staff have to be paid, the casino buildings kept up and a decent return paid to the casino's owners. In order to re-use the mathematics we have already developed, we decide to raise the money as follows. Each time a player lays a bet of 1 unit, we return 2 units to the player with probability p, we use the 1 unit to help pay running costs with probability r and we add the 1 unit to the casino's wealth with probability q. All the bets are independent and $p + q + r = 1$.

Exercise 10.4.2 *Suppose that the casino just described sets itself a target of owning N units and then stops. Let u_n be the probability that the casino avoids bankruptcy (more exactly never has zero wealth) and attains its goal if it starts with N units.*

(i) Explain why

$$u_n = pu_{n-1} + ru_n + qu_{n+1}$$

for all $1 \le n \le N-1$ and $u_0 = 0$ and $u_N = 1$.

(ii) Show that, if $p \ne q$,

$$u_n = \frac{1 - \left(\frac{p}{q}\right)^n}{1 - \left(\frac{p}{q}\right)^N}.$$

(iii) Find u_n when $p = q$.

(iv) Show that if the casino continues to take bets indefinitely it will go bankrupt with probability 1 *if* $p \geq q$. *If* $p < q$ *show that we can make the probability of bankruptcy arbitrarily small by starting with a sufficiently large fortune.*

After a large number n of bets we may expect, by the weak law of large numbers, that, of n units in bets that we have taken, we will have paid out about $2pn$ units to the bettors, raised about rn units to cover costs and that the sum retained by the casino will be about

$$(1 - 2p - r)n = (q - p)n.$$

This brings out a simple but important point. The greatly daring mathematician who decides to taste the joys of gambling by placing one bet of 20 dollars on red represents an expected profit to the casino of less than 1 dollar. This will not pay for very much of the Baroque[9] luxury that surrounds her. When you buy a book or a washing machine, the shop retains a large proportion of the price to cover its costs and can afford to let you take your time. The proportion r of each bet that the casino expects to cover its costs is quite small (5% for American roulette, $2\frac{1}{2}$% for European roulette and even less for some other games[10]), so the only way it can pay its way is to make many bets. The motto of any casino must be 'Speed, speed and more speed'. The more bets it can take in an hour, the more profitable it will be.

We have not yet looked at the surplus of roughly $(q - p)n$ retained by the casino. What purpose does it serve? In our analysis it simply serves as a reserve against a run of bad luck. But, as n increases, the expected value of the surplus increases indefinitely. It makes no economic sense to keep enormous sums unused against the remote possibility of a very long run of bad luck and eventually the investors will demand that it is returned to them.

Let us consider a simple model in which the casino pays out nothing (so we revert to $r = 0$) if its wealth is less than N but, when its wealth is N, pays out every extra unit to cover its costs and repay its investors. As before, players place one bet of unit value at a time. With probability p the bettor gets 2 units back and, with probability $1 - p$ nothing. All the bets are independent.

The first thing to observe is that the casino will ultimately go bankrupt with probability one. If it has not gone bankrupt after kN bets, its fortune will be no greater than N and so a run of N (or fewer) lost bets will ruin it. Thus

---

[9] Or Egyptian, or Roman, or what you please.

[10] Exercise 9.2.7 shows that the main bet in Craps yields only about $1\frac{1}{2}$%. The various side bets in this game are very much more advantageous to the casino.

the probability that it survives the next N bets is no greater than $1 - p^N$. The probability that it survives mN bets is no greater than $(1 - p^N)^m$.

Since we know that bankruptcy is certain, our first question must be how long the business can be expected to survive. Let e_n be the expected number of bets before bankruptcy if the casino starts with n units. Suppose that the casino has n units with $1 \le n \le N - 1$. If it takes a bet, then with probability p it will lose, have a new fortune of $n - 1$ and expect to survive a further e_{n-1} bets With probability $1 - p$ it will win, have a new fortune of $n + 1$ and expect to survive a further e_{n+1} bets. Thus

$$e_n = 1 + pe_{n-1} + (1 - p)e_{n+1}.$$

Exercise 10.4.3 *Explain why*

$$e_N = 1 + pe_{N-1} + (1 - p)e_N$$

and $e_0 = 0$.

We now seek to solve

$$(1 - p)e_{n+2} - e_{n+1} + pe_n = 1. \qquad\qquad ★$$

Suppose that $p \neq 1/2$. We know that, if

$$u_n = A + B\left(\frac{p}{1 - p}\right)^n,$$

then $(1 - p)u_{n+2} - u_{n+1} + ue_n = 1$ and we seek a particular solution x_n with

$$(1 - p)x_{n+2} - x_{n+1} + px_n = 1.$$

We try $x_n = Cn$ and obtain

$$C\big((1 - p)(n + 2) - (n + 1) + pn\big) = 1$$

whence $C(1 - 2p) = 1$ and $C = (1 - 2p)^{-1}$. Thus the general solution of ★ is

$$e_n = A + B\left(\frac{p}{1 - p}\right)^n + \frac{n}{1 - 2p}.$$

The condition $e_0 = 0$ gives $A + B = 0$ so

$$e_n = A\left(1 - \left(\frac{p}{1 - p}\right)^n\right) + \frac{n}{1 - 2p}$$

and the condition $e_N = 1 + pe_{N-1} + (1 - p)e_N$ gives

$$e_N - e_{N-1} = p^{-1}$$

whence

$$A\left(\left(\frac{p}{1-p}\right)^N - \left(\frac{p}{1-p}\right)^{N-1}\right) + \frac{1}{1-2p} = \frac{1}{p}.$$

Thus

$$A\left(\frac{p}{1-p}\right)^{N-1}\left(1 - \frac{p}{1-p}\right) = \frac{1}{1-2p} - \frac{1}{p}$$

and

$$A = \frac{1}{(1-2p)^2}\left(\frac{1-p}{p}\right)^N.$$

It follows that

$$e_n = \frac{1}{(1-2p)^2}\left(\frac{1-p}{p}\right)^N\left(1 - \left(\frac{p}{1-p}\right)^n\right) + \frac{n}{1-2p}$$

and, in particular,

$$e_N = \frac{1}{(1-2p)^2}\left(\left(\frac{1-p}{p}\right)^N - 1\right) + \frac{N}{1-2p}.$$

If $p < 1/2$ and N is sufficiently large, then the expected time to bankruptcy, if the casino starts with capital N and never returns money when its capital falls below this sum, is very long indeed.

The investors can choose N very large or they can decide to take the risk that from time to time the casino will run out of money and they will have to refinance it. They will be interested, not in the expected time before refinancing, but in how much the casino may be expected to pay out before it needs refinancing. Let us write f_n for the expected payout if the casino starts with a fortune n.

Exercise 10.4.4 *(i) Explain why*

$$f_n = pf_{n-1} + (1-p)f_{n+1}$$

for $1 \le n \le N-1$, $f_0 = 0$ and

$$f_N = (1-p)(1+f_N) + pf_{N-1}.$$

(ii) Let $p < 1/2$. Show that

$$f_n = \frac{1}{1-2p}\left(\frac{1-p}{p}\right)^N\left(1 - \left(\frac{p}{1-p}\right)^n\right)$$

and, so in particular,

$$f_N = \frac{1}{1 - 2p} \left(\left(\frac{1-p}{p} \right)^N - 1 \right).$$

We note, once again, the advantage to the casino of taking bets which are small compared with its total fortune.

Exercise 10.4.5 *Consider the purely theoretical version of Exercise 10.4.4 in which $p = 1/2$. Without making any calculations write down the value f_n for the expected payout if the casino starts with a fortune n. Explain your answer. (You are not asked to provide a proof.)*

Exercise 10.4.6 *Let us combine two of our models. Each time a player lays a bet of 1 unit the casino returns 2 units to the player with probability p, uses the 1 unit to help pay running costs with probability r and, with probability q, either adds the 1 unit to the casino's wealth, if that wealth is less than N before the bet, or pays it to the investors, if the casino's wealth before the bet is N. All the bets are independent and $p + q + r = 1$. We take $q > p$*

(i) Compute the expected number of bets until bankruptcy if the casino starts with N.

(ii) Compute the return to investors until bankruptcy if the casino starts with N.

Exercise 10.4.7 *The case $p = 1/2$ is of no interest to casinos but of great interest to mathematicians. Suppose that I play heads and tails with a fair coin and write $X_m = 1$ if the mth toss is head, $X_m = -1$ if the mth toss is tail. If I take*

$$Y_m(r) = r + X_1 + X_2 + \cdots + X_m,$$

and $Y_0(r) = r$, explain why Exercise 10.4.1 and the surrounding discussion show that, if $r \geq 0$, then the probability that $Y_n(r) = 0$ for some $n \geq 1$ is 1.

Now fix a large N and, if $0 \leq r \leq N$, let e_r be the expected number of tosses until the first time $Y_m(r)$ takes the value 0 or N. Explain why $e_0 = e_N = 0$ and

$$e_{n+1} - 2e_n + e_{n-1} = -2.$$

for $1 \leq n \leq N - 1$. By using a trial particular solution of the form Cn^2, or otherwise, show that

$$e_n = e_n(N) = n(N - n).$$

Show that, if $n \geq 1$, $e_n(N) \to \infty$ as $N \to \infty$. The appropriate interpretation is that, although a fair casino will go bankrupt with probability 1, the expected time until this occurs is infinite!

Exercise 10.4.8 *Two people are playing a standard game of heads and tails. They toss a fair coin repeatedly. If the total number of heads thrown ever exceeds the number of tails by N, the game stops and the second player pays one dollar to the first. If the total number of tails thrown ever exceeds the number of heads by N, the game stops and the first player pays one dollar to the second.*

In order to make the game more interesting, they introduce a doubling cube. *If player A thinks she has a good chance of winning the game she can 'turn the doubling cube'. Her opponent B must now either abandon the game, paying A one unit, or continue the game with doubled stakes. A now loses the right to double, but B retains the right to redouble at some later stage. If B redoubles, then A must either abandon the game, paying B two dollars, or continue the game with redoubled stakes (so the loser will pay 4 dollars to the winner). B now loses the right to double, but A regains the right to redouble at some later stage and so on.*

Suppose that the first player turns the cube for the first time at $Y_n = m$. If the expected value of the doubled game to the second player is less than -1, the second player will fold without hesitation. If the expected value of the doubled game to the second player is greater than -1, the second player will accept without hesitation. Since the first player wishes to make life as difficult as possible for the second player,[11] she should turn the cube when the expected value of the doubled game to the second player is exactly -1. Of course there may be no such value of m, but, if it exists, we call it the critical value and denote it by M.

(i) Explain why (if she has the right to double) the first player should always double when $Y_n = M$ and the second player should always double when $Y_n = -M$.

(ii) Show that (when M exists)

$$\Pr(\text{first player wins starting from position } r = M) = 4/5.$$

(iii) Show that M exists for the game described when N is divisible by 5.
[The use of the doubling cube in Backgammon is more complicated but this exercise is a good place to start.]

Exercise 10.4.9 *(i) To while away the time while they hang round the Danish court, Rosencrantz and Guildenstern each toss a fair coin. If both come down heads, Guildenstern gives his coin to Rosencrantz. If both come down tails, Rosencrantz gives his coin to Guildenstern. If neither event occurs, no coins change hands. The game ends when one or the other has no remaining*

---

[11] More sophisticated arguments are possible.

coins. If Rosencrantz starts with r coins and Guildenstern with g, show that the expected time until the game ends is 2rg.

(ii) To while away the time on their trip to England, Hamlet, Rosencrantz and Guildenstern decide to play a three-sided version of the game. They each toss a coin. If all three coins come down the same, no money changes hands. If they differ, the odd man out wins the coins of the other two. Let e_{hrg} be the expected time until one man runs out of coins if Hamlet starts with h coins, Rosencrantz with r and Guildenstern with g. Write down an appropriate equation and solve it by first guessing the form of the solution (or by any other method you wish).

We have seen that, from the point of view of the casino, the ideal client is one who spends a long time in the building making a succession of small bets until they leave having lost the maximum sum they can afford to lose easily. Fortunately for casinos, most of their clients wish to do exactly this. They visit the casino not to make money but to enjoy themselves.

The difficulty from the point of view of the casino is that the amount someone can afford to lose easily varies greatly from person to person. A rich gambler will derive no satisfaction from gambling for 10 dollars. A poor casino can do nothing about this and will not have any rich clients. However, since the key ratio is the size of a typical bet to the fortune of the casino, a sufficiently rich casino will be able to accommodate rich clients with larger bets. Since larger bets give correspondingly larger expected profits per bet, the casino can return some of these profits to its rich gamblers in the form of better odds or presents such as free drinks, cheap accommodation or more luxurious surroundings.[12]

Since glamour ceases to be glamour when shared with those poorer than oneself, rich gamblers have to be kept separate from poor ones. The upper and lower limits on the size of bets at the various games in a wealthy casino presumably have more to do with separating the different classes of gamblers than safeguarding the casino from bankruptcy. Nonetheless, the largest available bet in a casino, however grand, must be governed by considerations of the type discussed in this section.

Exercise 10.4.10 *(i) When Jack takes his cow to market, there is a probability p_j that he will be offered j magic beans in exchange $[0 \leq j \leq n]$. (We have $\sum_{j=0}^{n} p_j = 1$, since he can always give the cow away.) Jack decides to accept*

---

[12]   It's the same the whole world over.
       Isn't it a bloomin' shame.
       It's the rich what gets the pleasure
       And the poor what gets the blame.

any offer of m beans or more but otherwise to take the cow home and return to the market the next day, continuing in this way until he gets a satisfactory offer. Unfortunately, there is always a probability $1 - q$ that the cow will die overnight. Show that his strategy produces an expected number of beans e_m given by

$$e_m = \frac{\sum_{j=m}^{n} j p_j}{1 - q \sum_{j=1}^{m-1} p_j}.$$

Show that there is an integer m_0 with $0 \le m_0 \le n$ such that $e_{m-1} \le e_m$ when $1 \le m \le m_0$ and $e_m \le e_{m+1}$ when $m_0 \le m \le n - 1$. Advise Jack on the assumption that he wishes to maximise his expected number of beans.

(ii) Whilst Jack owns the cow, he has to feed it and go through the bother of taking it to market. He decides that it costs him the equivalent of k magic beans a day for each day the cow remains unsold. Advise him.

[No cattle were harmed during the production of this exercise.]

Exercise 10.4.11 *As I drive into work, I pass parking places*

$$\ldots, A_{n+1}, A_n, A_{n-1}, \ldots, A_2, A_1, A_0, A_{-1}, A_{-2}, \ldots$$

in order. I do not know whether a parking place is occupied until I reach it and, once I pass a parking place, I cannot go back. The probability that a place is occupied is p independently of what happens at the others and the distance from parking place A_j to my place of work is $|j|$ units.

Let V_m be the expected distance that I have to walk to my office if I follow the plan 'drive as far as A_m and then park in the first free parking place' [$m \ge 0$]. Explain why $V_0 = q/p$ where $q = 1 - p$. Find a recurrence relation for V_n and deduce that

$$V_n = n + \frac{(2q^n - 1)q}{p}$$

for $n \ge 0$. By looking at the sign of $V_{n-1} - V_n$, or otherwise, show that, if I wish to minimise the expected distance walked, I should take m to be the largest integer n with

$$n \le \frac{\log(1/2)}{\log q}.$$

10.5 A flutter on the lottery

It is said that the noted gambler and financier John Law[13] staked his last thousand pounds against a shilling in a wager that double sixes would not be thrown six times successively. He won, and repeated the experiment before the local authorities interfered.

Exercise 10.5.1 *(i) What is the probability that Law will lose his wager?*
(ii) How many seconds are there in a lifetime of 70 years?
(iii) There are 20 shillings in a pound. What is the expected value of the bet to the person who chooses to bet with Law?
(iv) Is the Kelly criterion of Section 2.6 relevant to this wager?

Law's bet is so favourable to him that few mathematicians would hesitate to join him in making the wager. However, it is genuinely difficult to make decisions which involve a very small chance of a very large loss. As usual, it is much easier to make satisfactory decisions when we are faced with a large number of similar cases.

In a lottery, a large number of bettors buy tokens which give them a very small chance of a very large prize. In the simplest lottery, the prize is announced in advance, participants buy tickets, one of the tickets is chosen at random and its owner receives the prize. The risk to the organiser lies in the possibility that insufficient tickets will be sold to cover the cost of the prize.[14] This risk can be avoided by taking the money paid for the tickets, removing a proportion for expenses and profit, and returning the remaining sum as the prize.

In the UK National Lottery, 50% of the cost of each ticket is returned to the buyers as prizes. Although it may seem irrational to enter such a lottery, our previous discussion suggests that, if you wish to become many times richer than you already are, you will be much better off buying a ticket in a lottery with a prize that will satisfy your desires rather than trying to reach that sum through a long succession of bets at low odds. It could also be argued that when you purchase a lottery ticket for £1 you get at least £1/2 of non-mathematical dreams in return[15] to add to the mathematical expected value of £1/2.

---

[13] The collapse of Law's banking schemes caused a French and European financial crisis. 'Çi gît cet Écossais célèbre, Ce calculateur sans égal, Qui, par les règles de l'algèbre, A mis la France à l'Hôpital.' Or, to give a free translation, 'This buried Scot beyond compare, At calculation showed unequalled flair, And by algebraic manipulation, Brought down ruin on the nation.' (See the first edition of the *Dictionary of National Biography*.)

[14] His enemies claimed that Horatio Bottomley avoided this risk by the expedient of not giving out the prize.

[15] This argument can be used to justify buying one ticket but not to justify buying two.

In many large modern lotteries, each participant buys a random number rather than a ticket.[16] A winning number is then selected randomly and anyone who has bought that winning number receives the prize.

Exercise 10.5.2 *Many national lotteries have the following form. Each player chooses six distinct integers r with $1 \leq r \leq 49$. Winners have to match a similar set which is randomly selected with due ceremony. Show that the chance of winning is $1/13\,983\,816$.*

Note that in our discussion of lotteries, we shall assume that the players do not *choose their numbers but are given randomly selected numbers. In Exercise 10.5.13 we look at why this makes a difference.*

In a lottery of the type described, each participant has a probability a of winning the prize, independent of what happens to the others and of the number of participants. However, it is now possible for more than one participant to win the prize or for nobody to win. We shall consider two possibilities for such a lottery. Either we have a *safe* lottery in which the prize winners (if any) share a fixed proportion of the money staked, or an *unsafe*[17] lottery in which the prize sum is fixed in advance and each winner receives the full sum.

It might be thought that the organisers of a safe lottery would have no interest in the number of winners, but, in the case of a weekly or monthly lottery, it is necessary to keep up the interest of possible participants. If too many people share the prize, the lottery will look less attractive and, although the use of 'roll overs' (in which the prize for a lottery in which there are no winners is added to the prize for the next lottery) may add excitement, people will become unhappy if the prize is won too infrequently.

Fortunately, there is a beautiful and remarkable theorem which gives a very close approximation to the probability that there are r winners.

Theorem 10.5.3 [**The Poisson approximation**] *Suppose that $a > 0$ is fixed. Suppose that we shoot at a target N times and there is a probability a/N that we hit the target at any shot independent of any other. Then*

$$\Pr(\text{hit target exactly r times}) \to \frac{a^r}{r!}e^{-a}$$

as $N \to \infty$.

---

[16] Of course you can place more than one bet, but the exposition is simplified by assuming that each participant makes one bet.

[17] That is to say, *unsafe* for the organisers.

Proof We use Lemma 2.2.2 and Exercise A.8 (or Exercise 10.5.10 (ii) at the end of this section) to obtain

$$
\Pr(\text{hit target exactly } r \text{ times}) = \binom{N}{r} \left(\frac{a}{N}\right)^r \left(1 - \frac{a}{N}\right)^{N-r}
$$
$$
= \frac{N(N-1)\cdots(N-r+1)}{r!} \left(1 - \frac{a}{N}\right)^{N-r}
$$
$$
= \frac{a^r}{r!} \times 1 \times \left(1 - \frac{1}{N}\right) \times \left(1 - \frac{2}{N}\right) \times \cdots
$$
$$
\times \left(1 - \frac{r-1}{N}\right) \times \left(1 - \frac{a}{N}\right)^{-r} \times \left(1 - \frac{a}{N}\right)^{N}
$$
$$
\rightarrow \frac{a^r}{r!} \times 1 \times 1 \times 1 \times \cdots \times 1 \times e^{-a}
$$
$$
= \frac{a^r}{r!} e^{-a}
$$

as $N \rightarrow \infty$.   ∎

Consider a safe lottery with a large number N of participants each of whom has a very small chance p of winning. If pN is very small compared with 1, then the probability that anyone wins is very small and, if pN is very large compared with 1, then the probability that there is at most one winner is very small. We are thus led to consider the case when $pN = a$ with a close to 1. Theorem 10.5.3 tells us that

$$
\Pr(\text{exactly } r \text{ winners}) \approx \frac{a^r}{r!} e^{-a},
$$

and, in particular, that

$$
\Pr(\text{exactly 1 winner}) \approx a e^{-a}.
$$

Exercise 10.5.4 *Show that ae^{-a} is maximised by taking $a = 1$.*

Thus the probability of exactly one winner is maximised when the number of participants is the reciprocal of the probability of a particular participant winning.

Exercise 10.5.5 *How can we maximise the probability that there are exactly m winners?*

We cannot control N exactly, but we can increase or decrease it by changing the interval of time between successive lotteries and we can choose the probability that a single participant will win. The choice of a method like that described in Exercise 10.5.2 reflects the organisers' view of the likely size of N.

As we said earlier, the organisers will not be too dismayed if there are no winners since this increases the size of the prize in the next lottery at no cost to themselves. A 'roll over' will greatly increase the number of participants in the next lottery and greatly decrease the chance that there will be no winner next time.

Now consider an unsafe lottery. We suppose that the organisers have a good idea of the likely number N of participants and wish to give out about half the entry money as prizes. They decide to choose an integer $m \geq 1$, to fix the probability of any single participant winning at m/N and to fix the size of the prize at $N/(2m)$.

Exercise 10.5.6 *Explain why, with these choices, the expected size of the sum the lottery pays out is indeed $N/2$.*

They also wish the probability that they have to pay out more than N in prizes to be extremely low.

Taking $a = m$ in Theorem 10.5.3 tells us that

$$\Pr(\text{exactly } r \text{ winners}) \approx \frac{m^r}{r!}e^{-m},$$

and so

$$\Pr(\text{pay out more than N}) = \Pr(\text{more than } 2m \text{ prize winners})$$
$$= 1 - \Pr(\text{at most } 2m \text{ prize winners})$$
$$= 1 - \sum_{r=0}^{2m} \Pr(\text{exactly } r \text{ prize winners})$$
$$\approx 1 - \sum_{r=0}^{2m} \frac{m^r}{r!}e^{-m}$$
$$= e^{-m}\left(e^m - \sum_{r=0}^{2m} \frac{m^r}{r!}\right)$$
$$= e^{-m} \sum_{r=2m+1}^{\infty} \frac{m^r}{r!}.$$

It is not as hard as it might look to get a reasonable estimate of $\sum_{r=2m+1}^{\infty} \frac{m^r}{r!}$.

Lemma 10.5.7 *If $m \geq 1$, then*

$$\frac{m^{2m+1}}{(2m+1)!} \leq \sum_{r=2m+1}^{\infty} \frac{m^r}{r!} \leq 2\frac{m^{2m+1}}{(2m+1)!}.$$

Proof Let us write $u_r = m^r/r!$. We observe that

$$\frac{u_{r+1}}{u_r} = \frac{m}{r+1} \le \frac{m}{2m+1} < \frac{1}{2}$$

for all $r \ge 2m+1$. Thus $u_r \le 2^{2m+1-r}u_{2m+1}$ for all $r \ge 2m+1$ and so

$$u_{2m+1} \le \sum_{r=2m+1}^{\infty} u_r \le \sum_{r=2m+1}^{\infty} 2^{r-2m-1}u_{2m+1} = u_{2m+1}\sum_{r=0}^{\infty} 2^{-r} = 2u_{2m+1}$$

as stated. ∎

Exercise 10.5.8 *Find a k such that*

$$\sum_{r=2m+1}^{2m+k} \frac{m^r}{r!} \le \sum_{r=2m+1}^{\infty} \frac{m^r}{r!} \le \frac{101}{100}\sum_{r=2m+1}^{2m+k} \frac{m^r}{r!}$$

(You are not asked for the best answer.)

Exercise 10.5.9 *Use a calculator to find $e^{-m}m^{2m+1}/(2m+1)!$ when $m = 5$ and $m = 10$.*

Exercise 10.5.9 shows that, if the organisers choose $m = 5$ and so offer prizes worth $N/10$ with each participant having probability $5/N$ of winning, there is a probability lying between 0.008 and 0.017 that they will have to give out more in prizes than they take in, but, if they choose $m = 10$ (so the prizes are worth $N/20$ but the chance of winning is $5/N$), then the probability that they will have to give out more in prizes than they take in is negligible.

We have chosen particular numerical values, but it is clear that, provided the organisers of an unsafe lottery choose odds which are rather favourable to themselves and make sure that each prize is fairly small compared with the total sum gambled, even an unsafe lottery is pretty riskless.[18]

If we wish to organise a single lottery, it makes sense only to offer large prizes. However, if we want to organise a sequence of lotteries, it is important to keep up the interest of the bettors. Since even the most optimistic individual becomes discouraged by a long sequence of total failures, regular lotteries also offer many small prizes. In effect, anyone who buys a ticket in such a lottery is actually buying two bets, one for a 'casino game' of the type discussed in the previous section with a reasonable probability of winning a small sum, and one for a 'safe' or 'unsafe' lottery with a very small probability of winning a very large sum.

Much casino betting is now mechanised. Expensive, troublesome and *slow* human croupiers and dealers are replaced by machines called variously 'one

---

[18] As usual, we assume that all involved on both sides are honest.

armed bandits', 'slot machines' (US), 'machines à sous' (France), 'fruit machines' (UK) and 'pokies' (Australia). In the old days, slot machines offered
a casino game with a reasonable probability of winning a small sum, but a series of bets with small prizes becomes rather boring. Modern electronics now
allows machines to offer very large 'jackpots'. Again, anyone who plays such
a machine is actually buying one bet for a 'casino game' and and one for a
'safe' or 'unsafe' lottery. Needless to say, both the game and the lottery will be
strongly advantageous and almost riskless for the casino.

Exercise 10.5.10 [The birthday problem]

(i) *Show that* $|x - \log(1 - x)| \le x^2$ *for* $1/2 > |x|$. *(If you need a hint, look at Exercise 10.2.6.)*

(ii) *If a is a fixed real number show that*

$$n \log \left(1 + \frac{a}{n}\right) \to a$$

as $n \to \infty$. *Conclude that*

$$\left(1 + \frac{a}{n}\right)^n \to e^a$$

as $n \to \infty$.

(iii) *Show, by induction or otherwise, that*

$$\sum_{r=1}^{n} r^2 \le \frac{(n+1)^3}{3}.$$

(iv) *Suppose that* $N(n) \to \infty$ *but* $N(n)n^{-1} \to 0$ *as* $n \to \infty$. *Show that*

$$\frac{2}{N(n)^2} \left[\log \left(1 + \frac{a}{n}\right) + \log \left(1 + \frac{2a}{n}\right) + \dots \log \left(1 + \frac{N(n)a}{n}\right)\right] \to a.$$

Deduce that

$$\left[\left(1 + \frac{a}{n}\right) \times \left(1 + \frac{2a}{n}\right) \times \left(1 + \frac{3a}{n}\right) \times \dots \times \left(1 + \frac{N(n)a}{n}\right)\right]^{2/N(n)^2} \to e^a$$

as $n \to \infty$.

(v) *Suppose we have a roomful of m people who are all equally likely to have birthdays on any of the 365 days of the year. (We ignore leap years and any other complications that the reader can think of.) Write down an exact formula for the probability that at least two have the same birthday. Use this, together with (iv), to get a simpler approximate formula for the probability that at least two people have the same birthday which will work if m is not too big.*

(vi) *Use your approximate formula to estimate the least r for which the probability that at least two have the same birthday is at least 1/2. Use the exact formula and some hard work to check or modify your answer.*

(vii) Roughly how many people must there be in the room so that the probability that two were born at the same hour of the same day of the year will be greater than $1/2$?

(viii) Consider the problem of issuing computer passwords where it is important that the same password should not be issued twice. To prevent people guessing passwords, or because several centres can issue passwords, it is decided to allocate passwords at random so that any of the n possible passwords is issued with probability $1/n$ in reply to any request, independent of what has gone on before. If the system is expected to issue m passwords, show that n should be chosen considerably larger than m^2.

Exercise 10.5.11 [More on birthdays] *What happens if the birthdays are not uniformly distributed throughout the year? The following ingenious argument due to Munford [45] shows that the probability of coincidences increases.*

(i) If x, $y > 0$ show that

$$xy \leq \left(\frac{x+y}{2}\right)^2$$

and that $xy = \big((x+y)/2\big)^2$ if and only if $x = y$.

(ii) Suppose that we have a year with N days, that we have n people in a room $[2 \leq n \leq N]$ and that the probability that any one of them has a birthday on the jth day of the year is $p(j)$, independent of any of the others. We assume further that $p(j) > 0$ for all $1 \leq j \leq N$. Show that if $j(1)$, $j(2)$, ..., $j(n)$ are distinct,

$$\Pr(\textit{together the n people have birthdays on days } j(1), j(2), \ldots, j(n))$$
$$= n! p_{j(1)} p_{j(2)} \cdots p_{j(n)}.$$

Deduce that

$$\Pr(\textit{the n people do not share birthdays})$$
$$= n! \sum_{1 \leq j(1) < j(2) < \cdots < j(n) \leq N} p_{j(1)} p_{j(2)} \cdots p_{j(n)}.$$

(iii) Show that

$$\sum_{1 \leq j(1) < j(2) < \cdots < j(n) \leq N} p_{j(1)} p_{j(2)} \cdots p_{j(n)}$$
$$= p_1 p_2 \sum_{3 \leq j(3) < \ldots < j(n) \leq N} p_{j(1)} p_{j(2)} \cdots p_{j(n)}$$

$$+ (p_1 + p_2) \sum_{3 \le j(2) < \cdots < j(n) \le N} p_{j(1)} p_{j(2)} \cdots p_{j(n)}$$

$$+ \sum_{3 \le j(1) < j(2) < \cdots < j(n) \le N} p_{j(1)} p_{j(2)} \cdots p_{j(n)}.$$

(iv) Use (i), (ii) and (iii) to show that, if we have a second group of people such that the probability that any one of them has a birthday on the jth day of the year is q_j, independent of any of the others and

$$q_1 = q_2 = \frac{p_1 + p_2}{2} \quad and \quad q_j = p_j \ for \ j \ge 3,$$

then

Pr(*first set people do not share birthdays*)

$$\le \text{Pr}(second \ set \ people \ do \ not \ share \ birthdays).$$

Show further that the two probabilities will only be equal if $p_1 = p_2$.

(v) Assuming, as usual, that there is a set of probabilities which minimises the probability of coincidence, show that this must be $p_j = 1/N$ for all j.

(vi) What does Exercise 2.5.19 tell you about the expected number of people in a room who share a birthday with you?

Exercise 10.5.12 *The management of the El Supremo holiday club has 1000 chalets which it lets out by the week. If someone who has booked arrives and there is no chalet for them, the club must pay compensation and loses £100. Otherwise each booking represents a profit of £100 whether the client turns up or not. The management knows from long experience that the probability that a client will not turn up is $1/500$ and that the pattern of cancellations fits the Poisson model (that is to say, the simple lottery model) of this section. How many bookings should they take to maximise their expected profit? Justify your answer.*

Exercise 10.5.13 *We return to the lottery described in Exercise 10.5.2 in which players choose their own numbers. Suppose that the participants never include the integers 1, 2, 3, 4 or 5 among their choices but otherwise make their choices at random. Show that about $1/2$ of the possible combinations will be unused. How will this affect the number of prize winners each week?*

A ticket for this lottery costs one pound and, in a moment of extravagance, you decide to buy one. We consider four cases.

(a) You do not know what the other participants are doing, so you choose your numbers at random.

(b) You choose the numbers 1, 2, 3, 4, 5 and 6.

(c) You choose the numbers 1, 7, 8, 9, 10 and 11.

(d) You choose the numbers 8, 15, 23, 32, 35, 46.

Discuss the effect on your expected winnings of your choices under the following four circumstances.

(A) There are about one million tickets sold. There is one prize of seven million pounds which is divided amongst all those who have matched the winning six numbers.

(B) There are about twenty eight million tickets sold. There is one prize of fourteen million pounds which is divided amongst all those who have matched the winning six numbers.

(C) There are about one million tickets sold. Everyone who matches the six winning numbers gets a prize of half a million pounds.

(D) There are about twenty eight million tickets sold. Everyone who matches the six winning numbers gets a prize of half a million pounds.

[It has been claimed that, during the first week of novel lotteries, it is possible to obtain close to favourable odds by using the fact that people do not choose numbers at random. As the population become more used to the lottery the effect wears off.]

10.6 Life is a lottery

Although historians of mathematics have traced versions of Theorem 10.5.3 back as far as de Moivre, they first entered the general scientific consciousness in 1898 with a book by Bortkiewicz entitled *das Gezetz der kleinen Zahlen* [66] (*The Law of Small Numbers*[19]) in which he gives several striking examples where real life observations conform closely to what we could expect from Theorem 10.5.3.

One example is the number of quadruplets born in Prussia in a period of 69 years. The total number of recorded quadruplets was 109, so we can imagine each mother in a particular year of N births holding a lottery ticket giving her a chance a/N of quadruplets with $a = 109/69$.

number of quadruplets	0	1	2	3	4	5	6	7+
years with that number	14	24	17	9	2	2	1	0
Poisson model	14.2	22.5	17.7	9.3	3.7	1.2	0.3	0.1

The bottom line labelled 'Poisson model' gives

---

[19] In spite of this, it might be better to talk about the 'law of rare events'.

expected number of years with r quadruplets

$$= \text{probability } r \text{ quadruplets according to Theorem 10.5.3}$$
$$\times \text{number of years considered}$$
$$= Ne^{-a}\frac{a^r}{r!}.$$

Thus our table says that, if we use the 'lottery' or 'Poisson' model, we would expect about 22.5 years with exactly one quadruplet birth and that, in fact, there were 24 such years.

Exercise 10.6.1 *Check the bottom line of the table just discussed.*

Bortkiewicz's most famous example concerns deaths from horse kicks in 20 Prussian army corps over a period of 20 years.

number of deaths	0	1	2	3+
army corps years	109	65	22	4
Poisson model	108.7	66.3	20.2	4.82

Exercise 10.6.2 *Check the bottom line in this table.*

In 1910, Rutherford and Geiger [18] counted the number of hits by alpha particles emitted by a sample of polonium recorded by their detecting apparatus over 2608 intervals of length 7.5 seconds. A total of 10 097 hits were recorded, so we apply our Poisson model with $a = 10\,097/2068$.

r	0	1	2	3	4	5	6	7	8	9	10	11	12	13	14+
H	57	203	383	525	532	408	273	139	45	27	10	4	0	1	1
P	54	210	407	525	508	393	254	141	68	29	11	4	1	0	0

Here r is the number of hits in an interval, H is the number of intervals with r hits and P is the number of hits predicted by the Poisson model. It is hard not to conclude that, at the level that we are observing, radioactive decay is governed by lottery rules.[20]

Returning to statistics in human affairs, we note the short paper [12] concerning the fall of flying bombs (a primitive cruise missile) on London towards the end of World War II. The author reports that 'frequent assertions were made that the points of impact of the bombs tended to be grouped in clusters. It was accordingly decided to apply a statistical test ...'. An area of south London was selected and divided into squares of area 1/4 square kilometres each and

---

[20] Einstein claimed that 'God does not play at dice', but a century of observation and theorising has not come up with anything better than lottery rules for radioactive decay.

the number of hits in each square counted. Since there were 537 bombs within the area the probability of a bomb falling within a particular square, *assuming no clustering*, was $a = 537/576$ and we can work out the probability $e^{-a}a^r/r!$ that r bombs fall in a particular square on this assumption. Clarke obtained the following table.

number per square	0	1	2	3	4	5+
actual number of squares	229	211	93	35	7	1
Poisson model	226.7	211.4	98.5	30.6	7.1	1.6

He suggests that this 'might afford material to future writers of statistical text-books'.

As part of his monumental study of the statistics of wars [56], J. F. Richardson tabulated the outbreak of the 59 moderately sized wars[21] that he could trace between 1820 and 1929. He then tabulated the number of years in which r such wars broke out. We can compare these numbers with the expected number of years that there would have been r winners in a yearly lottery in which the expected number of winners per year was 59/110.

numbers of outbreaks	0	1	2	3	4	5+
years with that number	65	35	6	4	0	0
Poisson model	64.3	34.5	9.3	1.7	.22	0

Looking at our data on horse kicks a suspicious reader might remark that, if four or five Prussian horses had missed four or five Prussian cavalry men, the match between the results for the model and actual deaths would not be quite so close. She may go on to ask two questions.

(1) How can one measure the match between the model for a series of random events and the actual outcome, particularly since probability theory itself predicts that there will rarely be a perfect match?

(2) Are the results quoted really typical? There must be many records of series of rare events. What is to prevent the author from choosing those which fit his thesis best and concealing the rest?

The answer to (1) is that this is indeed a difficult problem which we shall not tackle in this book. However, methods have been developed to answer (1) and they tell us that the examples chosen exhibit a good match.

---

[21] Those with an estimated death toll of between 3000 and 30 000. How many can the reader name?

'But what good came of it at last?'
Quoth little Peterkin.
'Why, that I cannot tell,' said he,
'But 'twas a famous victory.'

Southey The Battle of Blenheim

The answer to (2) must be a plea of guilty. It would be surprising if a collection of famous results cherished by generations of lecturers did not exhibit matters in their finest aspect. However, there are lots of data sets (number of births in a day in a medium town, number of goals in a football match, deaths from shark attacks each year in Australia, number of sentences per page whose 14th letter is t, ...) which the reader can obtain or construct and for which she can do her own checks.

Exercise 10.6.3 *Pick an appropriate data set and see what happens.*

It is important to note that I do not claim that the lottery model gives an exact description of any of the processes above. The flying bombs were indeed aimed at London,[22] but were sufficiently inaccurate that no pattern emerges at the scale considered. Presumably a death from horse kicks was followed by a period of increased caution and a decreased risk. All that is claimed is that *for particular processes* the lottery model may be good enough *for particular purposes*. The art of the mathematician and statistician lies in discovering when this is so and when it is not.

In the first half of the twentieth century, any conversation between two telephones required sole use of one telephone wire.[23] Thus the number of telephone calls possible at any one time between, say, London and Manchester was limited by the number of telephone wires between the two towns. If all the lines were in use, any further attempt to call would result in the message: 'All lines are engaged, please try later'. Since telephone lines were expensive to build and maintain, the telephone company would seek to install the minimum number of lines consistent with a reasonable standard of service.

At any time there would be a large number of telephones in London which might wish to be connected to Manchester, but only a small number that actually wished to be connected. Dialling Manchester was a low probability event for each telephone, but there were many telephones. The telephone company was in effect running an 'unsafe lottery' and the same ideas we used in the discussion centred on Exercise 10.5.9 to establish a number of prizes which was unlikely to be exceeded could be used to determine the number of telephone

---

[22] Double agents controlled by British Intelligence faithfully reported hits furthest from launching sites but did not report others, with the result that the Germans gradually moved their aiming point away from the centre of London and more flying bombs fell on less populated regions.

[23] As always, we simplify.

lines required to ensure that there was a low probability of a call failing to get through at a peak time.[24]

Mobile telephone companies divide their territories into cells. Only a limited number of phones can be active in each cell at one time but the smaller the company makes the cells the more cells it needs and the more expensive the system becomes. New technologies inherit old problems.

If we look once again at the vice of gambling and the virtue of insurance, we see that insurance belongs more to the lively but uncertain world of the race-track than to the controlled near certainties of the casino. However, our discussion does reinforce certain points. The law of large numbers tells us that no institution should insure against events which, from the point of view of that institution, have fairly high probability. Thus, the United States is so large that major natural disasters occur on a 'regular basis' and it makes no sense for the US government to do anything but pay out for disaster relief as it occurs. On the other hand, although New Zealand is under continual threat from destructive earthquakes, these events are sufficiently rare to justify the government of New Zealand in seeking insurance against such a disaster. We expect parts of the insurance business to deal with rare but expensive events, and combine the difficulties of running an unsafe lottery with those of running a bookmaking business.

Our discussion of lotteries confirms that no institution should offer insurance against events which involve a payout that is large compared with its available capital. Small institutions should take out insurance and large institutions should provide it. Just as in racing, bookmakers can bet with other bookmakers to cover the risk that some large bet may come off, so insurance companies can take out insurance with other insurance companies (reinsurance) and large risks can be shared.

---

[24] These ideas were first developed by Erlang who worked for the Copenhagen Telephone Company. Jensen, whose inequality we met in Theorem 3.5.3, was Chief Engineer for this company and Poulsen, who invented magnetic recording, also worked for them.

11

Prophecy

11.1 Coin tossing

In *The Napoleon of Notting Hill,* Chesterton wrote:

> The human race, to which so many of my readers belong, has been playing at
> children's games from the beginning, and will probably do it till the end, which is a
> nuisance for the few people who grow up. And one of the games to which it is most
> attached is called, 'Keep to-morrow dark,' and which is also named (by the rustics
> in Shropshire, I have no doubt) 'Cheat the Prophet.' The players listen very
> carefully and respectfully to all that the clever men have to say about what is to
> happen in the next generation. The players then wait until all the clever men are
> dead, and bury them nicely. They then go and do something else. That is all. For a
> race of simple tastes, however, it is great fun.

There is no headline journalists more enjoy writing or their readers more
enjoy reading than 'Experts get it wrong again'. But, although the future is
covered in mist, we may be able to glimpse vague shapes through that mist
and our actions should take account of those glimpses.

Consider, for example, the question of who will be president of the United
States of America in 9 years' time. Someone who believes that the future
is totally unknowable might offer odds of a thousand million to one against
us being able to name the future president. However, the Constitution of the
United States declares that

> No person except a natural born Citizen ... shall be eligible to the Office of
> President; neither shall any Person be eligible to that Office who shall not have
> attained to the Age of thirty-five Years, and been fourteen Years a Resident within
> the United States.

Thus, if we place a bet of 10^{-6} cents on each natural born Citizen over the age
of twenty-five, we are certain of profit.

Once this has been pointed out, our imaginary bettor might reduce the odds to a ten thousand to one against our success. But reflection suggests and history confirms that presidents will often come from the much more limited group of those interested by and already successful in the pursuit of political office. If we place a dollar on every serving or ex-congressman, senator and state governor, then we may lose our bet but it is a rational bet to place.[1]

If such simple considerations allow us to make a good bet at odds of ten thousand to one, it seems reasonable to suppose that careful research into the political scene of the United States would enable us to make good bets at odds of a thousand to one. On the other hand, it seems unlikely that, even with a great deal of knowledge and experience, we could make a good bet at odds of ten to one.

Most human organisations are constantly planning for the future. Our discussion suggests that, although certainty is impossible, careful study and thought may enable us to shave the odds in our favour. Under certain circumstances, mathematical techniques may be helpful and, because 'mathematical prophecy' and 'bookmaking' sound rather vulgar, we call such techniques 'statistics'.

As enemies of prophecy constantly point out, the real world is complicated and uncertain. However, history shows that the successful application of mathematics to the real world starts with the study of simple idealised models. Our brief excursion into mathematical prophecy will centre on the idealised model of coin tossing.

Suppose that you are playing a coin-tossing game with someone and he tosses 5 heads in a row. Should you stop playing with him? Perhaps, but remember that the probability of throwing 5 heads in a row with a fair coin is $(1/2)^5 = 1/32$ so, if you do a lot of coin tossing, you will soon have no one left that you trust.

Exercise 11.1.1 *(i) I toss r coins. If they are all heads, I stop. Otherwise, I toss the r coins again and continue until I get all r heads. Explain why the expected number u_r of tosses I make satisfies*

$$u_r = r + (1 - 2^{-r})u_r$$

and deduce that $u_r = r2^r$.

---

[1] Looking at twentieth-century presidents, we see that both Roosevelts, Taft, Wilson, Harding, Coolidge, Hoover, Eisenhower and Bush had not occupied such offices nine years before becoming president.

(ii) I toss r coins. If they are all heads or tails, I stop. Otherwise, I toss the r coins again and continue until I get r heads or all r tails. Find the expected number v_r of tosses that I make.

(iii) I toss a coin repeatedly until I get r heads in succession. Explain why the expected number e_r of throws satisfies

$$e_1 = 1 + 2^{-1}e_1$$

and

$$e_{r+1} = e_r + 1 + 2^{-1}e_{r+1}$$

for $r \geq 1$. Deduce that $e_r = 2^{r+1} - 2$.

(iv) I toss a coin repeatedly until I get r heads or tails. Find the expected number f_r of tosses that I make.

(v) Explain briefly why parts (i) and (iii) have different answers.

You are playing a coin-tossing game with someone and he tosses 10 heads in a row. Should you stop playing with him? The probability of throwing ten heads in a row are now $(1/2)^{10}$, so, perhaps, it would be good idea. But, presumably, you would have been equally suspicious if he had thrown ten tails. And, frankly, if he threw nine heads out of ten or nine tails out of ten it would look almost as bad. However, the probability of throwing nine or more heads or tails in ten tosses is $22 \times (1/2)^{10}$ which is almost one in fifty. And how about five heads followed by five tails or a perfect sequence *HT HT HT HT HT HT*? Once we start to suspect our fellow men where do we stop?

On the other hand, it does seem a little unwise to keep on playing with an opponent who throws 30 heads in a row.

In fact we are faced with two problems.

The golfer's fallacy A golfer hits the ball and it lands on a particular blade of grass. The odds on the ball landing on a particular blade of grass must be many tens of thousands to one, so the golfer must be incredibly skillful . . .

Obviously, the argument is fallacious.[2] We will only be impressed with the golfer's skill if she first nominates the blade of grass and then lands her ball on it. Only prediction counts and we should not re-use the data which led us to a particular hypothesis to test that hypothesis.[3]

---

[2] But difficult to avoid. Compare 'It still stands, a soaring testimony to one man's vision and his refusal to listen to the nay sayers' with 'The ruins are still visible, a mute testament to one man's hubris and his stubborn refusal to learn from others'.

[3] There may be circumstances when we have no choice. 'In a storm I will promise twelve candles to St Michael and a dozen to his dragon.' But such a path is fraught with danger, both philosophical and practical.

If I observe that my opponent has produced the sequence *HTHTHTHTHT* in his first 10 throws, then I may well make the prediction that his next 10 throws will be *HTHTHTHTHT*. If he is throwing a fair coin in a fair manner, then the probability that my prediction will be verified is 2^{-10}. If my prediction is verified, it is reasonable to conclude that he is not throwing a fair coin in a fair manner.

The cost of a mistake Let us continue with the coin-tossing example. I can make two sorts of mistake. I can decide that my opponent is cheating when he is not or I can decide that he is not cheating when he is.

Suppose that I intend to expose my opponent publicly as a cheat and a scoundrel if I believe him to be one. Then the cost of deciding that he is a cheat, when he is not, is very high. I shall have blackened the name of an innocent man and made an enemy for life. Surely, I should make an accusation only if my test prediction has very low probability indeed.

Suppose, on the other hand, that the game is one for high stakes and that the only action I intend to take is to cease playing (and that I can give a good explanation for withdrawing from the game). In this case, my test prediction need not have very small probability.

Prophecy should be directed towards action and should take into account how much action may cost.

11.2 A needle in a haystack

Traditionally, one of the main ways of looking for a new drug against a particular disease has been to try every chemical compound known to man to see if it has a biological effect which might make it a candidate for such a drug.

We may think of ourselves as given a roomful of coins. Most of them have very low probability of coming down heads (rarely show any effect in our test) and a few have fairly high probability of coming down heads (frequently show an effect in our test). We wish to reject any coin which has probability $1/10$ or less of coming down heads but to accept any coin which has probability $1/2$ or more.

Suppose we decide to throw each coin 5 times and reject every coin which comes down heads r times or less. We tabulate the probability that a coin, with probability $p = 1/10$ of coming down heads, will come down heads k times in 5 tosses.

number of heads out of 5	0	1	2
probability if $p = 1/10$	0.590	0.328	0.073

If we take $r = 0$, we shall let through over 40% of coins with $p = 1/10$ and, if we take $r = 1$, we shall let through over 8%.

Exercise 11.2.1 *Use a hand calculator to verify the table just given. You should check each table in this section as it appears.*

What about the probability of rejecting a 'good coin'? We tabulate the probability that a coin, with probability $p = 1/2$ of coming down heads, will come down heads k times in 5 tosses.

number of heads out of 5	0	1	2
probability if $p = 1/2$	0.031	0.156	0.313

Thus if we take $r = 1$ we shall be throwing away nearly 20% of coins with $p = 1/2$.

No choice of r is satisfactory. The reader may object that this is because we have chosen too lax a definition of a good coin. If we demand $p = 9/10$ in place of $p = 1/2$ we get the following result.

number of heads out of 5	0	1	2
probability if $p = 9/10$	0.000	0.000	0.008

A choice of $r = 2$ is then satisfactory. However, the clue to the formulation of a new drug may be found by looking at chemicals which have only moderate biological effects and so we cannot restrict consideration to those which are most potent.

The only way out of our problem is to perform more tests. If we double the number of tests, this will, more or less, double the cost of our preliminary trials, but we have little choice. Suppose we decide to throw each coin 10 times and reject every coin which comes down heads r times or less. We tabulate the probability that a coin, with probability $p = 1/10$ of coming down heads, will come down heads k times in 10 tosses.

number of heads out of 10	0	1	2	3
probability if $p = 1/10$	0.349	0.387	0.194	0.057

If we take $r = 1$, we shall let through over 1/4 of coins with $p = 1/10$ and, if we take $r = 2$, we shall let through over 7%.

What about the probability of rejecting a 'good coin'? We tabulate the probability that a coin, with probability $p = 1/2$ of coming down heads, will come down heads k times in 10 tosses.

number of heads out of 10	0	1	2	3
probability if $p = 1/2$	0.000	0.010	0.044	0.117

Thus, if we take $r = 2$ we shall turn down about 5%.

Should we be satisfied with a system of 10 tosses, throwing away those coins for which we record 2 heads or less? Perhaps. Remember that we are simply engaged in a preliminary sifting. Assuming that almost all the substances we investigate correspond to 'bad coins', we will have cut down the number of chemicals that we study further to less than a twelfth of the initial number. It is true that we shall have rejected about 5% of the 'good coins'[4] but drug development is a long and expensive process and we cannot investigate every lead.

Exercise 11.2.2 *Investigate what happens if we decide to toss each coin 20 times (doubling our costs once again).*

By spending more money on the first round of experiments, we can increase the number of tests on each chemical. By adjusting the 'acceptance level' we can use this increase

(1) to reduce the number of unlikely candidates going through to the next round and so reduce the cost of that round of investigations, or
(2) to reduce the chance of a likely candidate failing to go through to the next round, or
(3) some mixture of the two.

These benefits must be balanced against increased costs. Such decision making is aided by previous experience but can never be easy.

In the initial stages of drug development, it does not cost much to investigate any particular substance a bit further and, if we fail to investigate a particular substance further, it is unlikely to be a major mistake. However, as the process of development proceeds, the cost of mistakes rises very rapidly and we must use more and more expensive tests to reduce the chance of mistakes.

11.3 Tchebychev improved

It is clear from the previous section why mathematicians should be interested in obtaining estimates for

$$\Pr(a \leq Y_n \leq b)$$

where Y_n is the number of heads in n tosses of a coin with probability p of coming down heads.

---

[4] The reader may feel that this is slightly pessimistic since very good candidates are less likely to be rejected. However, in the absence of further information about the coins, we have no reason to draw more optimistic conclusions.

In order to deal fully with this, we would need the powerful but difficult central limit theorem of de Moivre.[5] However, Bernstein produced a beautiful argument which is often sufficient for the needs of mathematicians.

Theorem 11.3.1 *(i) Suppose that X is a real random variable with $|X| \leq 1$ and $\mathbb{E}X = 0$. Then*

$$\mathbb{E}e^{sX} \leq e^{s^2}$$

for all s.

(ii) Suppose that $X_1, X_2, \ldots, X_n$ are independent random variables with $|X_j| \leq 1$ and $\mathbb{E}X_j = 0$ for all j. Then

$$\mathbb{E}e^{s \sum_{j=1}^n X_j} \leq e^{ns^2}$$

for all s.

(iii) With the notation and assumptions of (ii),

$$\Pr\left(\left| \sum_{j=1}^n X_j \right| \geq a \right) \leq 2 \exp\left(-a^2/(4n) \right).$$

(iv) Suppose that $W_1, W_2, \ldots, W_n$ are independent random variables with $\mathbb{E}W_j = \mu$ and $1 \geq W_j \geq 0$ for all j. Then

$$\Pr\left(\left| \frac{W_1 + W_2 + \cdots + W_n - n\mu}{n^{1/2}} \right| \geq w \right) \leq 2e^{-w^2/4}.$$

(v) Let Y_n be the number of heads thrown in n tosses of a coin with probability p of heads. Then

$$\Pr(|Y_n - np| \geq 2n^{1/2}(\log \epsilon^{-1})^{1/2}) \leq 2\epsilon.$$

Proof (i) (We shall use infinite series, but anyone who knows enough to worry about our arguments should know enough to see that they are correct.) Observe that, using the Taylor expansion of exp (see Exercise A.10), we have

$$\mathbb{E}e^{sX} = \mathbb{E}\sum_{r=0}^\infty \frac{(sX)^r}{r!} = \sum_{r=0}^\infty \frac{s^r \mathbb{E}X^r}{r!}$$

$$= 1 + \sum_{r=2}^\infty \frac{s^r \mathbb{E}X^r}{r!} \leq 1 + \sum_{r=2}^\infty \frac{|s|^r \mathbb{E}|X|^r}{r!}$$

$$\leq 1 + \sum_{r=2}^\infty \frac{|s|^r}{r!}.$$

---

[5] It is unfair to call de Moivre's theorem difficult, since all it requires is the careful application of Stirling's formula. Still, we are reaching the end of a long book.

Now, if $|s| \leq 1$ and $q \geq 1$,

$$\frac{|s|^{2q}}{(2q)!} + \frac{|s|^{2q+1}}{(2q+1)!} \leq \frac{|s|^{2q}}{q!}$$

and so

$$1 + \sum_{r=2}^{\infty} \frac{|s|^r}{r!} \leq 1 + \sum_{q=1}^{\infty} \frac{|s|^{2q}}{q!} = e^{s^2}$$

while, if $|s| \geq 1$,

$$1 + \sum_{r=2}^{\infty} \frac{|s|^r}{r!} \leq \sum_{r=0}^{\infty} \frac{|s|^r}{r!} = e^{|s|} \leq e^{s^2}.$$

Thus $1 + \sum_{r=2}^{\infty} \frac{|s|^r}{r!} \leq e^{s^2}$ for all s and $\mathbb{E} e^{sX} \leq e^{s^2}$, as stated.

(ii) Since $X_1, X_2, \ldots, X_n$ are independent, it follows that $e^{sX_1}, e^{sX_2}, \ldots, e^{sX_n}$ are independent and so, using Lemma 2.4.11,

$$\mathbb{E} e^{s \sum_{j=1}^{n} X_j} = \mathbb{E} \prod_{j=1}^{n} e^{sX_j} = \prod_{j=1}^{n} \mathbb{E} e^{sX_j} \leq \prod_{j=1}^{n} e^{s^2} = e^{ns^2}.$$

(iii) Let us apply Exercise 2.5.4 with $X = \sum_{j=1}^{n} X_j, s > 0$ and $f(x) = e^{sx}$. Using (ii), we get

$$\Pr\left(\sum_{j=1}^{n} X_j \geq a \right) = \Pr(X \geq a)$$

$$\leq \frac{\mathbb{E} f(X)}{f(a)} = \exp(ns^2 - sa).$$

We now choose s to minimise $\exp(ns^2 - sa)$. With this choice, $s = a/(2n)$ and we have

$$\Pr\left(\sum_{j=1}^{n} X_j \geq a \right) \leq e^{-a^2/4n}.$$

A similar argument shows that

$$\Pr\left(\sum_{j=1}^{n} X_j \leq -a \right) \leq e^{-a^2/4n}$$

and the stated result follows.

(iv) Set $X_j = W_j - \mu$ and apply (iii).

(v) Let $W_j = 1$, if the jth toss is heads, and $W_j = 0$, if the jth toss is tails. Observe that

$$Y_n = W_1 + W_2 + \cdots + W_n$$

and apply (i) with $\mu = p$ and $w = 2(\log \epsilon^{-1})^{1/2}$. ∎

Exercise 11.3.2 *Check that $ns^2 - as$ attains its minimum when $s = a/(2n)$.*

Our proof of Theorem 11.3.1 is a little rough and ready. We polish it up in Exercise 11.3.3 which, I think, provides an excellent exercise for the reader. However, if she chooses not to do the exercise, she can be reassured that the main idea remains the same.

Exercise 11.3.3 [Hoeffding's inequality] *(i) By using the Taylor expansion of* exp *(see Exercise A.10), or otherwise, show that*

$$1 \leq \frac{e^t + e^{-t}}{2} \leq e^{t^2/2}$$

for all t.

(ii) By applying Exercise 3.5.9 (i) with $f(x) = -e^{sx}$, or otherwise, show that

$$e^{sx} \leq \frac{1-x}{2} e^{-s} + \frac{1+x}{2} e^{s}$$

for all s and all $|x| \leq 1$.

(iii) Suppose that Y is a real random variable with $|Y| \leq 1$ and $\mathbb{E}Y = 0$. By taking the expectation of both sides of the inequality

$$e^{sY} \leq \frac{1-Y}{2} e^{-s} + \frac{1+Y}{2} e^{s}$$

and applying part (i), show that

$$\mathbb{E}e^{sY} \leq e^{s^2/2}$$

for all real s.

(iv) Suppose that X is a real random variable with $|X| \leq a$ and $\mathbb{E}X = 0$. Show that

$$\mathbb{E}e^{sX} \leq e^{a^2 s^2/2}$$

for all real s.

(v) Suppose that X_1, X_2, ..., X_n are independent random variables with $\mathbb{E}X_j = 0$ and $|X_j| \leq a_j$ for all j. Show that, if we write, $A = \sum_{j=1}^{n} a_j^2$, then

$$\mathbb{E}e^{s \sum_{j=1}^{n} X_j} \leq e^{As^2/2}.$$

(vi) Continuing with the notation and conditions of part (v), show that, if
$y \geq 0$,

$$\Pr\left(\sum_{j=1}^{n} X_j \geq y\right) \leq \exp\left(-y^2/(2A)\right),$$

$$\Pr\left(\sum_{j=1}^{n} X_j \leq -y\right) \leq \exp\left(-y^2/(2A)\right) \text{ and}$$

$$\Pr\left(\left|\sum_{j=1}^{n} X_j\right| \leq y\right) \leq 2\exp\left(-y^2/(2A)\right).$$

(vii) Let Y_n be the number of heads thrown in n tosses of a coin with probability p of heads where $1 \geq p \geq 1/2$. Show that

$$\Pr\left(Y_n - np \geq Kn^{1/2}\right) \leq \exp\left(-K^2/2p^2\right),$$
$$\Pr\left(Y_n - np \leq -Kn^{1/2}\right) \leq \exp\left(-K^2/2p^2\right),$$
$$\Pr\left(|Y_n - np| \geq Kn^{1/2}\right) \leq 2\exp\left(-K^2/2p^2\right)$$

for all $K > 0$. What can you say if $1/2 \geq p \geq 0$?

If we put $K = 3$ and $p = 1/2$ in the inequality of Exercise 11.3.3, we see that, if we toss a fair coin n times, the probability that the number of heads differs from $n/2$ by more than $3n^{1/2}$ is less than $1/200$. If $n = 100$ this is not very striking, but if $n = 10\,000$ this gives us a rather clear test for fair coins.

Hoeffding's inequality overestimates[6] the probability that the number of heads can lie further than a certain distance from the expected number. This means that we have to toss our coin more times than we need in order to be sure that our result is unexpected. How can we improve things? Here are three possibilities.

(1) Use deeper mathematics. In the days before computers, this was the only possibility. However, as I have remarked before, a mathematical theorem, like a legal contract, means what it says rather than what we wish it to say. In particular, theorems are proved under certain assumptions and it is often difficult to check that these actually hold.[7]

---

[6] Moreover, this overestimate is substantial.

[7] The reader may feel that the widespread use of t-tests, F-tests, χ^2 tests and so on, casts doubt on this assertion. However, these tests are often only employed symbolically to decorate ritual objects like exam papers, Ph.D. theses and academic articles.

(2) Use brute force. In the previous section, when we discussed tossing a coin 5, 10 and 20 times, we calculated the probabilities explicitly and exactly. As Thorp realised (see page 59), the use of computers greatly extends the domain where brute force can be sensibly employed.

(3) Use brute force and ignorance. Suppose that we wish to find the probability that a fair coin will come down heads more than 60 times in a hundred throws. We can use a computer and a random number generator to simulate 10 000 experiments in which a fair coin is tossed 100 times. The weak law of large numbers tells us that (with high probability) the proportion of trials in which the imaginary coin comes down heads more than 60 times will be fairly close to the required probability.

In practice, a judicious mixture of these three methods, combined with common sense and understanding, may well produce the best result.

There is one further point I would like to make. Suppose that we toss our coin 1 000 000 times to check that it is fair. If the coin is fair, then the probability that the number of heads differs from 500 000 by more than 3000 remains less than 1/200. However, our test is now so sensitive that, if we toss a coin which has probability 51/100 of coming down heads, our test will almost certainly detect that it is unfair. We do not believe that a *perfectly fair* coin exists so we must believe that any sufficiently large trial will reject any coin that we test.

In order to produce an appropriate test, we must reconsider our purpose. We do not want to check whether our coin is fair but whether it is sufficiently fair.

Exercise 11.3.4 *Let $\epsilon > 0$ and $1 \geq p, q \geq 0$, $2q + 4\epsilon \leq 1$. Suppose we toss a coin n times and each toss has probability p of heads. Let Y_n be the number of heads we record. By using the results of Exercise 11.3.3, show that, if $|p - q| > 2\epsilon$,*

$$\Pr\left(|Y_n - nq| \leq n\epsilon\right) \leq 2\exp\left(-2n\epsilon^2\right).$$

11.4 A better needle?

Once a promising drug has been found, chemists must discover how to make it in quantity and in an appropriate form. If, as we hope, it has a powerful positive biological effect, it may also have powerful negative biological effects. It may be immediately toxic or it may have long term deleterious effects. Much can be learnt from test tube experiments on single cells, but the effect of the drug on different elements of a complex biological system may not tell us the effect on the whole, and eventually the drug must be tested on human beings.[8]

If the drug clears all these hurdles, then it has to be tested in a large scale trial to see if it is at least as good as the standard treatment. With so much at stake, the trials will use the best available mathematics.

This book does not provide the best available mathematics, but we can do something with what we have got. Fitting our trial to our mathematics, we suppose that the participants are paired and one member of each pair gets our new treatment while the other gets the old. If the patient with the new treatment is judged to have done better, then we register a success,[9] if not, we register a failure. In effect, we are tossing a coin with probability p of success (heads) and our new treatment will be at least as good as the old if $p \geq 1/2$.

A natural way to organise things would be to make n trials and record the number of successes Y_n. We then accept that the new treatment is better if $Y_n \geq a_n$ for some appropriate a_n.

Exercise 11.4.1 *Use Hoeffding's inequality to find an n and an a_n with the both following properties*
(i) If $p \geq 0.55$ the probability that we reject the new drug is less than 0.05.
(ii) If $p \leq 0.45$ the probability that we accept the new drug is less than 0.05.
More exact calculation using the central limit theorem shows that we can get n down to under 300.

If we were dealing with coins rather than people, our proposal would be entirely sensible. But we are dealing with people. What should we do if the new drug appears worse than the old in the first 20 comparisons? Would we continue to play against someone who throws 20 tails in a row? Surely not. Should we continue to give people a drug which appears inferior in 20 cases in a row?

Or suppose that our new drug appears better in the first 20 comparisons. Can we deprive people of what we are now fairly certain is a better treatment just to satisfy the requirements of the test? But, if we stop our trials if the first 20 results are all successes, should we not stop them if 19 out of the first 20 are, or 15 or 12 ... ?

The golfer's fallacy tells us that we cannot make up our tests as we go along. We must decide *in advance* what we are going to do.[10] We could, however,

---

[8] As this book was being written, the first human trial of a new drug nearly killed the six participants.
[9] It is sensible to leave this judgement to someone who does not know which treatment was used.
[10] Including a let-out clause which allows the trial to be abandoned completely in unforeseen circumstances.

change the form of the test so that we do successive comparisons as before, but accept the drug if the number of successes exceeds the number of failures by a certain fixed number a and reject the drug if the number of failures exceeds the number of successes by a.

Exercise 11.4.2 *Suppose that we accept the drug if the number of successes exceeds the number of failures by a and reject the drug if the number of failures exceeds the number of successes by a. We take the probability of success in each trial to be p.*

Let us write $q = 1 - p$ and let u_r be the probability that we will reject the drug if up to now the number of successes has exceeded the number of failures by r. Explain why

$$u_r = qu_{r-1} + pu_{r+1} \quad for \quad -a + 1 \le r \le a - 1,$$

$u_a = 1$ and $u_{-a} = 0$. *Find u_r using the methods of Section 10.3 and show that the probability that the suggested test ends with the acceptance of the new drug is*

$$\frac{1}{1 + \left(\frac{q}{p}\right)^a}.$$

Find the smallest value of a so that the probability of rejection is smaller than 0.05 if $p \ge 0.55$ and the probability of acceptance is less than 0.05 if $p \le 0.45$.

Why might one decide accept the drug if the number of successes exceeds the number of failures by a and reject the drug if the number of failures exceeds the number of successes by b with $b \ne a$?

As we have set it up, this test could involve arbitrarily many trials. In practice, there would have to be some fixed limit, but we will ignore this and ask instead for the expected number of trials involved. It is fairly clear that the worst case occurs when $p = 1/2$.

Exercise 11.4.3 *Consider our test when $p = 1/2$. Let e_r be the total number of expected trials from now on until the end of the test if up to now the number of successes has exceeded the number of failures by r. Explain why*

$$e_r = 1 + \tfrac{1}{2}e_{r-1} + \tfrac{1}{2}e_{r+1} \quad for \quad -a + 1 \le r \le a - 1,$$

and $e_a = e_{-a} = 0$. Show that

$$e_r = a^2 - r^2$$

and conclude that the expected number of trials required by our test is a^2.

What is the the expected number of trials for the a calculated in Exercise 11.4.2?

What happens in general?

Exercise 11.4.4 *Consider our test for general p. We wish to find the expected number of trials $R(p, a)$ required by our test. We set $q = 1 - p$.*

(i) Suppose that $p \neq 1/2$. Let $e_r(p)$ be the total number of expected trials from now on until the end of the test if up to now the number of successes has exceeded the number of failures by r. Explain why

$$e_r(p) = 1 + qe_{r-1}(p) + pe_{r+1}(p) \quad for \quad -a + 1 \leq r \leq a - 1,$$

and $e_a = e_{-a} = 0$. Show that

$$e_r(p) = \frac{2a}{p-q} \left(\frac{1 - \left(\frac{q}{p}\right)^{r+a}}{\left(1 - \frac{q}{p}\right)^{2a}} \right) - \frac{r+a}{p-q}$$

and conclude that the expected number of trial required by our test is

$$R(p, a) = \frac{a}{p-q} \left(\frac{1 - \left(\frac{q}{p}\right)^a}{1 + \left(\frac{q}{p}\right)^a} \right) = \frac{a}{p-q} \left(\frac{p^a - q^a}{p^a + q^a} \right).$$

(ii) Show that $R(p, a) \to R(1/2, a)$ as $p \to 1/2$.
(iii) Show that, if p is fixed with $p > 1/2$, we have

$$\frac{R(p, a)}{a} \to \frac{1}{p-q}.$$

as $a \to \infty$. Can you give an informal argument as to why we should expect this? What is the corresponding result when $p < 1/2$?
(iv) Compute $R(a, p)$ for $p = 2/3$ and a chosen as in Exercise 11.4.2.

However we design our test, it will only supply us with a limited amount of evidence. By increasing the number of trials we can obtain more evidence, but there is no sense in going beyond a certain point. Some side effects will take a long time to develop or will only occur in circumstances not covered by our test. Once a drug is made available, doctors may discover (or believe they have discovered) other uses for it. We have a duty to try and predict the future, but the strong limitations on our power of prediction place strong limits on our duties.

12

Final reflections

12.1 First the music, then the words

In 1981, the British Government commissioned a report on the teaching of mathematics. Part of the project involved interviewing a representative sample of the population.

> Both direct and indirect approaches were tried, the word 'mathematics' was replaced by 'arithmetic' or 'everyday use of numbers' but it was clear that the reason for people's refusal to be interviewed was simply that the subject was mathematics. . . . [The] apparently widespread perception amongst adults of mathematics as a daunting subject pervaded a great deal of the sample selection; half the people approached as being appropriate for inclusion in the sample refused to take part. . . . The extent to which the need to undertake even an apparently straightforward piece of mathematics could induce feelings of helplessness, fear and even guilt in some of those interviewed was, perhaps, the most striking feature of the study. . . . [These] feelings of guilt . . . appeared to be especially marked among those whose academic qualifications were high.
>
> *[[13], pages 6–7]*

Feelings of guilt can lead to excessive humility or to excessive hostility. Sometimes, hostility is expressed in the claim that any mathematical argument can be translated into non-mathematical terms. Thus, for example, the reason the moon can orbit the earth is shown by Figure 12.1,[1] in which a stone is projected with successively greater initial velocity. However, I know of no convincing 'non-mathematical' argument why the inverse square law of gravitation yields elliptic orbits.

---

[1] The diagram is taken from *The System of the World* [46], the Newtonian equivalent of *A Brief History of Time*.

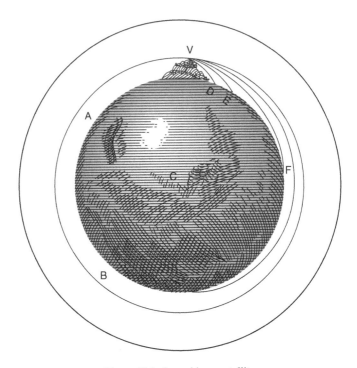

Figure 12.1. Launching a satellite.

The battle between believers in mathematical arguments and their opponents is, as might be expected, at its bitterest when it involves economists.

Let us take the proposition that, when the cost of something falls, people buy more of it. A mathematically inclined economist would consider a single 'household' with an income of €c which it can use to buy two sorts of things, say meat and bread. If meat costs €u per kilo and bread €v per kilo [u, $v > 0$], then the household can buy x kilos of meat and y kilos of bread subject to the conditions

$$ux + vy = c, \; x, \; y \geq 0.$$

The amount of 'satisfaction' that the household receives from its purchase is $f(x, y)$. The household will choose x and y, subject to the conditions stated, so as to maximise f.

Suppose that, when $c = c_0$, the household maximises its satisfaction by taking $x = x_0$, $y = y_0$ with x_0, $y_0 > 0$. To keep the mathematics simple, we suppose that, when δx and δy are small, we have

$$f(x_0 + \delta x, y_0 + \delta y) = f(x_0, y_0) + a\delta x + b\delta y +$$
$$\tfrac{1}{2}(A(\delta x)^2 + 2B\delta x \delta y + C(\delta y)^2),$$

but the general case runs in the same way. Note that, since we assume that the more meat or bread people have, the happier they are, we must take $a, b > 0$.

Observe that, if the household could have chosen $x = x_0 + \delta x$ and $y = y_0 + \delta y$, then

$$ux + vy = u(x_0 + \delta x) + v(y_0 + \delta y) = c$$

and so

$$u\delta x + v\delta y = 0.$$

It follows that

$$f(x_0 + \delta x, y_0 + \delta y) - f(x_0, y_0)$$
$$= (a - buv^{-1})\delta x + \tfrac{1}{2}(A - 2Buv^{-1} + Cu^2v^{-2})(\delta x)^2.$$

Since satisfaction is maximised by taking $\delta x = 0$, it follows that

$$a - buv^{-1} = 0, \quad (A - 2Buv^{-1} + Cu^2v^{-2}) \le 0$$

or, more symmetrically,

$$av - bu = 0, \quad Av^2 - 2Buv + Cu^2 \le 0.$$

We shall suppose that the maximum is unique and so

$$Av^2 - 2Buv + Cu^2 < 0.$$

Exercise 12.1.1 *Economists often demand that A, B and C be such that $Av^2 - 2Buv + Cu^2 < 0$ for all possible values of u and v $[(u, v) \ne (0, 0)]$. Show, by completing the square or otherwise, that this will be the case if and only if $A, C < 0$ and $B^2 < AC$.*

Having got this far, we now want to examine the effects of small changes in c, u or v. These will change the point of maximum satisfaction to some new (x_1, y_1) with

$$x_1 = x_0 + \Delta x, \quad y_1 = y_0 + \Delta y,$$

where we expect $\triangle x$ and $\triangle y$ to be small. Simple algebra gives

$$f(x_1+\delta x, y_1 + \delta y) = f(x_0 + \triangle x + \delta x, y_0 + \triangle y + \delta y)$$
$$= a(x_0 + \triangle x + \delta x) + b(y_0 + \triangle y + \delta y)$$
$$+ \frac{1}{2}(A(\delta x + \triangle x)^2 + 2B(\delta x + \triangle x)(\delta y + \triangle y) + C(\delta y + \triangle y)^2)$$
$$= f(x_0 + \triangle x, y_0 + \triangle y)$$
$$+ (a + A\triangle x + B\triangle y)\delta x + (b + B\triangle x + C\triangle y)\delta y$$
$$+ \frac{1}{2}(A(\delta x)^2 + 2B\delta x\delta y + C(\delta y)^2)$$
$$= f(x_1, y_1) + \tilde{a}\delta x + \tilde{b}\delta y + \frac{1}{2}(A(\delta x)^2 + 2B\delta x\delta y + C(\delta y)^2)$$

where

$$\tilde{a} = a + A\triangle x + B\triangle y, \quad \text{and} \quad \tilde{b} = b + B\triangle x + C\triangle y.$$

We can now investigate the effect of a small change in the price of bread from v to $v + \triangle v$ while leaving u and c unchanged. We must have

$$u(x_0 + \triangle x) + (v + \triangle v)(y_0 + \triangle y) = c$$

and so, since $ux_0 + vy_0 = c$,

$$u\triangle x + v\triangle y + y_0\triangle v + \triangle v\triangle y = 0.$$

By our previous results on behaviour at the maximum,

$$\tilde{a}(v + \triangle v) - \tilde{b}u = 0$$

and so

$$(a + A\triangle x + B\triangle y)(v + \triangle v) - (b + B\triangle x + C\triangle y)u = 0.$$

If, as we hope, all the quantities $\triangle x$, $\triangle y$ and $\triangle v$ are small, then 'second order terms' like $\triangle v\triangle y$ and $\triangle x\triangle v$ will be very small. If we ignore these very small terms, the equations obtained in the previous paragraph and this one reduce to

$$u\triangle x + v\triangle y + y_0\triangle v \approx 0,$$
$$(a + A\triangle x + B\triangle y)v - (b + B\triangle x + C\triangle y)u + a\triangle v \approx 0.$$

We know that $av - bu = 0$ so, after rearrangement, our equations take the form

$$u\triangle x + v\triangle y \approx -y_0\triangle v,$$
$$(Av - Bu)\triangle x + (Bv - Cu)\triangle y \approx -a\triangle v.$$

Substituting for $\triangle x$, using the first equation, we obtain,

$$\left(- v(Av - Bu) + u(Bv - Cu) \right) \triangle y \approx \left(- au + y_0(Av - Bu) \right) \triangle v.$$

Thus

$$\triangle y \approx \frac{-au + y_0(Av - Bu)}{-Av^2 + 2Buv - Cu^2} \triangle v. \qquad \bigstar$$

Mark Twain once compared speaking German to swimming the Atlantic with a verb in your mouth. What, if anything, can we say about the formula $\bigstar$ which we have obtained with so much effort? We first remember that we have

$$Av^2 - 2Buv + Cu^2 < 0.$$

Even so, we are faced with the worrying possibility that $-au + y_0(Av - Bu)$ might be negative so that a decrease in v (that is $\triangle v < 0$) would lead to a decrease in y (that is $\triangle y < 0$). This might happen if y_0 and v were both large.

Exercise 12.1.2 *By choosing explicit values for x_0, y_0, a, b, A, B, C, u, v and c, show that it is possible to satisfy all the constraints we have placed on them and still have*

$$-au + y_0(Av - Bu) < 0.$$

At this point, it is natural to search for some additional condition which prevents the possibility. The failure of such a search might force us to the conclusion first stated by Marshall.

> There are, however, some exceptions [to the statement that a decrease in price will lead to an increase in consumption]. For instance, as Sir R. Giffen has pointed out, a rise in the price of bread makes so large a drain on the resources of the poorer labouring families and raises so much the marginal utility of money to them, that they are forced to curtail their consumption of meat and the more expensive farinaceous foods: and, bread being still the cheapest food which they can get and will take, they consume more, and not less of it. But such cases are rare; when they are met with, each must be treated on its own merits.
>
> *([42] Book III, Chapter VI)*

Let us restate Marshall's argument at greater length. We are familiar with the idea that, as someone's income increases, they may buy less of a particular good. A rich family will eat more meat and less bread than a poor family.

Exercise 12.1.3 *Suppose that in our model we increase the total income by a small amount from c to $c + \triangle c$, and the household now sets $x = x_0 + \triangle x$, $y = y_0 + \triangle y$. Compute $\triangle y$ as a function of $\triangle c$. Under what conditions does $\triangle y$ decrease when $\triangle c$ increases?*

Now consider a poor family subsisting almost entirely on bread. If the price of bread drops, this is equivalent to an increase in their income. Since they much prefer meat to bread, they will use almost all of this increase to buy meat. Now that they have meat to eat, they can cut down on their bread consumption and use the saving to buy still more meat.

Notice that this argument takes place at the level of a single household. A real economy contains many households with different income levels and different desires. In a famine, the middling class will eat bread, the working class will eat potatoes and the poor will starve. When the famine ends, the price of bread will come down, the middling class will return to meat, the working class to bread and the poor to potatoes. The decline in the price of bread will not have lead to an *overall* decline in the consumption of bread although some households may now eat less bread.

If my readers are like me, they will feel much happier as to the possibility and meaning of the 'Giffen phenomenon' once they have read and understood Marshall's verbal explanation. However, it was the failure of our algebraic manipulation to deliver the expected result which alerted us to the possibility of such an effect.

Historians of economics have been unable to trace any unambiguous statement by Giffen of the phenomenon named after him, and the price of bread was stable during the period that Marshall refers to. Since Marshall was a first class mathematician, we may suspect that Marshall followed the same path[2] as we have done and followed the principles he laid out in a letter to Bowley.

> (1) Use mathematics as shorthand language, rather than as an engine of enquiry. (2) Keep to them till you have done. (3) Translate into English. (4) Then illustrate by examples that are important in real life (5) Burn the mathematics. (6) If you can't succeed in 4, burn 3. This last I did often.
>
> *(Quoted in [11])*

It seems plausible that the Giffen effect will mainly occur in very poor societies with a very limited number of commodities. If it does occur, then it reflects rational behaviour under certain unusual constraints. A much more troubling phenomenon, at least to mathematicians, may occur in rich societies if the satisfaction function f depends not only on the quantity of goods consumed but also on their price. (So, in the example above, $f(x, y)$ is replaced by $f(x, y, u, v)$.) In particular, the consumer may gain more satisfaction from

---

[2] Though, perhaps, geometrically rather than algebraically.

higher priced goods simply because they are higher priced so that increasing the price of a luxury might increase its sales.[3]

Biologists tell us that the peacock's tail is a signal to the peahen that here is a bird so healthy and successful that it can lavish energy on useless adornment. In the same way, the acquisition of expensive objects because they are expensive may serve goals invisible to the unsophisticated observer. Even so, this sort of behaviour suggests that there are limits to how far models based on rational behaviour can usefully explain human societies.

12.2 Mathematics and decision making

In this book we tried to study aspects of society using mathematics. To do this we ignored the question of how people actually behave. We acted like an astronomer who locks herself in a windowless room in order to think undisturbed.

It is not surprising that this approach only seems to work in a few instances, but it is remarkable how successful it is in the few cases when it does work.[4] Pulling the handle on our mathematical fruit machine rarely produces a winning combination, but when it does, the floor is covered with gold pieces.

One hundred years ago, a book was a book, a picture a picture and a telephone call a telephone call. Today, they are all strings of zeros and ones and so objects of mathematical study. In Section 4.5 we discussed one of the modern mathematical secret codes but, important though concealing information may be, it is the application of deep mathematics to storing, transmitting and treating information which is the foundation of much of the modern economy.

In the same way, although casinos and race-tracks represent very profitable applications of mathematics, their economic importance is dwarfed by the business of insurance and pensions which depend on the same ideas.

For generations, economists used to explain patiently to reckless rabble-rousers that there was no resemblance whatsoever between a stock market and a race-track. Today, they measure the efficiency of the market by the degree to which trading resembles betting at an idealised race-track. The mathematics involved is quite advanced, but the ideas follow those of the first part of this book.

The reader will not need to be persuaded of the importance of statistical theory or of that of many of the algorithms we discussed. She may need to be

---

[3] The classic work here is Veblen's witty *Theory of the Leisure Class*. Note, in particular, Chapter 14 on Higher Education.

[4] The reader will observe that much of this section consists of opinions. I would prefer the reader to reject these opinions out of hand rather than accept them uncritically, but I hope these are not the only possible outcomes.

alerted to the link between the noisy and quiet duels of Section 7.6 and auctions but, once the link has been made, she will not be surprised to learn that there is a useful mathematical theory of auctions.

In all the examples given so far, better mathematical methods lead (at least, if we act wisely) to better results. More advanced mathematics than that in this book, like the simplex method, 'continuous' probability, the central limit theorem, multivariable calculus, variational methods, compactness and measure theory, turn out to have striking applications.

Many of the remaining topics that we discussed, such as Arrow's theorem and n-person games, have not had such obvious success. In order to explain why I think they are still of interest to the decision maker, I will make a distinction between *explanatory* and *predictive* theories.

Most of my readers will know the elegant calculations which show that, if we neglect air resistance, the path of a shell will be a parabola. A little thought gives a rough idea of the way the path is modified by air resistance. It is clear that the theory *explains* the way that a shell flies, but it is also clear that, given a gun to fire, I have actually very little idea of where the shell will land. Up until the First World War, although professional artillery men knew roughly where a shell would land, they would not expect to hit their target first time but rather to use intelligent trial and error to hit it in a small number of attempts.

During that war, the need for surprise meant that gunners were asked to hit their target first time. Mathematicians and their assistants drew up tables which took into account the density of air at different heights, the weather, the barometric pressure and even the number of times a gun had been fired. The calculations were tedious, involved ad hoc adjustments and were generally unpleasant[5] but did their job of *prediction*.[6]

From our point of view, the most spectacular change from an explanatory to a predictive theory is given by weather forecasting. Anybody who thinks that it should be easy to replicate this success in other fields should bear in mind the following points.

(1) The laws that govern the weather are the same now as they were in 1900.
(2) The weather forecaster has clear goals. A correct forecast of the weather in two minutes' time is useless, but a forecast which is usually correct for 24 hours is immensely valuable.

---

[5] 'Even Littlewood,' wrote Hardy, 'could not make ballistics respectable'. [27]

[6] Second Lieutenant J. E. Littlewood 'devised a rapid and powerful method' for computing the trajectories of anti-aircraft shells. When this was put to the trial 'To the astonishment and joy of all concerned, the observed position of the shell bursts fell exactly on Littlewood's trajectories ...'. [43]

(3) There are agreed criteria for measuring the success of a series of forecasts.

(4) The data used in making the forecasts are available to known standards of accuracy and this accuracy improves as the years pass.

(5) The weather does not listen to our weather forecasts and adjust its future behaviour accordingly.

Viewed in this light, many of the topics discussed in this book are explanatory rather than predictive. The study of Prisoner's Dilemma or Hawks and Doves will not tell the decision maker what she should do but may help her understand the kind of decisions that she can make.

Thus the utilitarian argument for the study of voting systems or Morra is no different from the utilitarian argument for the study of classics ('the Roman Senate is not the British House of Commons but by seeing how Cicero swayed the one we may see how to sway the other') or history ('many things have changed in eighty years but a study of the Civil War will help us understand present day Irish politics'). The reader must decide how convincing she finds such arguments in each case.

The two great endeavours of mankind which depend most on mathematics are physics and engineering. But, on the whole, the best physics is done by physicists rather than mathematicians, and the best engineering by engineers. Chemistry, biology and medicine are slowly becoming more mathematical but, again, insight into the nature of the particular subject trumps mere mathematical ingenuity. In the same way, we should expect that, even when mathematics is important in decision making, mathematical skill will only be a small part of what makes a good decision maker.

Mathematics has made contributions to our understanding of decision making which could hardly have been arrived at using other modes of thought. No doubt mathematics will continue to make contributions, though the timing and nature of those contributions will be erratic and unpredictable. However, mathematicians should be modest in what they should expect to achieve.

> Fancy what a game at chess would be if all the chessmen had passions and intellects, more or less small and cunning: if you were not only uncertain about your adversary's men, but a little uncertain also about your own; if your knight could shuffle himself on to a new square by the sly; if your bishop, in disgust at your castling, could wheedle your pawns out of their places; and if your pawns, hating you because they are pawns, could make away from their appointed posts that you might get checkmate on a sudden. You might be the longest-headed of deductive reasoners, and yet you might be beaten by your own pawns. You would be especially likely to be beaten, if you depended arrogantly on your mathematical imagination, and regarded your passionate pieces with contempt. Yet this imaginary

chess is easy compared with the game a man has to play against his fellow-men
with other fellow-men for his instruments. He thinks himself sagacious, perhaps,
because he trusts no bond except that of self-interest; but the only self-interest he
can safely rely on is what seems to be such to the mind he would use or govern.

(George Eliot Felix Holt, The Radical*)*

Mathematicians should not be unduly worried by the limitations of mathematics as a tool for studying the real world. The object of mathematical study
is not power or even utility, but pleasure. Mathematicians are not to be found in
the council chamber where grave princes seek to decide the future of nations,
nor in the counting house where the merchant plans great commercial ventures,
nor in the sick chamber where the doctor wrestles against disease but outside
in the sunlight swapping riddles and tossing coins. If some of the games that
mathematicians play turn out to be useful, so much the better, but they remain
games and should be judged as such.

Appendix A

The logarithm

Since the logarithm function plays such an important role in this book, I include a series of exercises developing its properties. Even if the reader knows those properties, it may be useful to recall how they are derived. Solutions to most of the exercises in this book are sketched at the internet address given in the Introduction.

Exercise A.1 *(i) Sketch the function $g(x) = 1/x$ for $x > 0$.*

(ii) Suppose that $f(x)$ is a well-defined function of x for $x > 0$. Given the graph of $f(x)$, how would you obtain the graph of $F(x) = af(ax)$ by appropriate rescaling of the coordinate axes? What happens in the the particular case when $f(x) = 1/x$?

(iii) Show, by a suitable change of variable, that, if a, $b > 0$, then

$$\int_a^b \frac{1}{x}\,dx = \int_1^{b/a} \frac{1}{x}\,dx.$$

(iv) We define

$$\log t = \int_1^t \frac{1}{x}\,dx$$

for $t > 0$. Show, by writing

$$\int_1^{uv} \frac{1}{x}\,dx = \int_1^u \frac{1}{x}\,dx + \int_u^{uv} \frac{1}{x}\,dx,$$

that $\log uv = \log u + \log v$ for all u, $v > 0$.

Exercise A.2 *(i) If f is a well-behaved function, explain, by means of a diagram, why*

$$\frac{1}{h}\int_t^{t+h} f(x)\,dx \approx f(t)$$

345

and

$$\frac{d}{dt}\int_a^t f(x)\,dt = f(t).$$

(ii) Show that the function log *is differentiable with*

$$\log'(t) = \frac{1}{t}$$

for all t > 0.

Exercise A.3 *Prove the following results.*
  (i) $\log x$ *is a strictly increasing function of* x.
  (ii) $\log 1 = 0$.
  (iii) $\log 2^n \to \infty$ *as the integer* $n \to \infty$.
  (iv) $\log x \to \infty$ *as* $x \to \infty$.
  (v) $\log x \to -\infty$ *as* $x \to 0$ *through positive values of* x.
  (vi) The second derivative $\log'' x < 0$ *for all* $x > 0$.
  (vii) Sketch the graph of log.

Exercise A.4 *(i) Suppose that* f *is a well-behaved strictly increasing function. Show how, starting from the graph of* $y = f(x)$, *we can obtain the graph* $y = f^{-1}(x)$ *of the inverse function by swapping the x- and y-axes. (Recall that the inverse function* f^{-1} *is defined so as to satisfy the equations* $f^{-1}(f(x)) = x$ *and* $f(f^{-1}(y)) = y$ *wherever they make sense.)*
  Explain in terms of the resulting diagrams why

$$(f^{-1})'(y) = \frac{1}{f'(f^{-1}(y))}$$

wherever the equation makes sense.
  (ii) We define $\exp = \log^{-1}$ *(so* $\exp x$ *is defined for all real x). Sketch the graph of* exp.
  (iii) Show that exp *is everywhere differentiable with* $\exp' x = \exp x$.
  (iv) Why is $\exp x$ *always strictly positive?*

Exercise A.5 *Prove the following results.*
  (i) $\exp x \exp y = \exp(x + y)$ *for all* x *and* y.
  (ii) $\exp x$ *is a strictly increasing function of* x.
  (iii) $\exp x \to \infty$ *as* $x \to \infty$.
  (iv) $\exp x \to 0$ *as* $x \to -\infty$.

Exercise A.6 *Most of results of the next two exercises are not used in this book, but are included for completeness.*

(i) Show that, if $x > 0$ and n is a strictly positive integer,

$$\exp(n \log x) = x^n,$$

where x^n has its standard elementary meaning

$$x^n = \overbrace{x \times x \times x \times \cdots \times x}^{n}.$$

(ii) Show that, if $x > 0$ and n is an integer,

$$\exp(n \log x) = x^n,$$

where x^n has its standard elementary meaning.

(iii) Show that, if m and n are integers and $n \neq 0$, then

$$\left(\exp \left(\frac{m}{n} \log x \right) \right)^n = x^m$$

and so

$$\exp \left(\frac{m}{n} \log x \right) = x^{m/n}$$

where $x^{m/n}$ has its standard elementary meaning.

(iv) Since, as we have just shown, $\exp(a \log x) = x^a$ whenever $x > 0$ and a is rational, it is reasonable to define

$$x^a = \exp(a \log x)$$

for all $x > 0$ and all real a. Show that, if we write $e = \exp 1$, we recover the familiar equality

$$e^x = \exp x.$$

Exercise A.7 *In this exercise we check that the familiar index laws continue to hold for the extended definition of Exercise A.6 (iv). Verify that, if a, b are real and x, $y > 0$, then the following results hold.*
  (i) $(xy)^a = x^a y^a$.
  (ii) $x^{a+b} = x^a x^b$.
  (iii) $x^{ab} = (x^a)^b$.

Exercise A.8 *The result of this exercise is useful in probability theory.*
  (i) *Use the definition of differentiation to show that*

$$\frac{\log(1 + h) - \log 1}{h} \to 1$$

as $h \to 0$.

(ii) Deduce that, if a is real,

$$\frac{\log(1 + a/n)}{a/n} \to 1$$

as $n \to \infty$.

(iii) Deduce that

$$n \log\left(1 + \frac{a}{n}\right) \to a$$

as $n \to \infty$.

(iv) Conclude that

$$\left(1 + \frac{a}{n}\right)^n \to e^a$$

as $n \to \infty$.

(v) The 'rule of 72' says that, if you invest a sum of money €y at x% compound interest (so that, after n years, you will have €y$(1 + x/100)^n$), then it will double in approximately 72/x years. Where does this rule come from? For what values of x is it accurate?

[You may wish first to look at the matter theoretically and then try some values of x on a calculator.]

Exercise A.9 *In this exercise we obtain a Taylor series for* log.

(i) Show that, if $t \neq 1$, *and n is a positive integer,*

$$\frac{1}{1-t} = 1 + t + \cdots + t^n + \frac{t^{n+1}}{1-t}$$

and deduce that, if $|x| < 1$,

$$\int_0^x \frac{1}{1-t}\, dt = x + \frac{x^2}{2} + \cdots + \frac{x^n}{n} + R_n(x),$$

where

$$R_n(x) = \int_0^x \frac{t^{n+1}}{1-t}\, dt.$$

Conclude that

$$-\log(1 - x) = x + \frac{x^2}{2} + \cdots + \frac{x^n}{n} + R_n(x).$$

(ii) Show that, if $|t| \leq |x| < 1$, *then*

$$\left|\frac{t^{n+1}}{1-t}\right| \leq \frac{|x|^{n+1}}{1-|x|}$$

and deduce that

$$|R_n(x)| \leq \frac{|x|^{n+2}}{1 - |x|}.$$

Show that $R_n(x) \to 0$ *as* $n \to \infty$.

(iii) *Conclude that*

$$-\log(1 - x) = x + \frac{x^2}{2} + \cdots + \frac{x^n}{n} + \cdots$$

for all $|x| < 1$ *and so*

$$\log(1 + y) = y - \frac{y^2}{2} + \cdots + \frac{(-1)^{n+1} y^n}{n} + \cdots$$

for all $|y| < 1$.

Exercise A.10 *In this exercise we obtain a Taylor series for* exp.

(i) *Explain, by means of diagram, why, if f and g are well-behaved functions with* $0 \leq f(x) \leq g(x)$ *for* $a \leq x \leq b$, *we have*

$$\int_a^b f(x) \, dx \leq \int_a^b g(x) \, dx.$$

Deduce that if h and g are well-behaved functions with $|h(x)| \leq g(x)$ *for* $a \leq x \leq b$, *we have*

$$\left| \int_a^b h(x) \, dx \right| \leq \int_a^b g(x) \, dx.$$

(ii) *Show, by induction, or otherwise, that, if f is a well-behaved function with* $|f^{(n)}(x)| \leq A$ *for all* $|x| \leq X$ *and*

$$f(0) = f'(0) = f''(0) = \cdots = f^{(n-1)}(0) = 0,$$

we have

$$|f(x)| \leq A \frac{|x|^n}{n!}$$

for all $|x| \leq X$.

(iii) *Deduce that*

$$\left| \exp x - \left(1 + x + \frac{x^2}{2} + \cdots + \frac{x^{n-1}}{(n-1)!} \right) \right| \leq e^X \frac{|x|^n}{n!}$$

for all $|x| \leq X$.

(iv) *Conclude that*

$$\exp x = 1 + x + \frac{x^2}{2} + \cdots + \frac{x^n}{n!} + \cdots.$$

Exercise A.11 *If you know a standard method for proving that the Taylor expansion is valid, use it to obtain the result of the two previous questions.*

Let a, $x > 0$. It is sometimes useful to use the notation

$$\log_a x = \frac{\log x}{\log a}.$$

We say that '$\log_a x$ is the logarithm of x to the base a'.

Exercise A.12 *Let a, b, x, $y > 0$ and let k be a real number. Show that*
(i) $a^{\log_a x} = x$,
(ii) $\log_a xy = \log_a x + \log_a y$,
(iii) $\log_a x^k = k \log_a x$,
(iv) $\log_e x = \log x$,
(v) $\log_a b \log_b a = 1$.

Exercise A.13 *This question is to be done without a calculator (though you may wish to check your answer using one afterwards). You may use the fact that, to nine decimal places, $\log_{10} 2 = 0.301\,029\,996$ and $\log_{10} 3 = 0.477\,121\,255$.*

(i) Calculate $\log_{10} 5$ and $\log_{10} 6$ to three decimal places. By using logarithms, show that

$$5 \times 10^{47} < 3^{100} < 6 \times 10^{47}.$$

Hence write down the first digit of 3^{100}.
(ii) Find the first digit of the following numbers: 2^{1000}, $2^{10\,000}$ and $2^{100\,000}$.
(iii) (This has nothing to do with logarithms.) Find the last digit of 3^{100} and $2^{100\,000}$.

In book like this, where expressions like 2^n occur frequently, it is useful to use logarithms to base 2. Before pocket calculators were invented, logarithms to base 10 were in constant use.[1] However, when we do analysis, logarithms to base e work most smoothly.

---

[1] In a hangover from those days, some people and most pocket calculators reserve the notation log for $\log_{10}$ and use ln (from 'Naperian logarithm' or, possibly, 'natural logarithm') for $\log_e$. In my view, you should be aware of this convention but avoid it.

Appendix B

Cardano

Almost everyone who reads this book will have no difficulty with the following exercise.

Exercise B.1 *What is the probability that at least one of three ordinary dice show a* 1 *when they are thrown together?*

Who was the first person to find the answer to this question? So far as anyone knows, it was the remarkable doctor, author, mathematician and astrologer Cardano born in 1501.

In his lifetime, Cardano was chiefly famous as a doctor and astrologer. Later he was famous for such sixteenth-century best sellers as *De Subtilate Rerum* (On The Subtlety of Things) which could be considered one of the first popular science books. The observations and inventions, many due to others,[1] but some his own, that he reported earn him a place in histories of optics, hydrodynamics, geology, engineering and cryptography. In a more traditional vein, he wrote books on *Wisdom* and *Consolation* (believed, by some, to be the book Hamlet is reading when interrupted by Polonius).

Today, his claim to remembrance rests on three books. The first and most important was the *The Great Art* [9] on what we would now call the theory of equations. In 1500, algebra, in the sense that we know it, did not exist and all algebraic arguments had to be expressed verbally. People knew how to solve what we would call linear and quadratic equations, but those few who considered the matter held that further progress was impossible.

However, as Cardano relates:

> Scipione del Ferro, from Bologna, found in our time the rule for the cube and the unknown equal to a number,[2] something truly beautiful and admirable.

---

[1] It contained the first European description of the game of Chinese Rings which later inspired the Tower of Hanoi.

[2] That is to say, he found how to solve $x^3 + ax = b$ with a and b positive. The restriction to these coefficients appeared to be essential to his method which therefore did not provide a solution to other forms of the cubic equation.

> Such a discovery, a truly divine gift surpassing all human subtlety and the splendour or mortal ingenuity, is a proof of the virtue of the soul, a thing so marvellous that he who found it may have believed that there could be no difficulties he would not be able to surmount.

At the time, it was customary to hold mathematical contests in which the two opponents challenged each other to solve mathematical problems. The prizes were often substantial, but, even more importantly, the winner greatly increased his chances of appointment or reappointment to teaching positions. Del Ferro and his pupils would have considered his method a commercial secret. Cardano continues.

> Emulating this man, Niccolò Tartaglia from Brescia, our friend, who had entered into a contest with Antonio Maria Fiore, pupil of del Ferro, rediscovered this rule in order to win and he later confided it to me after I had made insistent requests . . .
>
> After I was in possession of this rule and had found a proof of it, I understood that many other things could be discovered, and with my confidence thus already increased I found such results, partly by myself and in part through the work of Lodovico Ferrarri, my former pupil.
>
> All that has been discovered by these men will be designated by their name and that which is not attributed belongs to me.

Ferrarri's[3] contribution included the solution of quartic equations

$$x^4 + ax^3 + bx^2 + cx + d = 0.$$

What Cardano does not say, is that he swore an oath to Tartaglia not to reveal his secret to anyone. Tartaglia had told Cardano that he intended to publish the result himself. Five years later, the result had not appeared and an impatient Cardano got permission to examine del Ferro's posthumous papers. Finding written proof that Tartaglia was not the first discoverer, Cardano decided that this released him from his oath.

Cardano's book contains much more than the solution of the cubic equation and represents a great leap forward in the study of equations. Of course most of his discoveries are now common property.

Exercise B.2 *(i) Show that, given an equation*

$$x^n + a_1 x^{n-1} + \cdots + a_n = 0,$$

we can find a c such that, writing $y = x + d$, we have

$$y^n + b_2 y^{n-2} + \cdots + b_n = 0.$$

---

[3] Ferrarri married Cardano's daughter. There is an old German academic saying: 'The law of heredity for mathematical talent is a little unusual, it passes from father-in-law to son-in-law'.

Explain how, if we can solve every cubic equation of the form

$$x^3 + ax + b = 0,$$

we can solve the general cubic equation

$$x^3 + ax^2 + bx + c = 0.$$

(ii) Show how, given one root α of an equation

$$x^n + a_1 x^{n-1} + \cdots + a_n = 0,$$

we can find an equation

$$x^{n-1} + b_1 x^{n-2} + \cdots + b_n = 0$$

having the same roots (apart, possibly, from α).
(iii) Show that the sum of the roots of the equation

$$x^2 + ax + bx + c = 0$$

is $-b$ and the product is c.
More generally, assuming that

$$x^n + a_1 x^{n-1} + \cdots + a_n = (x - \alpha_1)(x - \alpha_2) \cdots (x - \alpha_n),$$

show that the sum $\alpha_1 + \alpha_2 + \cdots + \alpha_n$ of the roots of

$$x^n + a_1 x^{n-1} + \cdots + a_n = 0$$

is $-a_1$ and the product is $(-1)^n a_n$.

On the other hand, if the reader does not already know how to solve a cubic, I claim that, in spite of 500 years of advances in mathematics, modern notation and, most important of all, the certainty that a general solution exists, the finding of such a method remains a challenging problem. (I shall give a solution in Exercise B.6 at the end of this section.)

The study of roots of polynomials leads inevitably to imaginary numbers. In an age when negative numbers were viewed with suspicion, Cardano explicitly considers the square roots of negative numbers, though he concludes his discussion with the words: 'This subtlety results from arithmetic of which this final point is, as I have said, as subtle as it is useless'.

Exercise B.3 *Cardano uses as an example the problem of finding two numbers whose sum is 10 and whose product is 40. Solve the problem.*

In old age, Cardano wrote what E. M. Forster called 'one of the great autobiographies of the world' *The Book of my Life* [10]. It tells us how, by intelligence and force of will, its author overcame poverty and illegitimacy to become one of the most famous men in Europe. (Not many autobiographies begin 'Although various abortive medicines – as I have been told – were tried in vain, I was born normally on the 24th day of September ...'.) However, it also tells of a man whose first and favourite son was beheaded for poisoning his wife and whose second son became a thief who stole even from his father. It passes in almost complete silence over the fact that, at the age of seventy, its author was arrested on the charge of heresy. (He had drawn up a horoscope for Jesus.) Although released after a few months, he had to make a formal recantation and was forbidden to teach or publish. Eventually, a new pope gave him an annuity, possibly because of his reputation as an astrologer, and he finished his life in Rome. A mixture of pride, melancholy and ruthless, though often deluded, self-examination gives *The Book of my Life* its unique tone.

Cardano was a great gambler and gamester, though he sometimes claims that he was only forced into such practices by poverty. Naturally, being Cardano, he wrote a book on the various games of skill and luck which were played at the time. His first version was written when he was young, but he seems[4] to have rewritten the part on games of chance and mixed chance and skill when he was about 65. This part of *The Book on Games of Chance* was found among his papers after his death and published a century later in his Collected Works.

The mathematical part of *The Book on Games of Chance* does not have the clarity of *The Great Art*. Incorrect reasoning in an earlier chapter will be followed by correct reasoning at a later point. The correct reasoning is then followed for the rest of the book but the earlier fallacious reasoning is left uncorrected to the confusion of the reader. However, there is no doubt that Cardano could calculate probabilities for any kind and any number of dice. The high point of the book occurs when he observes the result which, expressed in modern terms, states that if the probability of a particular outcome in one roll is p the probability that it is repeated n times in succession is p^n. He takes as his example the probability that, if we roll three dice three times in succession, each roll will contain at least one 1.

> Similarly, it has been stated that with three dice any one face, whichever one it be, has taken by itself 91 favourable cases in the whole circuit of 216. Therefore, if that face is required three times in a row, we shall multiply the whole circuit, and the result is 9 324 125. When the latter number is divided by the smaller of the

---

[4] Here, and in what follows, I rely entirely on the book [48] of Ore.

above numbers, namely 753 571, we get the odds determining the stake to be wagered, namely a little greater than 12 to 1.

Exercise B.4 *Verify Cardano's statement.*

Among Galileo's unpublished notes, there is one, written perhaps fifty years later, which shows that he, too, could handle dice problems.

Exercise B.5 *Suppose we throw three dice. Both the number 9 and the number 10 are produced by six different combinations. Galileo wrote an explanation for a puzzled friend of why, nonetheless, the number 10 appears more frequently in play than the number 9. Give your explanation.*

Apart from this, there is nothing until the explosive burst of activity described in Appendix C, another fifty years later, which marked the true birth of probability theory.

From the point of view of the progress of mathematics, the final publication of Cardano's work on probability came too late to have any influence. But it does raise the question why probability theory did not emerge earlier.

One answer, which I hope that the reader will consider, is that concepts like probability and expectation are not the products of common sense, but subtle ideas which it required great thinkers to uncover. Of course, this can be only part of the answer. Antiquity also contained great thinkers like Archimedes who combined subtlety of thought with technical brilliance and an openness to every question.

We cannot point to any single way in which the age of Cardano was unique. Most theories as to why probability theory did not appear earlier can be dismissed by paraphrasing Hacking [26]. 'Antiquity must have been full of impious and greedy men, equipped with excellent dice, gambling away like mad.'

Perhaps all we can say is this: Cardano's world was one in which some people earned a surplus to be spent on luxuries like gambling. It was one in which arithmetic was important for trade and which valued novelty for its own sake. It was an age with a fair amount of social mobility. All of these conditions had occurred before in human history (though not very frequently), but this time they produced something new. At the end of his long life Cardano wrote:[5]

It has been my particular fortune to live in the century which discovered the whole world – America, Brazil, Patagonia, Peru, Quito, Florida, New France, New Spain, countries to the North and East and South. And what is more marvellous than the human thunderbolt [gunpowder], which in its power far exceeds the

---

[5] The quotation is taken from [20]. The corresponding passage in [10] reads less smoothly.

heavenly? Nor will I be silent about thee, magnificent Magnet, who dost guide us through vast ocean, and night and storms, into countries we have never known. Then there is our printing press, conceived by man's genius, fashioned by his hands, yet a miracle equal to the divine.

It is true that, to compensate for these things, great tribulations are probably at hand; heresy has grown, the arts of life will be despised, certainties will be relinquished for uncertainty. But that time has not yet come. We can still rejoice in the flowering meadow of spring.

I cannot say that I regret my lot. I am the happier for having known so many things which are important and certain and rare. And I know that I have the immortal element within me and that I shall not wholly die.

Exercise B.6 *In this exercise we see how to solve the cubic. Remember that del Ferro did not use symbols but words and that he avoided (as we will not) the use of negative numbers. Cardano used geometric proofs where we would now use symbolic algebra.*

(i) Suppose that a and b are strictly positive. Sketch the graph of $x^3 + ax$ and explain why the equation $x^3 + ax = b$ has exactly one real root.

(ii) Suppose $x = u^{1/3} + v^{1/3}$ where u and v are real and we take the real cube root. Express $x^3 + ax$ in terms of u and v. Can you see what we should do next?

(iii) Suppose that

$$uv = -\frac{a^3}{27}$$
$$u + v = b.$$

Show that $u^{1/3} + v^{1/3}$ is a root of $x^3 + ax = b$. (Note the relevance of Exercise B.4.)

(iv) Show that u and v satisfy the conditions in (iii) if u and v are the roots of the quadratic

$$t^2 - bt - \frac{a^3}{27} = 0.$$

(v) Conclude that $x^3 + ax = b$ has the root

$$\left(\frac{b + \sqrt{b^2 + \frac{4}{27}a^3}}{2}\right)^{1/3} + \left(\frac{b - \sqrt{b^2 + \frac{4}{27}a^3}}{2}\right)^{1/3}.$$

(vi) The result of del Ferro was historically momentous since it represented the first time a modern mathematician had clearly outdone the ancients. In the rest of the question we leave del Ferro and Cardano and consider things from the perspective of complex numbers. You will need to know that 1 has three

complex cube roots 1, ω *and* ω^2. *In (v) we considered real roots. If we allow complex roots, while keeping a and b strictly positive as before, we seem to get nine possible solutions. Which of these are valid?*

(vi) Explain briefly why we can solve the general cubic

$$z^3 + Az^2 + Bz + C = 0$$

where all numbers are allowed to be complex.

(vii) If, having seen del Ferro's solution for the cubic, you think it all trivial, try and discover the solution for the general quartic. (If you fail, the result is in many algebra texts.)

(viii) If having discovered the solution for the general quartic, you think it all trivial (and, if you have genuinely done it by yourself, you have a right to this opinion) investigate the general quintic. (Two hundred years later, Galois explained why no similar trick will work. The ideas are explained in Galois theory. A excellent account of this theory will be found in any of the editions of [60].)

Appendix C
Huygens's problems

Lancelot Hogben's *Mathematics for the Million* [29] is such an excellent introduction to mathematics that mathematicians have forgiven him his evident dislike of their profession.[1] However, like many authors (including the present one), he finds it hard to distinguish between what ought to have been true and what actually happened.

Here is his account of the 'unsavoury origin' of probability.

> The first impetus came from a situation in which the dissolute nobility of France were competing in a race to ruin at the gambling tables. An algebraic calculus of probability takes its origin from a correspondence between Pascal and Fermat (about AD 1654) over the fortunes and misfortunes of the Chevalier de Méré, a great gambler and by that token *très bon esprit*, but alas (wrote Pascal) *il n'est pas géomètre*. Alas indeed! The Chevalier had made his pile by always betting small favourable odds, on getting at least one six in four tosses of a die and then lost it by always betting small odds on getting at least one double six in 24 double tosses.

Exercise C.1 *(i) Find the probability of getting at least one six in four tosses of a die and check that it is, indeed, greater than* 1/2. *Do you think it would be easy to make a fortune by betting even odds on getting at least one six in four tosses of a die?*

(ii) Find the probability of getting at least one double six in 24 *double tosses and check that it is, indeed, less than* 1/2. *Do you think it would be easy to make a fortune by betting even odds against getting at least one double six in* 24 *double tosses?*

---

[1] Witness his unwillingness to entrust 'the teaching of mathematics to people who put the head before the stomach, and who would tumble about the deep and high places of the earth if they had to teach another subject. Naturally this repels healthy people for whom symbols are merely the tools of organised social experience, and attracts those who use symbols to escape from our shadow world in which men battle for the little truth they can secure into a "real" world in which truth seems to be self-evident'.

(iii) Find the probability of getting at least one double six in 25 double tosses and check that it is greater than 1/2. Do you think it would be easy to make a fortune by betting even odds on getting at least one double six in 25 double tosses?

In fact de Méré was man of charm and good taste who divided his time between his small estate in Poitou and the French court.[2] His writings on how a gentleman ought to behave have given him a permanent place in the 'third division' of French authors. De Méré was one of those people who knows everybody and is interested (but not too interested) in everything. In this capacity, he knew and patronised Pascal who, he recalled, 'was then little known, but who later has certainly made people talk about him. He was a great mathematician who knew nothing but that. These sciences give little sociable pleasure, and this man, who had neither taste nor sentiment, could not refrain from mingling into all we said, but he almost always surprised us and often made us laugh' [49].

De Méré took pleasure in finding 'inconsistencies' in mathematics and asserted that the results of Exercise C.1 were a 'great scandal'. Some authors think that this reflects his own gambling experience, but I agree with those who think that the number of experiments and the exactness of record keeping required to obtain these results empirically go far beyond what can be reasonably be expected. This suggests that, although no records exist, some gamblers must have known rules for finding simple probabilities.

Exercise C.2 *Suppose that some event has small probability p of occurring in a given trial. Show that the number $N(p)$ of independent trials required for the probability of at least one success to exceed $1/2$ satisfies the approximate equation*

$$N(p) \approx \frac{\log 2}{p} \qquad \qquad ★$$

with the approximation improving as p becomes smaller.
  We thus expect

$$\frac{N(p)}{N(q)} \approx \frac{p}{q}$$

for p and q small.

---

[2] This account relies strongly on [49].

Ore suggests that this result was known as a rule of thumb to de Méré who expected

$$\frac{N(1/6)}{N(1/36)} \stackrel{?}{=} \frac{1/6}{1/36}$$

and cried scandal when this turned out not to be true.

Show that, nonetheless, ★ *gives a good estimate for* $N(1/6)$ *and a very good estimate for* $N(1/36)$. *(The approximation was first obtained rigorously by de Moivre.)*

More importantly, de Méré directed Pascal's attention to the problem of the interrupted game (called 'the problem of points' or 'the division problem'). A typical form of the question runs as follows.

Two teams play ball so that a total of 60 points is required to win, each innings counts 10 points and the stakes are 22 ducats. Due to circumstances, they cannot finish the game and one side has 50 points and the other 30. What share of the prize money belongs to each side?

This problem had been floating round Europe for 250 years. Various mathematicians had given various contradictory solutions, none of which seem satisfactory to us.

Exercise C.3 *(i) We would now restate the problem as follows. Two teams play ball so that a total of 60 points is required to win, each innings counts 10 points and the stakes are 22 ducats. The probability of either side winning an innings is 1/2. When one side has 50 points and the other 30, what are the expected winnings of each side?*

(ii) State and solve a more general version of the problem.

What Pascal did was to give the problem its modern interpretation and show how to solve it.[3] He then sought out the opinion of Fermat, the leading French mathematician, and, in a series of letters, the two worked out various methods for solving this and other problems in probability. 'I see,' wrote Pascal, 'that the truth is the same in Toulouse and Paris.'

Important as this correspondence seems to us, Fermat was mainly interested in problems in number theory, and Pascal's life was suddenly changed by an intense religious experience as a result of which he withdrew from the world and (with occasional relapses) from mathematics.

---

[3] The reader will not expect a single sentence to do more than act as a signpost to a major event in intellectual history. Note that the appropriate *interpretation* is at least as important as the actual *solution*.

When the young Huygens visited Paris a year later, he heard about the work of Fermat and Pascal, but Fermat was in Toulouse and Pascal in religious retreat. Huygens set out to recover their results and succeeded. He writes

> It should be said, also, that for some time some of the best mathematicians in France have occupied themselves with this kind of calculus so that no one should attribute to me the honour of the first invention. This does not belong to me. But these savants, although they put each other to the test by proposing to each other many questions difficult to solve, have hidden their methods. I have had therefore to examine and go deeply for myself into this matter by beginning with the elements, and it is impossible for me for this reason to affirm that I have even started from the same principle, but finally I have found that my answers in many cases do not differ from theirs.
>
> *[Quoted in [40]]*

Huygens circulated his results and received confirmation from both Fermat and Pascal that they were correct. He published his studies in a short work *De Ratiociniis in Ludo Alea* (On Reasoning in Games of Chance) concluding with five problems for the reader. Two of them had been sent by Fermat, while the fifth and hardest came from Pascal. The reader may enjoy solving problems of such a distinguished pedigree. (I have followed an early translation by Browne.)

Exercise C.4 *(i) A and B play together with a pair of Dice upon this Condition, That A shall win if he throws 6, and B if he throws 7; and A is to take one Throw [of both dice] first, and then B two Throws [of both dice] together, then A to take two Throws together, and so on both of them the same, till one wins. Shew that A's chance is to B's as 10 355 to 12 276.*

(ii) Three Gamesters, A, B, and C, taking 12 Counters, 4 of which are white, and 8 black, play upon these Terms: That the first of them that shall blindfold choose a white Counter shall win; and A shall have the first Choice, B the second, and C the third; and then A to begin again, and so on in their Turns. What is the Proportion of their Chances?

Jacob Bernoulli of pointed out that this problem could have several meanings including the following. (1) There is one set of 12 counters and the counters are not replaced after being drawn. (2) There is one set of 12 counters and the counters are replaced after being drawn. (3) Each player has his own set of counters and the counters are not replaced after being drawn.

Provide solutions for the three cases. The computations in (3) are particularly tedious, so just give reasonably explicit formulae from which the answer can be computed.

(iii) A lays with B, that out of 40 Cards, 10 of each different Sort, he will draw 4, so as to have one of every Sort. Show that the Proportion of his Chance to that of B, is as 1000 to 8139.

(iv) Having chosen 12 *Counters as before,* 8 *black and* 4 *white, A lays with B that he will blindfold take* 7 *out of them, among which there shall be* 3 *white ones. What is the Proportion of their Chances?*

Again Bernoulli points out that there are two interpretations. (1) A must take exactly 3 *white counters. (2) A must take at least* 3 *white counters. Solve both problems.*

(v) A and B taking 12 *Pieces of Money each, play with* 3 *Dice on this Condition, That if the Number* 11 *is thrown, A shall give B one Piece, but if* 14 *be thrown, then B shall give one to A; and he shall win the Game that first gets all the Pieces of Money. Show that the Proportion of A's Chance to B's is as* 244 140 625 *to* 282 429 536 481.

Huygen's little book was reprinted in various translations (Newton had his own annotated copy) and the exercises just given provided test problems for the next generation of mathematicians.

Exercise C.5 *One of the difficulties facing probabilists and statisticians in the days before computers was the sheer difficulty of generating and recording large numbers of random events. This final exercise is more of a project than an exercise and requires access to a computer, a source of random numbers and a little programming. Unless the reader is very fastidious, she will probably use the random number generator provided by her machine.[4]*

(i) That landmark of twentieth-century literature A Million Random Digits *[55] provides, in effect, a sequence Y_j of independent random variables such that*

$$\Pr(Y_j = k) = 10^{-1} \quad for \quad 0 \le k \le 9.$$

How would you use this sequence to produce a sequence corresponding the result of tossing a fair coin?

(ii) Use your random number source to produce a a sequence X_j of independent random variables such that

$$\Pr(X_j = 1) = \Pr(X_j = -1) = 1/2.$$

Graph the behaviour of

$$S_n = X_1 + X_2 + \cdots + X_n$$

as n runs from 1 *to* 10 000. *(You will need to experiment with methods of presentation and, in particular, with the scale of the vertical axis.) Repeat the experiment several times.*

---

[4] If the reader recalls the old political saying that 'the making of laws is like the making of sausages – the less you know about it, the more you respect the outcome', then a quick search of the internet will reveal several sources of unimpeachably random numbers.

(iii) Do the graphs of the various S_n cross the axis more often or less often than you expect? What is the relevance of Exercise 10.4.7?

(iv) We expect $|S_n|$ to have a tendency to grow larger as as n increases so it is reasonable to rescale in some manner. Graph the behaviour of S_n/n, $S_n/n^{3/4}$ and S_n/n^4 in a few experimental runs. (You should generate new random sequences for each run.) Do the results correspond to what you expect and why?

Guess how you expect $S_n/n^{1/2}$ to behave and then graph the behaviour in a few experimental runs.

There is beautiful treatment of coin tossing in Chapter III of the 3rd edition of Feller's masterpiece [19].

(v) Suppose that you play the game described in Exercise 2.6.1. Plot your fortune T_n if you follow the Kelly criterion for several runs. It is clear that you will have a rather bumpy ride. Why does the ride remain as bumpy as ever, even when your fortune has become large compared with its initial value?

If our only goal is to increase our fortune, then Kelly betting is best in the long run but, as Keynes remarked 'In the long run we are all dead'. Unless you have nerves of steel, near perfect judgement and can afford to take the very long view, you may prefer an α-Kelly rule where you bet α times the amount that Kelly suggests. Why should you never take $\alpha > 1$?

Plot the results for several runs with various choices of α. Many gamblers claim to use $1/2$- or $1/4$-Kelly systems.[5]

(vi) Let p be the probability of getting at least one six in four tosses of a fair dice (see Exercise C.1). Use your random number source to produce a a sequence Z_j of independent random variables such that

$$\Pr(Z_j = 1) = p, \ \Pr(Z_j = -1) = 1 - p.$$

Graph the behaviour of

$$T_n = Z_1 + Z_2 + \cdots + Z_n$$

over several runs. Plot de Méré's fortune U_n if he follows the Kelly criterion over several runs. Think about how many bets de Méré could make in real life. Plot de Méré's fortune if he follows other strategies.

---

[5] But what gamblers say they do, what gamblers think they do and what gamblers actually do are three very different things.

Appendix D

Hints on pronunciation

It may be helpful to know how certain symbols are read. Here are some hints.

- $\hat{a}$ 'a hat'.
- $\tilde{a}$ 'a tilde' (and 'tilde' is pronounced 'tilda').
- $\mathbf{a}$ 'bold a' or 'vector a'.
- $a \approx b$ 'a is approximately equal to b'.
- $\sum_{r=1}^{n} a_r$ 'the sum from 1 to n of a_r'.

$$\sum_{r=1}^{n} a_r = a_1 + a_2 + \cdots + a_{n-1} + a_n.$$

- $\prod_{r=1}^{n} a_r$ 'the product from 1 to n of a_r'.

$$\prod_{r=1}^{n} a_r = a_1 \times a_2 \times \cdots \times a_{n-1} \times a_n.$$

- $n!$ 'n factorial'.

$$n! = \prod_{r=1}^{n} r.$$

- $\binom{n}{r}$ 'n choose r' or 'binomial n, r'.

$$\binom{n}{r} = \frac{n!}{r!(n-r)!}.$$

- $A \cup B$ 'A union B', the set consisting of everything which is in at least one of A and B.
- $A \cap B$ 'A intersection B', the set consisting of everything which is in both A and B.
- $\varnothing$ 'the empty set', the set with no members.
- $A \setminus B$ 'A setminus B', the set consisting of everything which is in A but not in B.
- A^c 'A complement', 'the complement of A'. See page 45.

Table D.1. *Greek letters*

Lower-case	Upper-case	Name	Corresponds to	Note
α	[A]	alpha	a	
β	[B]	beta	b	
γ	Γ	gamma	c	
δ	Δ	delta	d	Often a small positive number.
ϵ, ε	[E]	epsilon	e	Often a small positive number. Do not confuse with $\in$ (belongs to).
ζ	[Z]	zeta	–	
η	[H]	eta	–	
θ	Θ	theta	–	
ι	[I]	iota	i	Often reserved for 'identity' objects.
κ	[K]	kappa	k	
λ	Λ	lambda	l	
μ	[M]	mu	m	Often used for the mean $\mu = \mathbb{E}X$.
ν	[N]	nu	n	
ξ	Ξ	xi	–	
[o]	[O]	omicron	o	Not used.
π	Π	pi	p	
ρ	[P]	rho	r	
σ, ς	Σ	sigma	s	We write $\sigma^2 = \text{var } X$ for the variance of X (page 63).
τ	[T]	tau	t	
υ	Υ	upsilon	u	Rarely used.
ϕ, φ	Φ	phi	–	
χ	[X]	chi	–	
ψ	Ψ	psi	–	
ω	Ω	omega	–	Used in connection with probability spaces (page 35).

- $f : A \to B$ 'the function f from A to B'.
- $\mathbb{R}$ 'the real numbers', $\mathbb{Z}$ 'the integers' and $\mathbb{C}$ 'the complex numbers'.
- $\text{Pr } A$ 'the probability of A'.
- $\mathbb{E}X$ 'the expectation of X'.

I conclude with a table of the Greek alphabet (see Table D.1). The Greek letters enclosed in square brackets are identical in form with ordinary (Roman) letters and are rarely used.

Bibliography

[1] A. Abdulkadiroğlu, P. A. Pathak and A. E. Roth. The New York City high school match. *American Economic Review, Papers and Proceedings*, **95**:364–367, 2005.

[2] R. M. Axelrod. *The Evolution of Cooperation*. Basic Books, New York, c1984.

[3] F. Bacon. *Essays*. John Haviland, London, 1625. Available on the web.

[4] W. W. Rouse Ball. *A History of the Study of Mathematics at Cambridge*. Cambridge University Press, Cambridge, 1889.

[5] D. Bernoulli. De la mortalité causée par la petite vérole, et des avantages de l'inoculation pour la prévenir. *Mém. Math. Phys. Acad. R. Sci. Paris*, **72**:1–45, 1766. There is an English translation by L. Bradley in *Smallpox Inoculation:An Eighteenth Century Mathematical Controversy* published by the Adult Education Department, Nottingham, 1971.

[6] J. Bernoulli. *The Art of Conjecturing*. Johns Hopkins University Press, Baltimore, 2006. English translation with notes by E. D. Sylla of the book *Ars Conjectandi* first published in 1713.

[7] J. L. Borges. *The Total Library*. Allen Lane, Cambridge, 2000. Edited by E. Weineberger; translated by E. Allen, J. Levine and E. Weineberger.

[8] S. J. Brams. *Mathematics and Democracy: Designing Better Voting and Fair-Division Procedures*. Princeton University Press, New Jersey, 2008.

[9] G. Cardano. *The Great Art or the Rules of Algebra*. The M.I.T. Press, Cambridge, Mass., 1968. Translated and edited by T. Richard Witmer.

[10] G. Cardano. *The Book of my Life*. The New York Review of Books, New York, 2002. A reprint of a translation by J. Stoner.

[11] J. W. S. Cassels. *Economics for Mathematicians*. Number 62 in LMS Lecture Note Series. Cambridge University Press, Cambridge, 1981.

[12] R. D. Clarke. An application of the Poisson distribution. *Journal of the Institute of Actuaries*, **72**:481, 1946.

[13] W. H. Cockcroft. *Mathematics Counts (The Cockcroft Report)*. HMSO, London, 1982.

[14] A. M. Colman. *Game Theory and its Applications*. Butterworth–Heinemann, Oxford, 2nd edn, 1995.

[15] I. G. Crumberry. *Why Things are not Otherwise*. U. U. Press, Ankh-Morpork, 1892.

[16] A. de Moivre. *The Doctrine of Chances.* H. Woodfall, London, 2nd edn, 1737. Versions available on the web.

[17] L. E. Dubins and L. J. Savage. *How to Gamble if You Must.* McGraw-Hill, New York, 1965. There is a Dover reprint.

[18] H. Geiger, E. Rutherford and H. Bateman. The probability variations in the distribution of alpha particles. *Philosophical Magazine, Series 6,* **20**:698–704, 1910. Reprinted in Rutherford's *Collected Works.*

[19] W. Feller. *An Introduction to Probability Theory and its Applications,* volume I. John Wiley, New York, 3rd edn, 1968.

[20] E. M. Forster. *Abinger Harvest.* Edward Arnold, London, 1936.

[21] D. Gale. *The Theory of Linear Economic Models.* McGraw-Hill, New York, 1960.

[22] D. Gale and L. S. Shapley. College admissions and the stability of marriage. *American Mathematical Monthly,* **69**:9–15, 1962.

[23] M. Gardner. *Mathematical Puzzles and Diversions.* Simon and Schuster, New York, 1961.

[24] I. Glynn and J. Glynn. *The Life and Death of Smallpox.* Profile Books, Hatton Garden, London, 2004.

[25] J. E. Gordon. *Structures, or Why Things Don't Fall Down.* Penguin, Harmondsworth, 1978.

[26] I. Hacking. *The Emergence of Probability.* Cambridge University Press, Cambridge, 1975.

[27] G. H. Hardy. *A Mathematician's Apology.* Cambridge University Press, Cambridge, 1940.

[28] C. A. R. Hoare. The emperor's old clothes. *Communications of the ACM,* **24**(2):75–83, 1981.

[29] L. T. Hogben. *Mathematics for the Million.* Allen and Unwin, London, 1936.

[30] D. Huff. *How to Lie with Statistics.* Norton, New York, 1954.

[31] H. Kahn. *On Escalation.* Princeton University Press, New Jersey, 1960.

[32] V. Kannisto, R. Thatcher, and J. W. Vaupel. *The Force of Mortality at Ages 80 to 120.* Odense University Press, Odense, Denmark, 1998. Available on the web.

[33] J. L. Kelly. A new interpretation of information rate. *Bell System Technical Journal,* **35**:917–26, 1956.

[34] D. M. Kilgour and S. J. Brams. The truel. *Mathematics Magazine,* **70**:315–326, 1997.

[35] D. E. Knuth. *Stable Marriage and Its Relation to Other Combinatorial Problems.* AMS, Providence, R. I., 1997. A translation by M. Goldstein of *Mariages Stables et Leurs Relations avec d'autres Problèmes Combinatoires* published by les Presses de l'Université de Montréal in 1976.

[36] M. Kraitchik. *Mathematical Recreations.* Allen and Unwin, London, 1944. A modified version of *La Mathématique des Jeux,* published in Brussels, 1930.

[37] F. Le Lionnais, editor. *Les Grands Courants de la Pensée Mathématique.* Hermann, Paris, 1998. First published in 1948.

[38] É. Lucas. *Récréations Mathématiques.* Gauthier-Villars, Paris, 1882–1896.

[39] R. D. Luce and H. Raiffa. *Games and Decisions.* Wiley, New York, 1957. There is a Dover reprint.

[40] L. E. Maistrov. *Probability Theory, A Historical Sketch.* Academic Press, New York, 1974. Translated from the Russian by S. Kotz.

[41] J. Manyard Smith. *Evolution and the Theory of Games*. Cambridge University Press, Cambridge, 1982.

[42] A. Marshall. *Priciples of Economics*. Macmillan, London, 8th edition, 1920. Available on the web.

[43] E. A. Milne. Ralph Howard Fowler. *Obituary Notices of Fellows of the Royal Society*, **5**:60–78, 1945.

[44] M. J. Moroney. *Facts from Figures*. Penguin, Harmondsworth, 1951.

[45] A. G. Munford. A note on the uniformity assumption in the birthday problem. *The American Statistician*, **31**:119, 1977.

[46] I. Newton. *Sir Isaac Newton's Mathematical Principles of Natural Philosophy and his System of the World*. University of California Press, Berkeley, 1934. Cajori's revision of Motte's translation. There is a Dover reprint.

[47] T. H. O'Beirne. *Puzzles and Paradoxes*. Oxford University Press, Oxford, 1965.

[48] O. Ore. *Cardano, the Gambling Scholar*. Princeton University Press, Princeton, 1953. With a translation of Cardano's *Book on Games of Chance* by S. H. Gould.

[49] O. Ore. Pascal and the invention of probability theory. *The American Mathematical Monthly*, **67**:409–19, 1960.

[50] H. Phillips. *The Week-end Problems Book*. Nonesuch Press, Bloomsbury, 1933. Republished in 2006 by Duckworth Overlook, London.

[51] H. Phillips. *Something to Think About*. Penguin, Harmondsworth, 1945.

[52] R. Porter. *English Society in the 18th Century*. Penguin, Harmondsworth, 2nd edn, 1991.

[53] W. Poundstone. *Prisoner's Dilemma*. Doubleday, New York, 1992.

[54] W. Poundstone. *Fortune's Formula*. Hill and Wang, New York, 2005.

[55] Rand Corporation. *A Million Random Digits with 100,000 Normal Deviates*. The Free Press, Glencoe, Illinois, 1955.

[56] L. F. Richardson. *Statistics of Deadly Quarrels*. Boxwood, Pittsburg, 1960.

[57] A. A. Rusnock. *Vital Accounts*. Cambridge University Press, Cambridge, 2002.

[58] T. C. Schelling. *The Strategy of Conflict*. Harvard University Press, Cambridge, Mass., 1960.

[59] B. Shaw. *Everybody's Political What's What*. Constable, London, 1944.

[60] I. N. Stewart. *Galois Theory*. Chapman and Hall, London, 1973.

[61] L. C. Thomas. *Games, Theory and Applications*. Wiley, New York, 1984. There is a Dover reprint.

[62] E. O. Thorp. A favorable strategy for twenty-one. *Proc. Nat. Acad. Sci. U.S.A*, **47**:110–112, 1961.

[63] E. O. Thorp. *Beat the Dealer*. Blaisdell, New York, 1962. There are later revised editions.

[64] S. Trybuls. On the paradox of three random variables. *Zastos. Mat.*, **5**:321–32, 1961.

[65] Voltaire. *Lettres Philosophiques*. E. Lucas, Amsterdam, 1734. Available on the web.

[66] L. von Bortkiewicz. *Das Gesetz der kleinen Zahlen*. Teubner, Leipzig, 1898.

[67] J. von Neumann and O. Morgenstern. *Theory of Games and Economic Behavior*. Princeton University Press, Princeton, 1944.

[68] H. Walpole. *The Letters of Horace Walpole, Earl of Orford*. Richard Bentley, London, 1857. Edited by P. Cunningham. Available on the web.

[69] D. Waltham. *Mathematics: A Simple Tool for Geologists.* Blackwell Science Ltd, Oxford, 2nd edn, 2000.

[70] J. D. Williams. *The Compleat Strategyst.* McGraw-Hill, New York, revised edn, 1966. This has been reissued by Dover.

[71] L. V. Williams. *Betting to Win.* High Stakes Publishing, 21 Great Ormond Street, London, 2004.

[72] L. L. Yong and Ang Tian Se. *Fleeting Footsteps.* World Scientific Publishing, Farrer Road, Singapore, 2004. Contains a translation of *Sun Zi Suanjing* (Sun Zi's Mathematical Manual).

Index

Printed in the United States
by Baker & Taylor Publisher Services